Landscape of Migration

FLOWS, MIGRATIONS, AND EXCHANGES

Mart A. Stewart and Harriet Ritvo, *editors*

The Flows, Migrations, and Exchanges series publishes new works of environmental history that explore the cross-border movements of organisms and materials that have shaped the modern world, as well as the varied human attempts to understand, regulate, and manage these movements.

BEN NOBBS-THIESSEN

Landscape of Migration

Mobility and Environmental Change
on Bolivia's Tropical Frontier,
1952 to the Present

The University of North Carolina Press *Chapel Hill*

© 2020 Ben Nobbs-Thiessen
All rights reserved
Set in Arno Pro by Westchester Publishing Services
Manufactured in the United States of America

The University of North Carolina Press has been a member of the
Green Press Initiative since 2003.

Library of Congress Cataloging-in-Publication Data
Names: Nobbs-Thiessen, Ben, author.
Title: Landscape of migration : mobility and environmental change
 on Bolivia's tropical frontier, 1952 to the present / Ben Nobbs-Thiessen.
Other titles: Flows, migrations, and exchanges.
Description: Chapel Hill : The University of North Carolina Press, [2020] |
 Series: Flows, migrations, and exchanges | Includes bibliographical references and index.
Identifiers: LCCN 2019044541 | ISBN 9781469656090 (cloth ; alk. paper) |
 ISBN 9781469656106 (paperback ; alk. paper) | ISBN 9781469656113 (ebook)
Subjects: LCSH: Migration, Internal—Bolivia—Santa Cruz (Department)—History—
 20th century. | Agriculture and State—Bolivia—Santa Cruz (Department)—History—
 20th century. | Indigenous peoples—Colonization—Bolivia—Santa Cruz
 (Department)—History—20th century. | Mennonites—Colonization—Bolivia—
 Santa Cruz (Department)—History—20th century. | Ryukyuans—Colonization—
 Bolivia—Santa Cruz (Department)—History—20th century. | Bolivia—
 Emigration and immigration—History—20th century. | Bolivia—Politics and
 government—20th century. | Bolivia—History—Revolution, 1952.
Classification: LCC HB2022.S26 N63 2020 | DDC 984/.305—dc23 LC record
 available at https://lccn.loc.gov/2019044541

Cover photo: *Pinwheel Squares in Bolivia,* NASA Photo ID ISS056-E-94529. Courtesy of the Earth
Science and Remote Sensing Unit, NASA Johnson Space Center (http://eol.jsc.nasa.gov).

To Laurie Nobbs and Linda Thiessen

Contents

Acknowledgments xi

Abbreviations in the Text xv

Introduction: The Meanings of Mobility in Bolivia's March to the East 1

CHAPTER ONE
Moving Pictures: Narrative, Aesthetic, and Bolivia's Frontier Imaginary 26

CHAPTER TWO
Military Bases and Rubber Tires: Okinawans and Mennonites at the
Margins of Nation, Revolution, and Empire, 1952–1968 65

CHAPTER THREE
Abandonment Issues: Speaking to the State from the Andes
and Amazonia, 1952–1968 102

CHAPTER FOUR
To Minister or Administer: Faith and Frontier Development
in Revolutionary and Authoritarian Bolivia, 1952–1982 139

CHAPTER FIVE
A Sort of Backwoods Guerrilla Warfare: Mexican Mennonites
and the South American Soy Boom, 1967–Present 187

Conclusion: Past and Present in the Bolivian Lowlands 230

Epilogue: From Abandonment to Autonomy 242

Notes 257
Bibliography 287
Index 305

Graphs, Illustrations, and Maps

GRAPHS

Jakob Knelsen's annual harvest 222

Jakob Knelsen's milk production in Mexico and Bolivia 226

ILLUSTRATIONS

Scenes from Jorge Ruiz's 1955 film, *Un poquito de diversificación económica* 28

Scenes from Jorge Ruiz's 1958 film, *La Vertiente* 35

Image from the pamphlet "Qué es el plan decenal?" 60

Images from a pamphlet titled "Como viviré y trabajaré mi nueva parcela?" 61

An aerial view of San Julián's unique settler nuclei 141

Sketch of San Julián's settler nuclei 171

Mennonite colonist Abram Wiebe sells jam at a busy intersection 231

MAPS

Administrative divisions of Bolivia 4

Infrastructure projects converge on Santa Cruz in the 1950s 38

The evolving Mennonite diaspora in the Americas 69

The evolving Okinawan diaspora in the Americas 70

Settler colonies near Montero 148

Mennonite colonies in lowland Bolivia 190

Acknowledgments

The migrants I interviewed over the course of this research told stories in which they traversed national, regional, and environmental boundaries as they moved from their places of origin—Mexico, Paraguay, Okinawa, the Andes—to the region they settled. Often their movements were not a singular, linear journey through space but a nimble series of comings and goings in which they continued to travel between, and draw support from, multiple locales and far-flung diasporas. As I developed a sense of this project, traced their narratives, and drafted and revised this manuscript over the past nine years, my growing family and I have engaged in a series of migrations of our own that brought us from Vancouver to Atlanta, Mexico, and Bolivia, back to Atlanta, onward to Arizona, and finally, back to the Pacific Northwest. So many friends, family, colleagues, and interlocutors made that intellectual and physical journey possible and deserve my sincere gratitude. I will surely only mention a fraction of them by name here.

A first book can (hopefully) be excused for delving further into the past than others. I have been shaped by many foundational educators including Chris Seppelt, Hilary Mason, and Bill French who led me on a five-month field study to Mexico in 2005. I returned to the University of British Columbia to complete an MA in Latin American history with Bill in 2007 and in large part went on to pursue a PhD after taking part in his Oaxaca Summer Institute in 2008. That five-week program was the first in an ongoing series of encounters with a wonderful community of fellow scholars including Derek Bentley, Nicole Pacino, Stephanie Opperman, Steve Allen, Brian Freeman, Jessica Fowler, and Nydia Martinez. In 2010, Karen and I drove across the continent with only the possessions that we could cram into the trunk and back seat of our car. In Atlanta we found a surrogate family of sorts. We were immediately taken in by Taylor and Caitlyn Mathes, Andre Domnigues, Ross and Bev Miller, Derek and Chelsea Bentley, Justin Barker, Alex and Jenny Baumann, Jim Ikemoto, and many more. We also reestablished connections with old friends Val and Brian Danin whose trajectory over the past decade in Georgia and Washington has closely matched our own. Along the way our family grew to include our cat Nora Jean, dog Skaha, and finally in July of 2015, our daughter Avery. I also arrived in Atlanta along with a cohort of emerging historians that included Jessica Reuther, Emma Meyer, Ashleigh Ikemoto, Colin Reynolds, Louis Fagnon, Scott Libson, and Rebekah Ramsey whose support and commiseration on the first leg of this long, strange odyssey was critical. The friendships of fellow Latin Americanists Chris Brown and Jennifer Schaefer extended from cycling to the seminar room to the soccer pitch. Angie Picone, Shari Wesja, and Jon Coulis were still generously offering help right up to the final days of preparing this manuscript.

At Emory my eternal thanks go to our triumvirate of Latin American historians: Jeffrey Lesser, Yanna Yannakakis, and Tom Rogers. Jeff has provided constant encouragement, exceptionally sound advice, and I benefitted greatly from Yanna's nuanced understandings of power and negotiation. I had no idea Tom would be joining the department when I decided to study at Emory, but I could scarcely conceive of this project without him. Our early conversations about development modernization, oral history, and agroenvironmental change were particularly informative, and he helped me piece together my earliest archival sources into a semblance of a narrative. Our Latin American subject librarian Phil MacLeod has been a great friend over the last decade while tirelessly tracking down even the most obscure materials on my behalf. I also received incredible mentorship and support from Karen Stolley, Peter Little, and Peggy Barlett.

In 2016, my family and I left Atlanta to begin a post-doc at Arizona State. In Tempe, our second child, Dylan, was born, and I met a fantastic group of scholars deeply engaged in the history and politics of migration who supported me as I prepared a book proposal—with invaluable assistance from Tore Olsson!—and revised this manuscript for publication. Special thanks to Alejandro Lugo, Lisa Magaña, Irma Arboleda, my fellow post-doc Henry Gonzalez, and the rest of the faculty and staff that made that academic unit feel like one big family. Anna Holian also offered generous support. In Tempe we were also lucky enough to settle alongside a couple of Emory expats— Colin Reynolds and Josh Robinson—who made our adjustment to the desert all the easier. In 2018, after those years in the southeast and southwest we returned home, settling into a small city only a day's drive from our families. It was here, fittingly, that I have drawn together this nearly decade-long endeavor. Throughout this time, I have appreciated the support of a fabulous group of Roots of Contemporary Issues (RCI) post-docs and supportive mentors. As I finalize this manuscript, I also want to thank those at the University of North Carolina Press whose faith in this project, and practical assistance, have made it a reality. This includes series editors Mart Stewart and Harriet Ritvo, acquisitions editor Brandon Proia, and editorial assistant Dylan White.

As I write these words, I am beginning research for a new project that received support from Conrad Grebel University College where I am serving as a J. W. Fretz Fellow for summer 2019. These last years of research and writing would not have been possible without the financial support of Emory's Laney Graduate School and the Social Sciences and Humanities Research Council (SSHRC) of Canada. I also received funding from the Conference on Latin American History, the Plett Foundation, and Drew University Archives. The project included collaboration with Royden Loewen's SSHRC-funded *Seven Points on Earth* project, and Roy's guidance has been critical to my development.

What about Bolivia? When I set out to study Latin America, I imagined myself a Paraguayan historian. I have Nicole Pacino to thank for my jump across the Gran Chaco into Bolivia. She introduced me to a vibrant community of national and foreign historians, anthropologists, sociologists, and journalists during my first visit to the semiannual Estudios Bolivianos Conference in Sucre in 2011. As I returned to Bolivia over the following years, I benefited from their wealth of knowledge about this country. I want to express particular appreciation to Jorge Derpic, Lesli Hoey (for inspiring chapter 4 and sharing time in the field together), Chuck Sturtevant, Carmen Soliz, Sarah Hines, Elena McGrath as well as Gabriel Hetland, Justin Blanton, Matthew Gildner, Hernán Pruden, and Gabi Kuenzli who helped me establish research contacts and navigate unfamiliar archives and government agencies. Special thanks are also due to the fantastic staff at the National Archives of Bolivia in Sucre, the Instituto Nacional de Reforma Agraria in Santa Cruz and the Ministerio de Desarrollo Rural y Tierras, and the library of the Inter-American Institute for Cooperation on Agriculture in La Paz, among others.

Living in La Paz and Santa Cruz I developed lasting friendships with a number of people who welcomed me into their lives and homes. I owe thanks to my friend Sara Shahriari (and her dog, Bell), Monica Flores, Nikki Evans, and Mariela Rodrigues in La Paz. In Santa Cruz I unexpectedly found myself living with two complete strangers, Sergio Reyes and Elena Méndez. They quickly became close friends and made the months I spent there—a time in which I would often stumble into the apartment exhausted and covered in mud from a long bike excursion in the colonies—ones I will remember fondly. Out in the farming communities of Santa Cruz so many settlers invited me into their homes to share food and stories it would be difficult to mention them all here. Special thanks to the Hamm, Buhler, and Enns families in Riva Palacio colony and the Fehrs, Falks, Brauns, and Ungers in Canadiense colony and all those that offered stories and *faspa* (a midafternoon meal) to a weary cyclist. Harry Peacock brought me to farm in San Julián colony and his house in Santa Cruz de la Sierra. Alejandro Araus and Jaime Bravo also shared their stories. Back in Santa Cruz, I drank *tereré* with Willmar Harder and his family on my frequent visits to Centro Menno. My research included two months in Mexico City. There I was lucky enough to live with two wonderful friends, Derek and Chelsea Bentley, and spent time in the city's archives and restaurants with Lance and Lauren Ingerwesen.

I want to close by thanking my family. As the first person in her family to attend college, my mother Linda Thiessen has been an enthusiastic supporter of my academic pursuits from my earliest days. She has also read and commented on every word of this book and has proven herself an effective, and at times ruthless, scourge of the passive voice. My in-laws Richard and Lexie Milton repeatedly visited Karen and me in Atlanta, Phoenix, and Pullman, traveled with us to Argentina, and welcomed us back to the Fraser Valley whenever we returned. My siblings, Andrea, Jesse, and Max, their partners Larry, Lisa, and Kathy-Ann, and my nieces and nephews have also been a constant source of pride, inspiration, and good-natured competition. Above all I wish to thank my wife Karen. This job has kept us on the move over the years from an eleven-month trip through Latin America in 2006–7 to Atlanta in 2010, Phoenix in 2016, and Pullman in 2018. Even as she has supported my academic pursuits, she has also reminded me that there are many other fine things in life, and I am very grateful for that. She has been a true friend and companion since the first day we met in September of 2001, and it comes as little surprise to me that since the birth of our children she has proven a wonderful and dedicated mother. A lot can happen in a decade. Over the course of researching this book I lost my father, and while writing and revising it, I became a father. Laurie Nobbs was a tremendously hard worker who held himself and others to a high standard but he was also a charmer who had a deep empathy for those around him. He set an example that I will strive to match as I raise my own children.

Abbreviations in the Text

ANAPO	Asociación Nacional de Productores de Oleaginosas
CAICO	Cooperativa Agropecuaria Integral Colonias Okinawa Ltda.
CAISY	Cooperativa Agropecuaria Integral San Juan de Yapacaní Ltda.
CAO	Cámara Agropecuaria del Oriente
CBF	Corporación Boliviana de Fomento
CECOYA	Central de Colonizadores de Yapacaní
CEN	Comité de Emergencia Nacional
CIDOB	Confederación de Pueblos Indígenas de Bolivia
CIU	Comité de Iglesias Unidas
COB	Central Obrera Boliviana
COMIBOL	Corporación Minera de Bolivia
CO	Conscientious Objector
EMBRAPA	Empresa Brasileira de Pesquisa Agropecuária
FES	Función Económico-Social (Social-economic Function)
FIDES	Fundación Integral de Desarrollo
FSB	Falange Socialista Boliviana
GRI	Government of the Ryukyu Islands
IBCE	Instituto Boliviano de Comercio Exterior
ICA	United States International Cooperation Administration
ICAIC	Instituto Cubano del Arte e Industria Cinematográficos
ICB	Instituto Cinematográfico Boliviano
IDA	Institute for Development Anthropology
IDB	Inter-American Development Bank
IICA	Inter-American Institute for Cooperation on Agriculture
INC	Instituto Nacional de Colonización

INRA	Instituto Nacional de Reforma Agraria
JICA	Japanese International Cooperation Agency
MAS	Movimiento al Socialismo
MCC	Mennonite Central Committee
MNR	Movimiento Nacionalista Revolucionario (Nationalist Revolutionary Movement)
MSC	Mennonite Service Committee
NADEPA	Núcleos Agrícolas de Producción Asociada
NEP	New Economic Policy
NGO	Nongovernmental organization
PCM	Pacto Campesino-Militar
ROEC	Ryukyuan Overseas Emigration Corporation
SPIC	Secretaría de Prensa, Información y Cultura
SRE	Secretaría de Relaciones Exteriores
TAP	Teacher Abroad Program
TCO	Tierras Comunitarias de Origen
TIPNIS	Isiboro-Sécure Indigenous Territory
UAGRM	Universidad Autónoma Gabriel René Moreno
UCAPO	Unión de Campesinos Pobres
UN	United Nations
UNESCO	United Nations Educational, Scientific and Cultural Organization
USAID	United States Agency for International Development
USCAR	United States Civil Administration of the Ryukyu Islands
USIS	United States Information Service
USOM	United States Operations Mission (Bolivia)
WGM	World Gospel Mission
YPFB	Yacimientos Petrolíferos Fiscales Bolivianos

Landscape of Migration

Introduction
The Meanings of Mobility in Bolivia's March to the East

[Mennonites] need to travel from time to time. We are accustomed to it. There are Mennonites in Canada, the United States and [here] in Mexico. We are now going to South America.

—Martin Dueck in an interview with Jorge Aviles Randolph in Mexico City, "A People Abandon Us," *Excelsior*, June 4, 1968

Most Okinawans, particularly the members of the young generation who feel frustrated by the lack of opportunities in the homeland consider emigration to be of supreme importance to their future welfare . . . as of September 1952, an estimated 172,000 persons had applied for permits to emigrate, chiefly to South American countries.

—Hoover Institute sociologist, James Tigner

With the road [to Santa Cruz] that Paz Estenssoro opened in the early 1950s, a great movement of people began, and I was among them. I am a migrant, that joined the wave of migration to the east of Bolivia.

—Aymara settler and Methodist director of Rural Colonies, Jaime Bravo

In mid-1968, horse-and-buggy Mennonite farmer Martin Dueck traveled 1,500 kilometers to Mexico City by bus from his home colony in the northern border state of Chihuahua to prepare his family's travel documents. Paperwork in hand, he exited the imposing Secretary of Foreign Relations (SRE) onto the Tlatelolco neighborhood's Plaza de Tres Culturas—a square that would become infamous as the site of a government massacre of student demonstrators only four months later. On that June day, flanked by the modernist SRE, a colonial Catholic Church, and a pre-Hispanic archeological site, Dueck gave an impromptu interview to journalist Jorge Aviles Randolph of *Excelsior*, one of the city's oldest and highest-circulating newspapers. Dueck explained to Aviles, in simple terms, why he and several thousand of his low-German speaking, pacifist coreligionists were leaving for Bolivia just shy of the fiftieth anniversary of their arrival in postrevolutionary Mexico from Canada. In contrast to the evocative idea of emigration as "abandonment" in

Aviles's subsequent article, Dueck conjured up a diasporic history of transnational Mennonite mobility as something so timeless, natural, and recurring, it bordered on the mundane.

A decade and a half earlier, U.S. sociologist James Tigner struck a more desperate tone when invoking the migratory impulse on Okinawa. He worried that the "frustrated" youth of the Ryukyuan archipelago, where existing population pressure had become catastrophic due to postwar repatriation and U.S. military land expropriations, might turn to leftist political parties if emigration—a venerable Okinawan survival strategy with roots in the late nineteenth century—was not once again offered as a safety valve. In Cold War East Asia, it was a claim that surely resonated with nervous U.S. officials. Tigner had just completed a grueling circuit of thousands of kilometers on both sides of the Pacific on their behalf. His travels took him from Washington D.C., to Japan, Okinawa, and Hawaii and then through the sites of the Japanese Latin American diaspora in Brazil, Peru, and Bolivia. In his final report, Tigner advised his military sponsors to support the mass migration of displaced and impoverished Okinawans to the Bolivian frontier.

In mid-2014, Jaime Bravo adopted a different perspective when reflecting on a lifetime of mobility that had carried him from his birthplace in Bolivia's western Andes to the country's tropical and semitropical lowlands, and later, through study, exile, missionary, and nongovernmental organization (NGO) work, to Argentina, Peru, the United States, and Canada. "We all heard that the future of Bolivia was in the east," Bravo began, "that there were fertile lands, that there were opportunities to expand oneself, to grow, opportunities to improve life, improve the economy and for us, the youth, we had the hope of looking upon a new horizon."[1] For Bravo, as with the hundreds of thousands of indigenous Aymara and Quechua settlers that joined him, individual migration was inextricably bound to the unfolding of personal development *and* national control on the frontier.

What unites these disparate narratives in which migration figured as a diasporic imperative, a geopolitical reality, or the fulfillment of a frontier myth? Ultimately the destination was the same. The lowlands of eastern Bolivia appeared the answer to problems originating across the Pacific, in Northern Mexico, and in the nation's own Andes. In the 1950s and 1960s, Mennonite, Okinawan, and Andean trajectories converged on Bolivia's forested frontier. But settlers are not the sole migrants in this narrative. As agrochemical giant Syngenta would ominously proclaim in an infamous 2003 advertisement, in which a sea of green spilled across the national frontiers of South America, "soybeans do not recognize borders."[2]

That humble bean was one of the many nonhuman and human migrants—from aging U.S. tractors and Indo-Brazilian cattle to a globetrotting cast of filmmakers, missionaries, agronomists, and development anthropologists—that also crossed borders to take part in one of the largest per capita tropical colonization initiatives of the twentieth century. In the process, these transnational migrants recast Bolivia's tropical lowlands as a *landscape of migration.* Their intertwined histories make the region an exceptionally fertile terrain for capturing the intersection of transnational, national, and local histories in a defining element of twentieth-century modernization—namely, the desire to transplant people, plants, ideas and technologies across the globe. The result was an unprecedented alteration of landscapes in Bolivia along with much of the tropical world.

The Nature of a Revolution:
National Visions and Global Currents

This "March to the East" (*marcha hacia el oriente*) as it became known, emerged out of one of Latin America's transformative midcentury revolutions. In April of 1952, a Nationalist Revolutionary Movement (MNR) composed of middle-class politicians, students, workers, farmers, and miners overthrew the mining oligarchy that had dominated Bolivia—one of the globe's four large tin producers—for most of its independent existence. In the wake of that revolution, Bolivia's new government, led by President Víctor Paz Estenssoro, implemented a series of radical and well-documented reforms. These included the nationalization of the mining sector; land redistribution in the highlands; and the extension of voting, education, and public health to the country's indigenous majority which was, and remains, proportionately the largest in the Americas.

Although often missing from a burgeoning new historiography of the National Revolution of 1952, the MNR also began to reimagine their small, impoverished, and landlocked nation at the heart of South America in spatial terms. The problems they faced were considerable. By 1952, Bolivia imported many of its food staples; had lost significant portions of its western, northern, and eastern frontiers to Paraguay, Chile, and Brazil; and registered 80 percent of its population concentrated in less than one-third of national territory in the western highlands. Worryingly, an increasing number of impoverished Andeans were engaging in annual midcentury pilgrimages to Argentina to work as field laborers or *braceros* (as their Mexican contemporaries in the United States were also known). These officials envisioned a *revolution in*

Administrative divisions of Bolivia. Created by TUBS, Wikimedia Commons.

nature as a fundamental aspect of their political revolution. By encouraging indigenous Bolivians to instead migrate from the overcrowded Andes to colonization zones along the nation's undeveloped lowland frontiers, the MNR hoped to solve these interlocking problems with a single blow, ensuring food security, territorial integrity, and demographic symmetry in the balance. Their gaze swept across the entirety of the Bolivian lowlands, an area larger than the state of Texas, ranging from humid Amazonia in the north to the semiarid bushlands of the Gran Chaco in the southeast. The linchpin in their plans was Santa Cruz Department. The largest of Bolivia's nine territorial divisions, Santa Cruz occupied a unique ecotone where those two lowland landscapes (Chaco and Amazon) converged while the Chiquitano dry forest extended out to the east.

In its impressive scope the March to the East was not unlike the contemporary projects of reformist and newly independent nations across the global south. Swept to power by revolutionary fervor and anticolonialism, high modernists everywhere looked to relocate bodies and reorder landscapes. They would call new publics into being through, and in the service of, development. In postcolonial Africa this was evident in a series of mass resettlement and *villagization* projects led by independence leaders like Julius Nyerere in Tanzania and Kwame Nkrumah in Ghana.[3] A similar logic was at play in Sukarno's attempts to sedentarize the Meratus people of Indonesia and shift surplus populations between islands through transmigration.[4] In postcolonial India, dam-building projects to supply power and irrigated agriculture became a favored tool of the state that also displaced, and resettled, millions.[5] The environmental and the cultural were equally apparent in Bolivia where the MNR paternalistically envisioned mobility as the dual refashioning of indigenous subjects into settler-citizens and a fugitive frontier landscape into a site of intensive agricultural production.

Closer to home, the March to the East finds strong parallels in Latin American republics engaged in similar projects of internal colonization in regions that had long resisted the control of the state. Already in the 1930s, Brazil's populist leader Getúlio Vargas had championed the country's own March to the West, proclaiming in one 1940 speech on the banks of the muddy Amazon River near Manaus, that the "greatest task for civilized man," lay in, "transforming its blind force and extraordinary fertility into disciplined energy."[6] In 1948 the very borders of the "Legal Amazon" were expanded to create an administrative unit that included most of Matto Grosso state. In the 1950s, president Juscelino Kubitschek continued this project with the construction of a new capital—Brasília—closer to the geographical center of the country. In the following decades, Amazonian expansion (encompassing directed colonization and road building) became a favored policy of Brazil's military government. President Emílio Médici famously promoted the transmigration of excess drought-stricken populations as "the solution to two problems: men without land in the Northeast and land without men in the Amazon."[7] Brazil's military leaders also funneled investment into the *cerrado* region of Matto Grosso leading to a booming soybean and cattle industry on Bolivia's eastern frontier.[8] By building infrastructure and encouraging indigenous migrations to the tropical lowlands, other Andean nations like Bolivia also turned inward over the second half of the twentieth century, linking ideas of citizenship with environmental change in the process.[9] In Peru, architect and president Terry Belaúnde imagined a transnational "Marginal Highway of the Jungle" that would integrate new settler colonies and link the Peruvian and Bolivian Amazon in the south to Ecuador and Colombia in the north.[10] These contemporary territorial imaginings led to new rounds of displacement for lowland indigenous communities, development, and environmental change across the Amazon and the rest of the South American interior.

Postcolonial and midcentury state-building projects in Bolivia and elsewhere were also intertwined with Green Revolution science in which U.S. technical funding promised to help nations convert marginal lands into breadbaskets to feed growing populations.[11] The spread of crops and production technologies was part of an emerging geopolitics that drove waves of investment in selected Cold War battlegrounds. The corresponding growth of an international development industry produced yet another set of migrants—an emerging transnational class of practitioners—that circulated through Latin America, sub-Saharan Africa, and Southeast Asia. Bolivia's revolutionary leaders, just socialist enough to be alarming to U.S. officials, became the largest per capita recipient of Point Four and Alliance for Progress funding through the 1950s and 1960s as U.S. planners came to view the nation—and Santa Cruz Department in particular—as a perfect laboratory for the implementation and export of rural modernization.[12] The connections between Santa Cruz and places like Brazil, Tanzania, the Ivory Coast, and Indonesia are more than parallel examples. The Cold War logic of development led the Methodist Mission Board to proclaim Bolivia, along with Korea, Malaysian Borneo, and the Belgian Congo as "lands of decision" in the 1950s and principal sites of missionary activity. The globetrotting members of SUNY-Binghamton's Institute for Development Anthropology that prominently included Bolivia's San Julián Project in their 1981 state-of-the-art global evaluation of new land settlement provide another example of the proliferation of experts that physically linked those disparate locations as they passed through Bolivia before translating their skills to similar projects across the global south.[13]

Recasting Bolivia from the Margins

Although the vision of state planners looms large in past and contemporary discussions of development, this is not simply a story of how the center reshaped the margins as the familiar logic of internal colonization might suggest. Far more than its original architects imagined, the reverse was equally true as the frontier came to reorder Bolivia's regional balance and national political order. According to one estimate, 63,738 Andean families—nearly a quarter of a million individuals—had migrated to Santa Cruz and other lowland regions by 1980.[14] Even greater numbers of seasonal laborers and spontaneous settlers went uncounted, while over the next two decades of economic turmoil and neoliberal reform, highland–lowland migration (to city and frontier) increased exponentially.

Among those hundreds of thousands of settlers was none other than the future president of Bolivia. In 2006, less than a half-century after Jaime Bravo had left the nation's Andean heartland to start a new life in the tropics, Juan "Evo" Morales Ayma returned from the Amazon frontier to take his place in La Paz's *palacio quemado* (presidential palace) as the nation's first indigenous leader. His victory may have been improbable—flying in the face of centuries of entrenched racism—but his trajectory will be all too familiar to readers of this book. Born in an impoverished region of the highland department of Oruro, a major sending region for indigenous settlers, he had traveled, like so many other Aymara, to Argentina in the 1960s where his father worked as a seasonal laborer in the sugar cane fields. Facing extreme drought back home in the highlands in the early 1980s, the Morales family had again relocated, joining tens of thousands of settlers in expanding colonization zones in the Amazon basin portions of La Paz and Cochabamba departments. It was in the tropical Chapare region of Cochabamba, whose colonies were first established by the MNR and the International Development Bank in the 1960s, that Morales rose to prominence as a union organizer for the region's coca growers before ascending to the national political scene during protests over water and gas privatization in the early 2000s.

Indigeneity figured prominently in Morales's 2006 inaugural address, but he opened with a striking paean to his personal mobility. "I salute the place where I came from, Orinoca [in the highland department of Oruro]," he began, ". . . [and] the Federation of the Tropics of Cochabamba . . . which is my place of birth in the union fight and in the political fight . . . these two lands taught me about life."[15] Evoking the windswept treeless altiplano at nearly 4,000 meters and the lush forests of the Amazon at less than 300, Morales embraced his dual identity as highlander and lowlander. His words were reminiscent of propaganda films produced by the MNR in the 1950s to promote the March to the East, in which personal migrations bound together Andean and Amazonian Bolivia. His address also echoed the founding manifesto of the National Federation of Colonizers. At that raucous 1971 meeting in La Paz, delegates from across the lowlands proclaimed the colonist to be the vanguard of the revolution because, "the act of migration has conditioned important cultural changes that gave them an accelerated awareness of the Bolivian drama."[16]

Morales's speech may have offered a comforting narrative resolution to Bolivia's bifurcated territorial identity but ironically—as president—he faced an immediate challenge from the nation's lowland elite centered in the city of Santa Cruz de la Sierra. After half a century of state driven integration, *cruceños* (as residents of Santa Cruz are known) met Morales's election with calls for regional autonomy. Their demands were often framed in explicitly racialized terms separating *kollas*—indigenous highlanders like Morales—from *cambas*—ostensibly "whiter" *mestizo* lowlanders. The 2006–8 autonomy movement—discussed further in the epilogue to this book—figured prominently in international reporting and academic studies of Santa Cruz over the last decade. Yet, as with Morales's unlikely trajectory from settler to president, it was also one more indication of how much the March to the East had reordered Bolivia from the margins. On the eve of the 1952 revolution, Santa Cruz was a dusty frontier town of 40,000 with no paved streets nor all-weather roads to the nation's highlands. By the time of Morales's victory, the city—buoyed by a steady stream of internal migration from the Andes and connected to national and global markets by road, rail, and air—had surpassed highland centers like La Paz and Cochabamba to become the largest and wealthiest urban area in the country. That year a global organization of city mayors ranked Santa Cruz as the fourteenth fastest growing city in the world and second fastest in the Americas.[17] Notably, agricultural service centers in the region's settler zones were expanding at an even more rapid pace. National agricultural censuses show that cultivated land in Santa Cruz jumped fivefold from 1950 to 1984 and more than that from 1984 to 2013—expanding from roughly 60,000 to 300,000 to nearly 1.7 million hectares of farmed land.[18]

The emergence of Santa Cruz—*lowland ascendency* as it has been termed—was a novel phenomenon for Latin America where, despite twentieth-century regional growth in many countries, economic power and demographic dominance remained with highland and coastal capitals established in the pre-Colombian, colonial, and republican eras.[19] Like a twenty-first century variant of gilded-age Chicago, the upstart Santa Cruz had truly become *nature's metropolis*, mobilizing the latent wealth of the surrounding soil and subsoil to sustain its unprecedented growth. The refining of coca and natural gas are well-known elements in this narrative. But critically, the city's expanding rural hinterland had also given way to dozens of settler colonies and agroindustrial operations as fields of sugar cane, corn, rice, sesame, peanuts, sunflowers, and soybeans as well as pastures teeming with Zebu and Holstein cattle covered much of its once-forested plains. The former frontier had become the center of national wealth.

A (Trans)National Revolution

As Tigner and Dueck's explanations betray, what could be written as a national project of internal colonization or a regional story of rapid development, was, in practice, a remarkably transnational affair. In the immediate wake of the 1952 revolution, pacifist Mennonites, an Anabaptist group with a centuries-long history of migration and frontier farming were wandering the halls of newly created government ministries petitioning for rights to settle the frontier. With official legal sanction protecting their peculiar habits and customs, they would flock to Bolivia from Canada, Paraguay, Belize, and Mexico.[20] In the wake of Tigner's report, several thousand Okinawan and Japanese colonists also arrived on Bolivia's tropical frontier, while, like Morales, many indigenous Andeans came to Santa Cruz after years spent as migrant laborers in neighboring Argentina. The transnational underpinnings of local agroenvironmental change were clear to U.S. consul William Dietrichs who was stationed in the city in the early 1970s. In his words, "meeting the daily plane from La Paz at the Santa Cruz airport was an experience in diversity, although I don't think we used that word yet." Remembering difference through dress, he recalls that, "on a good day" alongside "Santa Cruz natives in *guayaberas* and sport shirts" he would encounter "groups of highland Indians in their bowler hats and ponchos . . . Japanese with a young girl in a kimono carrying a bouquet of flowers [and] overalled, poke-bonneted Mennonites."[21]

In bringing the stories of Andeans, Mennonites, and Okinawans into a single narrative, *Landscape of Migration* departs from the bulk of Latin American scholarship on migration that typically follows one migrant group (Italians in Argentina, Germans in Brazil) in isolation from others while treating internal migration and immigration as separate objects of study. A similar division exists within the Bolivian case, for which historian Royden Loewen and anthropologists Taku Suzuki, Carolyn Stearman, and Lesley Gill provide compelling but individualized discussions of Mennonite, Okinawan, or Andean migrants where the presence of neighboring settler communities is reduced to footnotes.[22] Conversely, as Julian Lim, Elliot Young, and Lara Putnam have shown, borderlands and boomtowns throughout the Americas were points of overlap for disparate migrant routes.[23] Such connections demand transnational histories of the frontier that attend to cross-cultural and multiracial comparisons.[24] It is in this process of forging these comparisons while moving between the local, national, and transnational spaces created by migration that a new borderlands history exposes the contested nature of foreignness and belonging, and the relational ironies of inclusion and exclusion.

In following this course, I also seek to push the boundaries of traditional Bolivian historiography which, as if mirroring the nation's landlocked geography, has largely eschewed transnational narratives. This is particularly true of migration history, which is seen as something foreign to Bolivia and more relevant to major receiving societies such as Brazil and Argentina. Yet as Jürgen Buchenau points out, while Argentina and Brazil are often treated as central to the history of Latin American immigration, their experiences are not paradigmatic but in fact atypical for a region that generally did not receive massive numbers of immigrants.[25] In nations of relatively low immigration like Mexico and Bolivia, immigrants remain important objects of study particularly when they occupied key industries, economic niches or remote frontiers as both Mennonites and Okinawans came to do. Extensive public debates and policies related to migration (even small-scale immigration and "immigration that never was") also cast implicit understandings of race, citizenship, and national identity into high relief.[26] Furthermore, if we broaden our framework we can recognize that, in the second half of the twentieth century, intense mobility defined nations like Bolivia that missed out on the earlier era of mass migration. Rapid urbanization, colonization, seasonal migration, and emigration took place on an unprecedented scale. In the late 1960s, anywhere between 400,000 and 700,000 Bolivians (an astounding 10 to 20 percent of Bolivia's total population) lived in Argentina where they worked as laborers in the harvest as well as in urban occupations.[27] Internal migration also reshaped Bolivia. These processes are most evident at the extremes as they simultaneously drew indigenous migrants to the highest and lowest points of the nation, the sprawling high plains suburbs of El Alto sitting on the valley rim overlooking La Paz, and the frontiers of the semitropical lowland department of Santa Cruz.[28]

In addressing the long legacies of Bolivia's 1952 revolution I find good company in a vibrant new scholarship that explores revolutionary citizenship through land reform, cultural policy, resource allocation, and public health.[29] Yet such discussion, like much of Bolivian historiography, largely focuses on events in the Andes at the expense of lowland history and eschews migration. I engage with this rich highland-focused literature by taking it in an unexplored direction. How did notions of citizenship emerge in the process of lowland colonization in Bolivia after 1952? This approach draws on provocative work in other national contexts. Heidi Tinsman's study of the Chilean

Agrarian Reform explores its relationship to gendered notions of citizenship, a process by which the state sought to create *new men* and in which, just as in Bolivia, family stability and masculine honor emerged as privileged dynamics.[30] As in Allende's Chile, Bolivians who took part in settlement schemes experienced them as a form of citizenship continually deferred by a government that withheld clear title to the land they worked.[31] I complicate that question by juxtaposing the struggles of recent citizens (indigenous Andeans) with the noncitizens (Mennonites and Okinawans) they settled alongside in what I argue, at least in the lowlands, was an increasingly (trans)national revolution.

Bringing internal migration and immigration, often treated as separate phenomena, into the same frame of analysis sharpens the contradictions of deferred citizenship for Bolivia's indigenous majority while revealing surprising parallels in migrant strategies for engaging the state, local actors, and the land. Whether foreign or national, all three groups occupied a noncitizen status—albeit with different implications. Indigenous Bolivians won full legal citizenship rights through the 1952 revolution but were still seen as social and cultural outsiders by revolutionary leaders who held that the primary, and paternalistic, goal of the revolution was to "make a citizen out of the Indian." Suspect citizens at home, they became full-fledged noncitizens once again as they crossed the border to work as cane cutters in Argentina. Like indigenous Andeans, Okinawans had long been treated as culturally deficient citizens in the Japanese empire, by turns subject to economic exploitation and cultural assimilation. They became legal noncitizens in the postwar period when the United States took sovereignty of Okinawa from Japan without extending citizenship to its new wards. Mennonites, while possessing a plethora of passports (Canadian, Mexican, Paraguayan, Belizean) remained noncitizens by choice and law, having secured state exemptions that freed them from participation in many crucibles of modern citizenship from the classroom to the barracks.

Environmental History and Migration

Faced with a tenuous claim to legal and cultural citizenship, Andeans, Okinawans, and Mennonites each invoked a form of agrarian citizenship based on their role as transformers of a tropical frontier and producers for region and nation. This varied discourse, and the practices it entails, sheds critical light on the tight relationship between the twin subjects of this narrative: environmental history and migration. As Marco Armiero and Richard Tucker point out, with a few notable exceptions, "environmental historians have not been significantly active in studying the history of mass migration, nor have the historians of migration ever been interested in the environment."[32] Slowly, a new wave of agroenvironmental historians has begun to address this gap.[33] Carol MacLennan makes this case for Hawaii whose "environment bears the imprint of the successive waves of migrants and their agricultural pursuits," and Eunice Nodari has done so for settlers in South America's Atlantic Forest, while Linda Nash has undertaken this layered approach to California's Central Valley—situating human and nonhuman migrant actors in the creation of a "hybrid landscape."[34] In linking *belonging*, a core concept for migration historians, to the production of migrant landscapes through agrarian citizenship, this book charts a similar path.

What did agrarian citizenship look like in practice? The substance of settler claims could range from the telluric, describing patriotic communities rooted in the soil, to the technological—in which new machinery, improved seed, and stock were paramount. Agrarian citizenship was also about performance. Mennonite delegations took regular trips to La Paz to negotiate with government officials while emphasizing their nonthreatening religious identity as the "quiet in the land." Japanese and Okinawans might make symbolic gifts of colony rice to the Mayor of Santa Cruz while holding periodic cultural events displaying their agricultural prowess to which they pointedly invited Bolivian officials. Andeans, for their part, could engage in similarly comforting displays and official commemorations or break from those scripts altogether. By the late 1960s, they would take public officials hostage, blockade roads, and occupy the plazas of Santa Cruz and regional centers as part of radical assertions for agrarian rights guaranteed by the revolution.

Under the logic of agrarian citizenship, Mennonites, Okinawans, and Andeans secured space by claiming to be "feeding the nation" or rendering vacant or abandoned lands productive. This resonated with nineteenth century physiocracy—the belief that all wealth is rooted in agriculture—as well as the slogan of twentieth-century Latin American agrarian reform, "the land belongs to those who work it," which, under Bolivian law applied to foreigners and nationals alike. The assertion took on new meaning in the post-1952 era when Bolivian officials targeted food security as a major policy objective in line with other developing nations. Like dwarf-wheat farmers in India's Punjab state or growers of IR8 miracle Rice on Luzon in the Philippines, migrant farmers in Santa Cruz cast themselves as central actors in an emerging global Green Revolution that would sustain rapidly growing populations. Such confident positioning often jarred with the reality of frontier settlement.

As Eunice Nodari documents in an earlier wave of colonization in Brazil, migrant-driven environmental change in the South American interior was as tenuous as it was profound. Settlers experienced new landscapes as chaotic, failed to produce favored foods, misread ecological conditions, and often developed production strategies that were economically and environmentally unstable.[35] Successive agricultural boom and bust cycles and widespread landscape alteration are testament to similar dynamics in Bolivia's March to the East.

While making a common claim to agrarian citizenship embraced by the nation-state, the experiences of Mennonites, Okinawans, and Andeans often fractured along racial and ethnic lines. In the Bolivian lowlands, expectations of foreignness and belonging were complicated by a pronounced regional identity separating lowlander (camba) and highlander (kolla). Racial, environmental, and regional identities intermingled in unexpected ways. In one of the more surprising ironies of the March to the East, lowland elites ultimately accepted Mennonite and Okinawan immigrants as "our colonists" while denouncing their fellow Bolivian farmers from the nation's highlands as "foreign invaders." Ultimately, crossing a racialized environmental barrier separating highland and lowland proved more difficult for Andeans than traversing international boundaries for Okinawans and Mennonites.

In addition to troubling understandings of foreignness and belonging, exploring agrarian citizenship within a comparative frame complicates our expectations about tradition and modernity. Mexican Mennonites like Martin Dueck came to Bolivia intent on maintaining old ways represented by steel-wheeled tractors and horse-drawn buggies but soon came to exemplify the modern, market-oriented, small family farms the revolutionary government had envisioned for its own citizenry. Yet even as their demographic, economic, and environmental impact grew, Mennonites remained cultural outsiders for most Bolivians who were often unsure of how to square the prodigious production and conspicuous privileges of these traditionalists with national narratives of modernization. In contrast, Japanese and Okinawan settlers encountered outspoken xenophobia when they first arrived in Bolivia in the 1950s but ultimately combined agroindustrial expertise with prominent public roles in the political and economic life of the nation to mitigate hostility. Even more surprisingly, they recast their threatening foreignness as a critical asset for a globalizing regional economy eager for development assistance from Japan.[36]

The aims of the March to the East, and the rhetoric of agrarian citizenship, underwent a profound change in the latter decades of the twentieth century with the arrival of a new transnational migrant. As tropical forest gave way to farmland, settlers experimented with transplanted crops and new livestock breeds while Santa Cruz's nascent agroindustry became part of a global Green Revolution. Among these unfamiliar flora and fauna, none had a greater impact than the soybean. Transplanted from Asia to the United States and then again to Brazil, soybeans were adapted for a semitropical climate. Along the way their meaning changed, profoundly reshaping the economy and ecology of the South American interior in the process. A food crop in Asia became a versatile agroindustrial cash crop in the Americas. By the time agrochemical behemoth Syngenta was boasting that soybeans "did not recognize borders" in 2003, a broad swath of newly cleared lands, stretching from western Brazil, into eastern Paraguay, Uruguay, Argentina, and eastern Bolivia produced the majority of the world's soy. The rapid conversion of more than 60 million hectares of land into a vast transborder *soyscape* constitutes one of the most significant agroenvironmental changes in Latin American history. Yet the soy boom has only slowly crept into academic writing on South America. Soy expansion provides a speculative conclusion to Greg Grandin's earlier history of Brazilian Amazonian rubber development and exists as a menacing specter, literally consuming the region's past, in Gastón Gordillo's exploration of ruins and placemaking in the Argentine Chaco.[37] Yet half a century removed from its origins in Brazil, the history of South American soy is meager in comparison to the exhaustive literature on earlier Latin American agroexport booms in sugar, coffee, or bananas. The existing body of work understandably tends to focus on disembodied corporate capital (Archer Daniels Midland, Bunge, Monsanto) and technology rather than the cultures of agrarian production that pioneered and sustained the boom in distinct local contexts.[38]

In Bolivia, soy first emerged as a star nontraditional export crop as the country's traditional mineral exports reeled in the face of economic crisis and neoliberal reform at the tail end of the twentieth century. The transition seemed fitting, nearly poetic, given the MNR's long-standing goal of balancing the mineral dependency of an extractive Andean state with a cultivated or agrarian state in the east. Indeed, after two decades of military and authoritarian rule, the MNR's original leaders (first Hernán Siles Zuazo and later Víctor Paz Estenssoro) returned to power and were personally on hand to oversee this transition. Something strange happened along the way. As soy supplanted tin, farming came to look a lot more like mining. Chemical dependent and highly mechanized, the process of soy farming marked a serious break from earlier and continuing modes of agrarian production such as corn, rice, and even commercial sugar cane. Soy's low labor requirements have come to constitute what anthropologist Ben McKay and Gonzalo Colque refer to as "productive exclusion."[39] By 2013, fewer than 2,500 producers farmed two-thirds of the cultivated land in Santa Cruz on large farms of more than 1,000 hectares while representing 2 percent of the department's 113,000 producers.[40] Economically, as soy's physical and geographical footprint has expanded in the twenty-first century, soy has joined expanding gas and struggling mineral production in what anthropologists Nicole Fabricant and Bret Gustafson characterize as a form of "neo-extractivism."[41] The rhetoric of agrarian citizenship shifted accordingly. Although some small farmers could, and still did, claim to be feeding the nation, many, particularly expanding Mennonite populations producing soy on relatively small farms of 50–100 hectares, could also now assert their role as mechanized cash croppers and producers for a growing export economy. This new discourse was not without its discontents. One eager Bolivian *soyero* (soy farmer), may have fondly described it as "the best harvest that I know, clean, easy, good" and most tellingly, "almost done without labor." But other informants rejected soy agriculture as the antithesis of agrarianism, decrying it as little more than mining the soil, a new form of pulling wealth from the earth.[42]

Visions of Territoriality and State Sovereignty in a Truncated Nation

Although this book focuses on the period after 1952, plans, dreams, and hopes for the development and colonization of lowland Bolivia predate the national revolution. For decades, commentators decried the abandonment of a region of great potential they claimed had been neglected or ignored by the state. In doing so they forged a frontier discourse that finds parallels with other regions and time periods.[43] In their writings, classic liberal theories of territorial integrity and state sovereignty merged with the unique geographic and political realities of late nineteenth- and early twentieth-century Bolivia. It is more than coincidence that many of these commentators were foreigners. In Bolivia as elsewhere, national integration relied on many non-nationals who—as surveyors, planners, missionaries, and public health officials—served as proxies for an absent state.[44]

"I shall never forget the impression produced on my mind by my introduction to Eastern Bolivia," observed geographer J. B. Minchin in his 1881 report to the Royal Geographical Society in London.[45] Looking down from the imposing heights of the Andes onto the seemingly endless forested expanse of unbroken lowlands, he experienced the allure and promise of one gazing upon the "the heart of the continent." Like explorer James Orton, who traveled through the region several decades earlier, Minchin agreed that "only the margin of the continent is developed" while the core was "as wild and prolific as ever."[46] In contrast to the densely populated Andean mining regions of the west, the vast lowland landscape appeared a barren zone to Minchin. Yet he acknowledged the incredible potential that existed along this eastern Andean frontier. Imagining the expanding Río de la Plata nations across the Chaco, he speculated, "What may we not expect when her vast interior becomes opened up, and when the tide of immigration, spreading over and utilizing her boundless pampas and inexhaustible forests reaches the Andes?"[47] Minchin, a foreigner, worked previously as a proxy for the Bolivian state demarcating its contested boundary with Brazil. He understood the problem these fugitive landscapes posed for state sovereignty in gendered terms.[48] It would only be once Bolivia had penetrated the Chaco and transformed its idle potential into productive use that the nation's cartographic fictions would become meaningful claims and that Bolivia could "take the place that is expected of her in the scale of nations."[49]

Nearly half a century later, a beleaguered party arrived in Santa Cruz de la Sierra. The group, led by Bolivian diplomat and future president Mamerto Urriolagoitía (1949–51) included British writer Julian Duguid, cinematographer John Charles Mason, and Russian "tiger-hunter" Alexander Siemel. As expedition leader, Urriolagoitía had recruited the three foreigners to assist him in making an official commercial inspection of the eastern lowlands on behalf of the Bolivian government. Beginning in Buenos Aires with a 3,000 kilometer voyage by river steamer up the Río de la Plata and Paraguay River, the journey took several months. At La Gaiba Lake on the Bolivian-Brazilian border the party disembarked and, on foot, completed the nearly 600 kilometer crossing of the densely forested eastern lowlands to Santa Cruz de la Sierra. "Bolivia is in a curious state of development," reflected Duguid, the British member of the team.[50] Although the lack of modern transport crossing the eastern frontier was hardly surprising, he marveled at the complete absence of permanent roads between "the forests of the east and the plateau of the west." After a few days in Santa Cruz, the party ascended a distance into the nearby mountains for an expansive view of this ecotone—where Andean, Amazonian, and Chaco landscapes merged—not unlike the vantage enjoyed by Minchin a half-century earlier. Looking over the land that Duguid would popularize as a *green hell* in his 1931 travel memoir, Urriolagoitía admitted his disappointment. The former Jesuit missions they had passed through appeared "dead," "dusty," and "decayed" and the great eastern forest contained, in his opinion, little more than "a few tortoises and mudholes."[51] Attempting to offer words of encouragement, Duguid reminded the despondent Urriola-goitía, that "Argentina was like this once." Gesturing at the plains, Siemel agreed, suggesting that, "in fifty years . . . this may well be populated."[52]

The hopes of Minchin and Duguid would find a belated realization in the post-1952 era, but the intervening years were defined by territorial loss rather than frontier development. Bolivia had already ceded portions of its eastern and northern frontier to Brazil when Minchin surveyed the Santa Cruz plains and would lose another territory (Acre) during the turn of the century Amazon rubber boom. The country's territorial integrity was also challenged from the west. As Minchin looked east in 1880, Bolivian forces were suffering their final defeat before Chilean forces at the battle of Tacna. The War of the Pacific (1878–83) would cost Bolivia its nitrate and copper-rich Pacific coast. With the loss, Bolivia's leaders looked east to the Paraguay River. Running along the eastern edge of the Gran Chaco before entering the Paraná and Río de la Plata, this was the route that Urriolagoitía's expedition had taken to reach Santa Cruz in 1930. That same year, Bolivian engineer Miguel Rodríguez was proposing a massive navigable canal that would redirect the Río Grande—which ran past Santa Cruz on its way to the Amazon—into the Paraguay River, giving the city access to the sea. Bolivia's defeat in the Chaco War (1932–35) forestalled that plan, resulting in another territorial loss for the now truncated nation.[53]

For cruceños the Chaco War served a bitter but valuable lesson proving that "the desert, the uncultivated bush, the absence of roads, railways and population [in the east]" were a dire threat to the sovereignty of the nation.[54] Cruceño topographer Constantino Montero Hoyos, completed an exhaustive tour of the lowlands in the post–Chaco War era and insisted that with *vías y brazos* (roads and labor), Santa Cruz would flourish.[55] But until U.S. cultural attaché Erwin Bohan toured Bolivia in 1942, this and similar calls went largely unanswered. Bohan's visit was a product of the U.S. Good Neighbor Policy. U.S. officials had identified Bolivia—one of the world's four largest tin producers—as a key strategic ally in the war against fascism.[56] In his subsequent report on the state of Bolivian national development (the Plan Bohan), he advocated for U.S. funding to build an all-weather highway linking Santa Cruz to the nation's highlands. Additionally, Bolivia independently signed agreements with both Brazil (1938) and Argentina (1945) to build railways that would connect Santa Cruz de la Sierra to their borders.

In 1952, as the MNR's revolutionary coalition toppled Bolivia's mining oligarchy, all three of these projects were nearing completion. In the oft-cited Marxian equation, road and rail promised to "annihilate space with time" at the very moment when the revolutionary government was initiating the first concerted effort to colonize the frontier.[57] Over the following half a century, Bolivia—a small, landlocked Andean nation at the heart of South America—embarked on a dramatic program of self-fashioning along its Amazonian frontier.[58] In contrast to the unrealized hopes of previous decades, waves of infrastructure, directed colonization, and international financing accompanied and supported the post-1952 March to the East.

Sources and Methodology

This book draws from films, letters, pamphlets, diaries, newspapers, oral histories, and the archives of migration authorities, as well as agricultural and land ministries. This varied source base responds to the challenges of studying the frontier and researching migrants. Both frontier and migration history sit in tension with national narratives, and relevant records often slip in and out of national archives. Operating in another Amazonian context, Hugh Raffles has emphasized the importance of drawing from just such a heterogenous source base not only out of necessity but to highlight the conflicting ways in which the frontier was imagined, experienced, and recast by writers, laborers, large-property owners, and government officials.[59] I have similarly drawn from the records and memories of settlers, filmmakers, missionaries, local elites, and government officials who physically and discursively shaped the landscape of Santa Cruz.

Migration historians encourage us to follow the migrants as they cross national borders, forge and sustain diasporic communities, and reorder local environments.[60] This book follows this transnational turn, searching for the "connections—and sometimes the disconnections" which at times has required me to move across borders as well.[61] I explore the history of displacement that brought Okinawan migrants to Bolivia through the files of the United States Civil Administration of the Ryukyu Islands (USCAR) housed in the U.S. National Archives. I trace the origins of large-scale Mennonite migration to Bolivia in archives in Mexico City. National archives in Bolivia also provide a range of information on the cross-border movements of Bolivian braceros to neighboring Argentina, while Mennonite Central Committee and Methodist archives in the U.S. contain detailed records of their activities in Bolivia during a period when national archives are incomplete or altogether missing. The latter two archives also capture the remarkable mobility of a transnational class of faith-based development practitioners (discussed in chapter 4) who often stood in as proxies for the absent Bolivian state in secular services such as public health, agronomy, and education. Their movements in turn linked colonization and development sites across the global south, global faith communities, international financing organizations, and university campuses in the decades after 1952.

Much of the history I document escaped the public archive. Personal collections and oral history figure prominently in the latter two chapters. Because Bolivian Mennonites do not participate in the military, public education, or politics, their archival imprint, in comparison to Andean migrants, is particularly light. Unlike Okinawan and Japanese settlers, Bolivian Mennonites have also not engaged in an extensive documentation of their own history. I was able to uncover aspects of Mennonite history in the archives of the Agrarian Reform and the Ministry of Agriculture, but I also conducted more than thirty interviews with Mennonite settlers. Together these unarchived histories proved critical to situating the large but relatively silent history of Mennonites within the narrative of the March to the East.

This history is comparative in nature, but it is not rigidly so. In tracing initial migrant routes, I focus on Okinawan migration to the exclusion of Japanese migration given the unique trajectory of the former (displacement and resettlement by the U.S. military), whereas in Bolivia, where both groups were frequently conflated by locals, I discuss both Okinawan and Japanese responses to xenophobia. Those two migrant streams never reached the lofty numbers forecast in the 1950s and 1960s. More importantly, Okinawan and Japanese history has been well documented by visiting anthropologists and sociologists as well as by colony members themselves. I thus give greater attention to Andean and Mennonite settlers in the post-1968 era as their presence across the lowlands expanded exponentially while new Japanese and Okinawan migration definitively ground to a halt. Despite this imbalance, I sustain a comparison with Japanese and Okinawan communities throughout. Although few in number, they remain key agroindustrial producers in Santa Cruz, much like the recent small-scale wave of Brazilian expatriate farmers who have settled in the region, and continued to appear as points of comparison in newspaper editorials, government reports, and oral histories concerning Andean and Mennonite settlers.[62] Indirectly their presence has also led to large-scale investment from the Japanese International Cooperation Agency (JICA) in regional development projects.

Organization of the Book

The second half of the twentieth century was a tumultuous period across much of the developing world. Revolutionary regimes gave way to authoritarian ones, midcentury booms were followed by economic crises, state institutions strengthened and then collapsed by the close of the century amid a turn to neoliberalism. Financing from global superpowers—so critical for modernizing cash-strapped nations across the global south—proved highly contingent. Bolivia was no exception to these dizzying trends. Yet one thing has remained remarkably consistent through the twentieth century and into the twenty-first. Whether revolutionary or reactionary, statist or laissez-faire, an unbroken succession of Bolivian governments maintained a fundamental consensus: the future of national development lay in eastern expansion.

This frontier imaginary was highly contested—as settlers, natives of the region, departmental and federal officials, large-scale farmers, and peasant producers voiced competing claims to the land—but it proved remarkably durable. Such developmental continuity amid political discontinuity is hardly surprising to environmental historians who remind us that environmental policy routinely transcends ideological affiliation and structures of governance.[63] Evo Morales's 2005 election, accompanied by a strong environmentalist rhetoric, did little to question this premise. Plans for lowland development continued apace—and even accelerated—under his long-standing regime. While his government was opposed by supporters of lowland autonomy, cruceño concerns were ultimately mollified and racialized resistance diminished—though hardly disappeared—in favor of a grudging acceptance of Morales's pro-lowland stance. Conversely, those same extractivist policies, including support for export-focused agriculture, led to mounting criticisms from indigenous communities that were initially strong supporters of Morales but witnessed firsthand the environmental consequences of accelerated frontier development. These trends, evident in Bolivia over the last decade in protests over road construction, intensified in the wake of large-scale fires in the Bolivian lowlands on the verge of Morales' controversial fourth campaign for office. The embattled president found his environmental policy, compared to his far-right counterpart, Jair Bolsonaro, in Brazil and faced renewed regionalist resistance in Santa Cruz, this time framed in environmental terms largely absent in earlier struggles for regional autonomy.

Bolivia's conflictive but enduring frontier consensus gives shape to *Landscape of Migration* which is divided into five chapters each of which explores the role of an intertwined set of migrants. While the first three chapters are largely focused on the revolutionary period in Bolivia (1952–64), chapters four and five bring the narrative of the March to the East forward in time, respectively covering periods of authoritarian rule and neoliberal transition. The conclusion and epilogue carry us into the plurinational present.

Chapter 1, "Moving Pictures: Narrative, Aesthetic, and Bolivia's Frontier Imaginary," explores the role of images and imagemakers in the imagination of the frontier. I approach the March to the East as a national romance that needed to first be visualized to be enacted. The chapter draws from an impressive corpus of visual and written materials that sought to define the meanings of mobility in human and environmental terms. This includes the work of a different sort of migrant, filmmaker Jorge Ruiz, who traveled throughout the Bolivian lowlands documenting colonization for the Bolivian state and its U.S. sponsors. In his films, Ruiz played with the striking visual distinction between the arid highlands of Bolivia and the tropical lowland jungle. His camera followed the lives of several fictional characters whose personal migrations through unfamiliar landscapes would overcome profound regional differences and unify a fractured national body. Shown in cinemas across the country, Ruiz helped the state consolidate an enduring frontier imaginary that the future of the country lay in the east. Yet residents of Santa Cruz responded to such films in conflicting ways. Lowland elites had long nurtured a discourse that their region had been abandoned and at first eagerly embraced new highways and railways that would link Santa Cruz to the rest of the nation. Yet they harbored a deep-seated fear of the Andean indigenous bodies that would accompany these new forms of mobility. Their cries of abandonment soon shifted to ones of invasion when confronted with a wave of incoming Andean migrants. Ruiz and his images also circulated far beyond Bolivia and I conclude this chapter by following his transnational trajectory from Bolivia to Ecuador, Guatemala, Peru, and back again, in the 1950s and 1960s. Ruiz's success at transplanting his aesthetic repertoire highlights the flow, pervasiveness, and flexibility of midcentury development ideology and the role of film as a powerful vehicle for representing those changes.

In chapter 2, "Military Bases and Rubber Tires: Okinawans and Mennonites at the Margins of Nation, Revolution, and Empire," I weave together the geopolitical and environmental forces—military occupation and midcentury drought—that led the Okinawan and Mennonite diasporas to Santa Cruz. In the first half of this chapter I return to postwar Okinawa where the U.S. military displaced farmers as it constructed bases on expropriated lands across the Ryukyuan archipelago. From political protests and blockades to performances of model agrarian citizenship, Okinawans contested removal. These actions failed to achieve their stated aim yet in an emerging Cold War climate, Okinawan resistance led U.S. officials to seek relocation options. I follow several thousand Okinawan settlers as they moved to Bolivia in the mid-1950s where they often faced racial and nativist attacks from local authorities. As I show, Okinawan colonists employed the same strategy of model agrarian citizenship they had used to contest U.S. removal on the Ryukyuan islands to successfully counter local xenophobia in Santa Cruz. The second half of this chapter begins with the small-scale migration of Paraguayan Mennonites to Bolivia in the mid-1950s. Although scarcely noticed in Santa Cruz, news that Bolivia had extended special exemptions rippled through the broader Mennonite diaspora including in Mexico where a prolonged midcentury drought was devastating farming communities in Chihuahua. In the face of drought many Mexican Mennonites turned to transnational migration. They first traveled north to work as harvesters on Canadian farms but returning to Mexico these braceros brought modern goods and evangelical missionaries back to their traditional colonies. The result was a bitter conflict that centered on the use of rubber tires, rather than steel wheels, on Mennonite tractors and pushed forward an exodus of conservative Mennonites to Bolivia in 1968.

The chapter concludes by placing Okinawan and Mennonite migrations in explicit comparison. Even as a large-scale wave of Mexican Mennonites emerged in the late 1960s, a planned mass migration of Okinawans to Bolivia was dwindling due to the postwar Japanese economic miracle and the return of Okinawan sovereignty to Japan between 1968 and 1972. Okinawans soon eschewed the remote Bolivian jungle in favor of factory labor in Toyota and elsewhere in the Japanese industrial heartland, limiting their presence in Santa Cruz to three small but dynamic colonies. In contrast, with a high birth rate and continued migration, Mexican Mennonites quickly became the largest, and most conspicuous, immigrant farmers in lowland Bolivia over the next half-century (the subject of chapter five).

In chapter 3, "Abandonment Issues: Speaking to the State from the Andes and Amazonia, 1952–1968," I examine the intertwined movement of indigenous letters and bodies in the March to the East. In contrast to visions of development produced by the state—what James Scott has famously termed "seeing like a state,"—I privilege ways of *speaking* to one. Andeans actively articulated their own imaginings of the March to the East apparent in an array of letters written by individuals and groups who demanded to take part in colonization in the 1950s and then denounced its shortcomings in the following decade. I follow their petitions as they traveled from highland hamlets and humid settlement zones to the halls of government and the office of the president. Letters produced in the Andes in the 1950s and 1960s paired a provocative portrait of desperate situations in their home communities with the promise and allure of the tropical environment of the lowlands. Writers attempted to shame the state by emphasizing their struggles as migrant laborers or braceros in neighboring Argentina and demanded land as part of the state's commitments to its own revolutionary legacy. Along the lowland frontier, the reality of colonization often failed to match the harmonious human experiment depicted in state propaganda. Government officials blamed a high rate of settler abandonment in new colonization zones on the "backwards" cultural practices of indigenous migrants, but settlers flung this accusation back on the state. In a litany of denunciations that streamed into government offices, they claimed that it was the MNR that had abandoned them by failing to provide even the most basic of services. In this chapter failure looms large—as it does in many critiques of development modernization. However, for both planners and colonists, failure served as impetus for new forms of intervention (on the part of the state) and radicalism (on the part of settlers) as colonization increased over the following decades.

Chapter 4, "To Minister or Administer: Faith and Frontier Development in Revolutionary and Authoritarian Bolivia, 1952–1982," inserts a new migrant actor in this state–settler dynamic. Across the global south, missionary and religious organizations became active intermediaries and state proxies in "secular" modernization projects. In Bolivia, Protestants flocked to colonization zones at the invitation of the MNR. Some, like the decidedly evangelical World Gospel Mission, aggressively sought "uncontacted" tribes along the margins of new settlement zones while the Methodist Mission Board and the Mennonite Central Committee (a North American relief agency), began working extensively with new colonists. Those latter two organizations soon made Bolivia a center of their global operations and were ideally positioned to expand their role in the face of intertwined environmental and political disasters on which this chapter pivots. The first was a massive flood that devastated colonization zones in the lowlands in 1968. The Methodist Church and the MCC joined several Maryknoll nuns in an improvised United Church Committee (CIU) to support the resettlement of flood victims.[64] This ad hoc response soon transitioned into a new model of settler "orientation" and in 1970 the CIU contracted with the National Institute of Colonization to administer San Julián—the largest colonization program in Bolivian history. The second disaster took place the following year when General Hugo Banzer staged a military coup definitively ending Bolivia's revolutionary period and ushering in a period of authoritarian right-wing rule. The CIU weathered this transition and over the following decade, these religious actors became key "go-betweens," able to work on the ground with colonists, gain the confidence of Banzer, and channel international funding from North American religious and secular sources. During that time, San Julián also attracted a range of academics and planners, who were drawn to its unique orientation program and its radical spatial design. Organizations like the Institute for Development Anthropology and individuals like Thayer Scudder helped place San Julián in dialogue with African "villagization" and the resettlement of "displaced peoples" in South Asia. I follow the trajectories of these mobile actors who translated their work in Bolivia to new roles with international agencies and NGOs across the global south as they crossed boundaries separating the revolutionary and the authoritarian, the secular and the sacred, and the frontier and the academy.

Chapter 5, "A Sort of Backwoods Guerrilla Warfare: Mexican Mennonites and the South American Soy Boom," explores the intertwined migration and expansion of two temperate zone transplants—Mennonites and soybeans—in semitropical Santa Cruz. The transnational history of Bolivian Mennonites offers several compelling and interrelated ironies that drive home the paradox of national development in lowland Bolivia. A revolutionary nation-state that sought to use colonization to transform traditional indigenous subjects into citizens welcomed foreign horse-and-buggy Mennonites and granted them special exemptions that explicitly freed them from the central domains of modern citizenship (the barracks, the ballot box, the classroom). Seeking to develop modern, market-oriented agribusiness on its eastern frontier, the MNR invited a communitarian, traditionalist agricultural community that shunned a wide range of technological innovations. Yet, surprisingly, horse-and-buggy Mexican Mennonites emerged over the following fifty years as exactly the sort of model, mechanized farmers the Bolivian state hoped to create of its own citizenry. Their success in negotiating with the state and transforming the land rested on a further paradox: the highly mobile nature of a handful of Mennonite go-betweens. Moving between Mexico, Paraguay, Argentina, Bolivia, and the United States, enterprising Mennonite traders established a thriving dairy industry with imported Holstein cattle and farmed land and cleared bush with secondhand U.S. machinery. These nonhuman transplants underpinned Mennonite success in an unfamiliar environment as select Mennonites allowed a colony that appeared closed, agrarian, and immobile to rapidly expand the frontier.

In the second half of this chapter I situate Mennonites at the center of the defining agroenvironmental change of late-twentieth-century South America: the dramatic expansion of soybean production that has converted the forested heart of the continent into the world's preeminent soy region. Soybean production emerged from a confluence of regional and global factors—from agroindustrial technology in Brazil and the collapse of fish stocks in Peru to personal relationships between Mennonites and an immigrant entrepreneur in Santa Cruz. Ironically, soybeans boomed just as Bolivia and the rest of Latin America was entering a profound economic crisis. I conclude by exploring the expansion of Mennonite settlement in the turn to neoliberalism in the 1990s. By then, the logic of the March to the East had definitively shifted from national self-sufficiency to the export of profitable cash crops. Mennonites stood at the center of this neoextractivism even as they continued to produce dairy within an earlier logic of food security.

The Epilogue of this book extends the history of the March to the East to the present. I return to the personal history of Bolivian President Evo Morales—who like Mennonite and Okinawan settlers followed a transnational route (through Argentine cane fields) to become a lowland colonist. I link his personal trajectory in the March to the East to his administration's plans to extend the agricultural frontier. I also examine the ways that transnational and regional dynamics continue to unfold in this national state-building project. Just as ideas of abandonment provided a key framing narrative for the body of this work, conflicting notions of autonomy help us understand Santa Cruz at the beginning of the twenty-first century. During the well-publicized autonomy movement of 2008, residents of Santa Cruz challenged state authority emanating from the Andes and lashed out at the visible presence of highland indigenous migrants. This occurred even as lowland indigenous peoples voiced a very different set of demands for autonomy. Long silenced in the March to the East, the Guaraní, Chiquitano, Sirionó, Ayoreo, and other indigenous communities recast the narrative of settlement as one of displacement and organized to demand the return of their traditional lands. In the process they also often described indigenous Andean settlers as "invaders," while surprisingly pointing to forms of legal autonomy exercised by neighboring Mennonites as a model for maintaining their own customs, language, and land.

CHAPTER ONE

Moving Pictures

Narrative, Aesthetic, and Bolivia's Frontier Imaginary

In the highlands I was searching and searching, but I found what
I was looking for in the east.

—Santos, the miner-turned-colonist and protagonist of the 1955 film
A Little Bit of Economic Diversification

Jorge Ruiz's 1955 film, *A Little Bit of Economic Diversification*, opens with a scene
that appears quintessentially Bolivian. A whistle pierces the air in a highland
mining town surrounded by the craggy peaks of the Andes. Workers emerge
from shacks and file into the mine to begin the dangerous task of drilling, laying charges, and hauling material. At the mine's office a letter awaits a foreman. It is from his brother-in-law Santos, a former coworker who had left the
mines to travel to the country's eastern lowland frontier. The other miners,
eager for news of their compatriot, gather around. One man criticizes Santos's decision to abandon his work in gendered terms. Another argues that,
"No, Santos was *macho* too," affirming that the hard life in the mines is simply
not for everyone. Turning to their foreman they encourage him to read the
letter aloud, a request that precipitates a series of flashbacks tracing Santos's
journey.

The scope of this coming-of-age story is both personal and national. It begins in the back of an open truck as the letter's author passes under the com-
memorative arch of the first all-weather road to connect the nation's highlands
with the rapidly expanding lowland frontier department of Santa Cruz. The
new highway, the product of "Bolivian force and Yankee money," notes Santos,
is transforming the country in its path. He celebrates *macho* road construction crews and his gaze is drawn to dams and irrigation projects alongside the
highway. As the truck struggles over one final, foggy, highland pass before
beginning the dramatic descent into the lowlands, the tires themselves seem
to say to him "go, go, never come back."

Arriving in Santa Cruz de la Sierra, Santos is overwhelmed by the heat, the
humidity, the chaotic muddy streets, and the mountains of unfamiliar tropical fruit. A stranger in this hedonistic city, his first encounter is with a pair of
gregarious, hard-drinking *cambas* (lowlanders). Masculinity is once again at

play. Although he is unaccustomed to drink, he writes of a need to "stand up for *kolla* [highlanders] honor." Soon he is intoxicated, penniless, and alone. The city clearly has nothing to offer the highland migrant, but Santos soon discovers a booming agricultural center a short distance to the north. In the town of Montero at the terminus of the new highway, he returns to cataloging the region's many transformations including endless cane fields and a cathedral-like modern sugar refinery. "Let all our friends come" he writes to his brother-in-law. "There is land for everyone and our government will help." In this landscape of possibility, Santos falls in love with a local camba woman named Pilar. With financing and technical support courtesy of the U.S.-funded Inter-American Agricultural Service (SAI), the two receive their own parcel of land. In a final flashback, a mature Santos is the proud father of what could be described as Bolivia's new *mestizo*. Moving beyond the colonial designation of the offspring of indigenous and European, their child represents the inter-regional union of a masculine race and a feminine environment, or as Santos bluntly puts it, "kolla blood and camba land."

Despite its innocuous title and light-hearted script, *A Little Bit of Economic Diversification* made a radical claim. Personal migrations would bind together a bifurcated nation and bridge racial, environmental, and economic divides separating Andean and Amazonian Bolivia. These were the ideas at the heart of Bolivia's revolutionary March to the East, and filmmaker Jorge Ruiz played an active role in representing that process. He was tasked with creating documentaries depicting infrastructure, colonization, health, and sanitation efforts for the state's Instituto Cinematográfico Boliviano (ICB) and Bolivia's U.S. sponsors. In a country with a tiny film industry, Ruiz was one of few well positioned to do so. Recognizing documentary filmmaking as an effective didactic medium while studying agronomy in northern Argentina in the early 1940s, he made his first films with support from Kenneth Wasson, the U.S. embassy's cultural attaché and had produced the Aymara-Spanish documentary *Bolivia Seeks the Truth* about the 1950 census prior to the revolution.[1] In films like *A Little Bit of Economic Diversification*, Ruiz utilized Bolivia's divergent landscapes and the striking visual distinction between west and east in the service of his migrant narratives. Juxtaposing the contrasting environmental aesthetics of sweeping mountain vistas and arid high plains with dense jungle and sinuous rivers, Ruiz presented the highlands as a place mired in a hopeless past while depicting the lowlands as a transformative space of limitless possibility. Like other midcentury Latin American filmmakers, he melded drama and documentary, his camera capturing the movements of fictional teachers, roughnecks, and ex-miners. These characters' personal migrations and

Scenes from Jorge Ruiz's 1955 film, *Un poquito de diversificación económica*. Santos' former coworkers in the mines (left) read of his new life and family in the tropics (right).

boundary-crossing romances overcome profound regional differences and thus unify a fractured national body.

Ruiz's films emerged alongside an eclectic body of images and narratives that sought to give structure and meaning to the unprecedented mobility that characterized midcentury Bolivia. These include government reports and the Bolivian Congress's annual, florid "Homage to Santa Cruz," editorials in Santa Cruz's *El Deber* newspaper and crude pamphlets produced for potential migrants. Those materials reveal the visual and narrative repertoire that accompanied Bolivia's attempt to reimagine itself in its eastern lowlands. Along with images that present the March to the East as a national project are regional understandings of mobility that both mesh with and challenge this vision. Moving beyond Bolivia reveals a transnational orbit in which such images also circulated. Ruiz and others imagined happy resolutions to problems ranging from technical obstacles to regional hostilities, land tenure, and cultural change. In practice these tensions were often exacerbated rather than overcome in the process of eastward expansion. The meanings of mobility, so clear in some films, were often contested by their imagined subjects (and viewers) in the lowlands. The image of the east as a land of unlimited promise may seem simplistic, but it remained an enduring ideal for many of the hundreds of thousands of migrants that entered Santa Cruz over the following half century, and continues to shape government policy today. It is thus crucial to understand the formation of this frontier imaginary in the postrevolutionary period.[2]

Although these materials spoke to a particular Bolivian reality, they find predecessors and contemporaries among of the wealth of literary and visual material that accompanied neighboring Brazil's frontier mythmaking. As early as the 1920s Brazilian director Silvino Santos produced films like "*No paiz das amazonas*," depicting natural wealth and nascent modernization in the Amazon. By the late 1930s and early 1940s, as the March to the West became state policy, Brazilian periodicals like *Revista Oeste* (West Magazine) would support this symbolic construction of the frontier's importance by linking demographic considerations, border anxieties, and developmental imperatives to calls for Brazilians to become modern-day frontier *bandeirantes* (seventeenth-century settlers, slavers, and fortune seekers).[3] In Brazil, as in Bolivia, this relatively limited literary representation of frontier development would give way to a plethora of popular imaginings in films and pamphlets as the March to the West grew in scope.[4]

Filming a National Romance

Ruiz's presentation of the country's vast lowlands as a transformative space for both individual and nation gave visual representation to revolutionary policy. Those ideas were at the heart of the *Plan inmediato de politica economica del gobierno de la revolución nacional* written by a prominent intellectual in the MNR, Wálter Guevara Arze, in 1953. His proposal, directed at fellow party members and U.S. funders, encouraged self-sufficiency and reduction of imports through the development of transport corridors and large-scale, mechanized farming. Succinctly expressing the logic of import-substitution, Guevara Arze noted that Bolivia derived nearly all its export income from tin while spending two-thirds of those earnings importing food. Bolivia's challenge, he argued, was to "produce more minerals at lower cost when possible and stop importing what the country can produce itself."[5] Guevara Arze combined this economic argument, which would emerge as economic gospel for midcentury Latin American nations, with a spatial sensibility unique to Bolivia. "The majority of the population of Bolivia lives in the departments that encompass the Altiplano, the Cordillera and the valleys whose extension does not even reach 33% of [national territory]. In this space, constituted in its majority by sterile and poor land, lives something more than 72% of the population of Bolivia. The lowlands, semitropical and tropical, plains that are generally apt for agriculture and reach 67% of the total extension [but] do not even contain 28% of the population."[6] Population transfer would not only balance the nation and drive agricultural expansion, it would also reduce the strain of overcrowding and unemployment in the highlands. Guevara Arze targeted two surplus populations for relocation. He was primarily concerned with the excess of labor in highland mining camps. Overstaffing was promoted by the tin oligarchy to suppress wages and exploited miners were key to the success of the revolution. However, this notably Communist population was an increasing liability for the more moderate and conservative members of the MNR coalition. Guevara Arze hoped to fire 10 percent of miners and "take them to new productive occupations such as agriculture."[7] He also looked to Bolivia's rural majority, worried that the rapid destruction of the *latifundia* had created a generation of small subsistence farmers (*minifundia*).[8] These frustrated agriculturalists migrated to cities, mines, and the cane fields of northern Argentina in search of work. Redirecting this stream to Santa Cruz promised to convert them "into providers of essential products instead of consumers."[9] Ruiz's film closely follows this model with lines from the *Plan inmediato* appearing nearly verbatim in *A Little Bit of Economic Diversification*.[10] But in another sense, the technocratic language of the *Plan inmediato* differs fundamentally from the coming-of-age film. It was good policy, Ruiz remembered years later, but at heart, it "lacked a certain aesthetic seduction."[11] Ruiz translated Guevara Arze's statistics into sweeping vistas, his arithmetic of population transfer into the heroic movement of machines and bodies, his projections for increased production into swaying fields of cane and a "cathedral-like" refinery. Ruiz and scriptwriter Oscar Soria provided a narrative hook for the audience by inserting the fictional *bildungsroman* of Santos into an otherwise straightforward documentary.

This formula, which Ruiz employed in other productions, drew from a long tradition in Latin American popular culture resonating with nineteenth- and early twentieth-century foundational fictions, from *Martín Rivas* to *Doña Bárbara*. In these novels that quickly became national romances for modernizing Latin American nations, the serious conflicts produced by export booms, regional integration, and frontier development are mediated through the exploration of interpersonal, familial, and generational tensions in which hopes for sexual and national union intersect.[12] Characters from the periphery, be it the plains of Venezuela or the mining districts of northern Chile, interact and intermarry with those from urban centers as race, regionalism, and class intermingle. In *A Little Bit of Economic Diversification*, a similar process unfolds. The dense logic of import-substitution is reframed through a personal migration in which the reconstitution of the national body is reduced to the movement of the individual. The fecundity of Santa Cruz's land is inextricably linked to the fertility of cruceño women and the potentially conflictive east-meets-west encounter between lowlander and highlander is reconciled in the new mestizo child of Santos and Pilar.

As the MNR attempted to execute a socialist-inspired nation-building project in the mid-twentieth century, it turned, much like other revolutionary states, to film rather than the novel.[13] The Soviet Union immediately embraced the medium as did the postrevolutionary Mexican government whose leaders attempted to "forge a sense of patriotism and national identity based on its own evolving version of modernity."[14] The social message film had been viewed as a powerful tool of colonial governance in Africa. In 1957, Ghanaian independence leader Kwame Nkrumah, about to embark on a dramatic resettlement campaign of his own, nationalized film production, "aware of the potential role of the mass media in nation-building."[15] Beginning in 1959, the Cuban Institute of Cinematography and Art (ICAIC) also produced documentaries that sought to "personalize narratives," giving viewers, "immediate and spontaneous access to individuals and their stories."[16] As with these revolutionary movements, the MNR spoke to the creation of a new public or "the incorporation of the mass of Bolivians into national life," by broadcasting evidence of progress in agriculture, education, and health."[17]

Bolivian filmmakers also drew inspiration from Anglo-American nations. During WWII, the United States had promoted film by sending Hollywood icons from Walt Disney to Orson Welles to Latin America to act as cultural ambassadors. U.S.-financed audiovisual centers sprang up throughout the region, providing financial and technical support to nascent national film industries. In 1958, British filmmaker John Grierson visited Bolivia as part of a South American tour during which he was a guest of honor at an international film festival sponsored by Uruguay's State Broadcasting and Entertainment Service (SODRE).[18] A pioneer of the documentary, Grierson produced numerous interwar and wartime propaganda films with the authoritative narration that would characterize Ruiz's work. In the postwar era he advocated for the creation of national film boards, notably in Canada, that, like the ICB, supported cultural and nationalist production and served as UNESCO's director of mass communications and public information.[19] While in Bolivia Grierson toured the country with Ruiz—a formative moment of professional validation for the young filmmaker.[20]

The MNR was highly attuned to the value of propaganda. Under the direction of José Fellman Velarde, the Secretary of Press, Information, and Culture (SPIC) produced printed material, broadcast on government-controlled Radio Illimani, and published articles in the government-run newspaper *La Nación*.[21] In 1953, SPIC's Institute for Political Capacitation produced a pamphlet titled, "Lessons of Propaganda, Organization and Agitation," revealing how the government imagined production, dissemination, and reception. The authors did not shy from paternalism. Propaganda was to adopt "a tone that corresponds to the level of culture of the masses to whom it is directed [which] would naturally seem vulgar for educated people but has a surprising effect on multitudes and collectives."[22] A pragmatic response to high illiteracy rates, visual propaganda also drew inspiration from Bolivia's colonial past.[23] As the Spanish "inculcate[d] the Indians with religion, [they] presented religious ideas by means of almost perfect graphics that were truly works of art of the age." For SPIC, evangelization permanently altered indigenous "aesthetic sensibility, [producing] the love for the image" and the Andean landscape, with its "strong colors and the sharpness of the atmosphere," had made indigenous people "aficionados of color."[24] The pamphlet further addressed the intense regionalism that characterized midcentury Bolivia and many other Latin American nations. Propaganda would be tailored to "the fundamental difference of a ranching or farming region, of forms of life that are more or less primitive, of climatological conditions," going so far as to determine "if a people have a better disposed spirit for the humorous if they are from a tropical climate than from a mountain one."[25] It would also break the provincialism of the receiver and translate regional value for a disparate public based in mines, fields, and cities. The paradoxical result was an intensification of regional tropes in much of the MNR's early propaganda. By offering stock figures that a diverse national audience could easily identify, Ruiz's films simultaneously elevate regional distinctiveness before collapsing it in the romantic unions of their central characters.

On March 20, 1953, Bolivian president Víctor Paz Estenssoro signed Supreme Decree 3342 founding the ICB and named Walter Cerruto as its first director. Mirroring the language of the SPIC pamphlet, Paz Estenssoro's decree affirms the need to amplify the "installations and the sphere of activity with the goal of having an organism that better constitutes a means of propaganda for the understanding of national reality and serves to educate the popular classes."[26] The decree also mandated the distribution and display of ICB films in all movie theaters providing the Institute with a captive, if perhaps unwilling, audience. While exploring many aspects of the revolution, a significant number of ICB films were set in the eastern lowlands.[27] *Traveling Through Our Land* was a recurring short that introduced audiences to unfamiliar eastern locales—the show severing as a space for the ritualized encounter between Bolivia's diverse peoples and landscapes. The series attempted for Bolivia, what Vargas's 1940s newsreels like "March to the West," or "In the Trail of the Pioneers" had done for Brazil, the latter transporting urban

Brazilians to regions "where Brazil still maintains its primitive jungle."[28] In their "endless pilgrimage" across the country, the producers of Bolivia's *Traveling Through Our Land* invited highlanders to board a riverboat on the Beni or stroll the streets of tropical Riberalta. One episode opens with an aerial shot of a sinuous river winding through the jungle. Viewers were welcomed to Trinidad, "one of the most picturesque small cities in Bolivia." The "fecundity of this tropical land" was captured in a verdant plaza adorned with towering palms. The narrator balances the city's "enchanting" tradition with an insistence that even in this faraway place, "day-by-day new industries are born that contribute to the greatness of the patria."[29]

The liminality of lowland Bolivia comes across in other episodes of *Travelling Through Our Land*. In one, the camera travels to a mission station in Santa Cruz "where men of good will incorporate [hunter-gatherer communities] to civilization." The Sirionó, recipients of this civilizing mission, demonstrate traditional hunting techniques for the camera even as the narrator assures viewers that their children are "learning to love and serve the patria."[30] When the program passes through the city of Santa Cruz in a subsequent episode, the narrator finds a place "typified by its exuberance," in which unpaved streets "show the contrast between traditional oxcarts and modern vehicles." In what would become a persistent trope in lowland representation, the program blends tropical fertility with alluring feminine beauty, with scarce concern for subtlety. "In its gardens, under a serene and warm sky, the most beautiful perfumed flowers, just as beautiful as its women, open their petals," the narrator intones suggestively as images of tropical flowers fade to a closing scene of a couple dancing in the plaza to a song with lyrics extolling "the beautiful eastern land, . . . Santa Cruz, my love."[31] Yet while frequently emphasizing the femininity of its exquisite nature, the ICB also produced images of Santa Cruz that were just as likely to fetishize modern infrastructure and goods from water towers and roads to imported Zebu cattle. In a news short on the construction of an electrical plant, the ICB reminds viewers that "Santa Cruz has been one of the regions of the country in which the government of the national revolution has put its greatest interest . . . this *city of the future* has received one of its most basic necessities [emphasis added]."[32]

It is unclear how the ICB's early films were received among a public that may have watched grudgingly while waiting for the feature to begin. Government officials, however, were thoroughly impressed. The simple images and didactic language of these first films provided a stable representation of policy-in-action for the high modernists of the MNR just as much as for the film's imagined illiterate public. In September of 1956, the Bolivian Congress heard a motion to commend ICB director Walter Cerruto for his work. Deputy Mendoza-López led the charge, expressing a sentiment of "logical and just gratitude," given that before the creation of the ICB, "it had not been possible to understand, in this country or abroad, the diverse activities of the *campesino* sector." In a frank admission of their own bias, the deputies acknowledged that the Institute had done no less than transform rural Bolivia, the vast majority of the nation's population, into "an element worthy of consideration where before it was not."[33]

Mendoza-López found the eastern vistas of programs like *Travelling Through Our Land* important for another reason. "For the first-time sectors of the east have seen themselves, their panoramas, their landscapes and activities exhibited for the contemplation of the nation." As with postrevolutionary Cuba and Russia, where citizens had "rarely seen themselves on screen" prior to the revolution, film rendered the Bolivian east accessible and produced a broadened self-awareness.[34] Yet Mendoza's comments hint at the potential awkwardness of this Lacanian moment in which lowlanders were assumed to be watching ICB film while imagining themselves as highlanders watching lowlanders. "National contemplation," in this formulation involved the judgment of La Paz in relation to a peripheral Santa Cruz.[35] This uniquely Bolivian *orientalism* hinged on a display of regional types, from crocodile hunters to wandering troubadours, that was not always flattering. Despite their land's great promise, lowlanders, just as much as highland transplants, were seen to suffer from serious limitations often cast in explicitly gendered terms that implied cultural and racial difference. In ICB productions lowland men appeared as both hypermasculine and deficient family providers. In contrast, the highlands were often spoken of as a conflictive place with a troubled future. Out of place in the lowlands, that volatile highland energy was not a political liability but a perfect counterpart to the repeated assertions of lowland complacency. Development discourse often hinges on the construction of a deficient subject in need of intervention, a perspective that characterizes two Ruiz films of the late 1950s during a period in which he took over directorship of the ICB from Cerruto.[36]

In 1958's *La Vertiente* (*The Spring*), Ruiz's camera travels to the village of Rurrenabaque. Like *A Little Bit of Economic Diversification*, the film drew on actual events while incorporating a fictional hook for the audience. A young schoolteacher from La Paz is the agent of change and conflict is once again written as an ethnic, interregional romance, this time between the impetuous *paceña* and a rugged crocodile hunter. While completing an obligatory year in the provinces, the teacher acts as a transformative force in a lowland envi-

Scenes from Jorge Ruiz's 1958 film, *La Vertiente*. Regional tropes are presented in gendered form through the romance between a young schoolteacher from La Paz and a crocodile hunter from Rurrenabaque.

ronment characterized by drunkenness and a lack of interest in change. She repeatedly encourages the town to build a modern water system instead of drinking directly from the river. After a young student dies of a waterborne disease, she leaps into action enlisting her own class to build a water line through the jungle before breaking her leg in a dramatic tree-falling accident. Shamed into action by the gender transgression of this martyr for development, the hunter abandons drink and rallies the support of the town, the government, and the military. The project is heroically completed, and order and sanitation arrive in the village just as the safely convalescing schoolteacher's injuries are healing. The film concludes with the happy embrace of hunter, masculine honor restored, and tearful schoolteacher bound more than ever to the lowland town.[37]

Although lacking a highland-lowland romance, a similar relation between reconstituted gender order, family, and regional development is evident in *Los Primeros* (*The First Ones*). This 1956 ICB film, centers on Camiri in the south of Santa Cruz. In the *Plan inmediato* Guevara Arze described the area "on the eastern flank of the last ridges of the Andes [as] one of the greatest regions of petroleum possibility in the western hemisphere."[38] Once again, the film recasts abstract policy as personal transformation. *Los Primeros* open with an elderly woman darting through dense bush. Doña Ramona knows "the secret of the jungle" the narrator reveals. Arriving at a hidden oil seep in the forest, she scoops a bucket of the precious liquid and scurries back to town. The home that Doña Ramona returns to typifies the disordered gender structure of lowland life. She sells the oil to cover necessities and the vices of her adult son. Like the crocodile hunter from *La Vertiente*, he is a "vagabond," who "only thinks of jokes and drinks." Heading to the bush to play guitar with friends he insists that, "life is more beautiful without worries." The next scenes of dramatic state intervention offer a striking contrast to Doña Ramona's traceless gathering practices. In the process, the unrealized promise of her undeveloped well and her unproductive son are brought forth. Men from the state petroleum company cut through the bush with machetes, survey the land by air, and set off dynamite to test wells. Soon a massive derrick rises out of the forest and Doña Ramona finds a fence blocking access to her well. Her neighbors, benefiting from a steady supply of refined gas, no longer want her "dirty petroleum." As opportunities close for Doña Ramona they open for her son who finds work as a roughneck thereby fulfilling his filial obligation to mother and patria. "Living in the jungle a long way from their family, these workers have taken their responsibility to the great Bolivian family," the narrator proclaims, casting oil work as a form of patriotic masculinity akin to settlers or the road crews building the new highway. This dramatic change to the natural landscape surrounding Camiri results in a corresponding change to the town's interior landscapes. Doña Ramona now has a comfortable domestic space with electricity where she serves coffee to her laboring son. Not entirely removed from the sphere of production, she continues to sell oil as a special tonic, a false panacea (the narrator chuckles), for the ills of her neighbors. In contrast, state-controlled oil offers the "dignification of men [like her son] through work" and is also integrated in a global capitalist order. The film ends with a shot of the Chilean port of Arica, where Bolivian oil is exported by tankers, the final step in transforming Doña Ramona's "secret of the jungle" into Guevara Arze's model for national development.

In *A Little Bit of Economic Diversification, La Vertiente,* and *Los Primeros*, the March to the East is presented as a project that would transform landscapes, move bodies, and thus unify nation and family. As Doris Sommer argues, in the oscillation between "epic nationalism and intimate sensibility," the distinction is ultimately collapsed.[39] The gender order of the family appears as both effect and cause of national unity. Rather than masking the conflicts inherent in this process, Ruiz's films work to resolve them. Yet the nationalist discourse of eastern expansion, and the mobility that it engendered, was not without its critics, particularly in the promising yet sleepy regions and cities of the future that the MNR sought to transform.

Rivers, Roads, and Rail: Regional Understandings of Mobility

Santa Cruz produced its own series of popular and literary images of mobility that both meshed and conflicted with that of the central government. This counterveiling regionalist discourse emerged through the Santa Cruz-based newspaper *El Deber* founded in 1953 just a few months after the creation of the ICB. Currently the periodical with the largest circulation in Bolivia; at the time *El Deber* was a humble broadsheet of four pages and the city's only regular newspaper. Under the directorship of Lucas Saucedo Sevilla, *El Deber* styled itself as the authentic voice of cruceños, a forum for a limited but influential public to reflect on regional modernization. The paper embraced the MNR's March to the East as a *modernizing vision*, celebrating new forms of transport, industry, and agriculture. But its authors had an ambivalent relationship to the March to the East as a *migratory project* that would bring thousands of Andeans to the region. They also chafed at the representation of the people of Santa Cruz as *deficient subjects* in need of intervention rather than the authors of their own transformation.[40] The paper soon developed a characteristic regionalist critique which reached a fever pitch during a series of conflicts between Santa Cruz and the national government in the late 1950s. Saucedo and his team of writers gathered in a long tradition of elite cruceño reflections on their region's abandonment by an absent state.[41] Anthropologist Penelope Harvey describes a similar ambiguity for frontier elites in the neighboring Peruvian Amazon in which, "the state appears in another guise as an object of desire and fantasy and people's fears are as likely to focus on abandonment as on control."[42] Likewise cruceños often claimed two rhetorical positions, clamoring for, and then questioning the results, of modern infrastructure. Mobility would break the isolation of the east and provide access to new markets, but it would also render Santa Cruz fundamentally accessible to the mass migration of Andeans and other groups that cruceños deemed foreign. Over the following years, their initial articulations of difference, as experienced through infrastructure, would harden into a racialized reaction to outsiders.

Of the major changes that Santa Cruz experienced in the 1950s, few resonated more strongly with elite cruceño dreaming than the arrival of the railroad. The city had been fighting since WWI for rail links with Cochabamba, Brazil, and Argentina. In one 1942 pamphlet they bitterly lamented their continual abandonment before envisioning a glorious future when they would "gather before the image of the arriving locomotive . . . [and] salute it with the euphonious and vibrant call of 'Long live the immortal Bolivia, long live the great Bolivia of the future.'"[43] More than a decade later that image of triumphant mobility was brought to life. *El Deber* reported on the festivities as the Brazilian President João Café Filho met Bolivian President Paz Estenssoro in Santa Cruz on January 5, 1955, to inaugurate the new railway as a binational initiative. The city's unpaved streets were decked with flags of both nations,

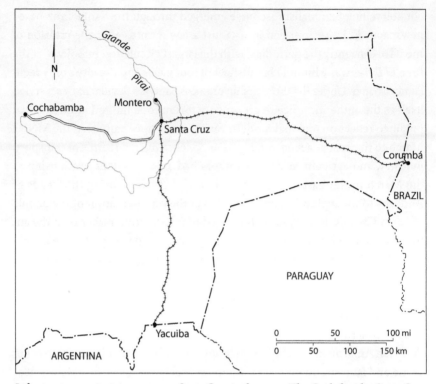

Infrastructure projects converge on Santa Cruz in the 1950s. The Cochabamba-Santa Cruz highway, the Santa Cruz-Corumbá railway, and the Santa Cruz-Yacuíba railway connected the frontier city to the nation's highlands and to the borders of Brazil and Argentina. Map adapted from *UN Map of Bolivia*, Wikimedia Commons.

and a large crowd gathered at the airport to wait for the arrival of the Brazilian presidential plane flanked by a Bolivian escort. Café Filho disembarked to a marching band playing the national anthems of both countries. The two presidents fraternally embraced and traveled a short distance outside of town to board an incoming train. Here the diplomatic, religious, and technological merged. As they arrived in Santa Cruz, the Archbishop was on hand to bless the train. Brazilian and Bolivian engineers made speeches and the presidents planted a friendship tree sent by Belo Horizonte's Rotary Club. That after-noon the railway commission served lunch along the Cochabamba-Santa Cruz highway, deliberately highlighting the link between international rail and interregional road. In the evening the city's most prestigious club hosted a banquet where the respectable families of Santa Cruz danced the night away alongside foreign and national delegations as fireworks exploded over the new train station.[44]

El Deber's account reads like something from the height of the late nineteenth century railway building frenzy. For regions at the margins of Latin America's export boom (1850–1930) such spectacles were long-deferred, and their belated realization carried an air of both modernity and antiquarianism. Their choreographed displays often concealed as much as they revealed.[45] The scene in Santa Cruz on the fifth of January seemed to offer the realization of the municipal government's vision of 1942, in which they inextricably linked modernity with mobility, and the MNR's March to the East. Yet the carefully orchestrated display bore little resemblance to the subsequent experience of riding the rails. Over the following years, mounting frustration with the train, ranging from the tragic to the mundane filled the pages of *El Deber* as editorialists, journalists, and users grappled with the meaning this new rail line would have for Santa Cruz and the nation. Discussions of mobility in *El Deber* slipped easily from anger at a missed connection to reflections on food security, sovereignty, and modernity. While fetishizing technology, these authors found transportation's promises to be fundamentally incomplete.

Already in the first month of the train's operation, *El Deber* noted a dramatic increase in passengers. The distance between the city of Santa Cruz and the Brazilian border at Corumbá—formerly, weeks of mule train away, had truly been collapsed. Cruceños now had access to modernizing Mato Gross and industrial São Paulo. Yet it is telling that the presidents began their ceremonial ride at a point just outside the city limits beyond which this vision of modern mobility quickly unraveled. A short distance to the east, the Río Grande remained a broad expanse of mud and sand capable of diminishing to a narrow channel in the dry season and extending to more than a kilometer with the summer rains. No bridge had been built despite the inauguration of the line. Travelers had to disembark from the train, unload their possessions onto barges and, as they had done in decades past, make a precarious and costly crossing of the river before boarding a waiting train on the other side. Scarcely a month after inauguration, five passengers drowned when an overloaded boat overturned in flood waters.[46] Denouncing the disaster, *El Deber* demanded the creation of a Port Captaincy to remedy the "uncontrolled situation" provoked by the dramatic increase in traffic.[47]

The river crossing was only the most extreme of a series of environmental challenges that railway engineers struggled with. On January 30, 1955, a writer broke down a typical train trip in an article titled "The Railway Service Is Painful." The author traveled less than half the line's total distance—from the Bolivian town of San José in Chiquitos west to Santa Cruz—at the height of the rainy season. The train left on January 22 at 10:00 A.M. and arrived at the Pozo del Tigre station at 6:30 P.M. where it was forced to make an unscheduled stop due to a washed-out track at KM 522. Passengers and cargo remained in place until the following morning at 7:30 A.M. when they resumed their journey to Pailón on the eastern side of the Río Grande. They only advanced as far as KM 590 where they found the track was washed out again. Machinery sent to fix the rails was also paralyzed because of the unrelenting rains. The conductor considered returning to the previous station but the track behind them at KM 581 had also collapsed. They had no choice but to disembark, leave their cargo, and board a locomotive on the other side of the washout. On the twenty-fourth, the train finally arrived at the Río Grande around 4:00 P.M. where the passengers disembarked and crossed the swollen river on barges. They spent that night on the other side of the river in Puerto Pailas waiting for a train to Santa Cruz. That train only took them a short distance to KM 609 where they switched trains again around 6:00 P.M. Arriving in Cotoca at 6:15 P.M., they waited at the station until 8:30 P.M. because of a lack of water for the boiler, a delay repeated only an hour later at Guaracachi station. At 10 P.M., more than seventy-two hours after departing San José, the exhausted passengers arrived in Santa Cruz thereby completing a journey of scarcely 270 kilometers. Their luggage, the writer reminded readers, had still not arrived.[48]

The grueling saga is worth recounting in all its excruciating detail precisely because the kilometer-by-kilometer and minute-by-minute account best expresses the painful experience, as opposed to the abstract logic, of mobility. Time and space, far from being annihilated, had become agonizingly apparent. Yet the frustrated punctuality of the rider-writer also indicates the degree to which expectations (if not realities) of mobility were in flux in Santa Cruz's belated entrance to the railway era. For all its inconsistencies, formerly isolated cruceños flocked to the new service. Even tragic accounts demonstrate the allure of new Brazilian markets. With demand surpassing the train's limited capacity, the service took to placing surplus passengers on open decks without railings to prevent falls. On July 2, *El Deber* noted the deaths of nine such passengers due to frequent violent stops and failure to signal departures.[49] In December, during the feast of the Virgin of Cotoca, another rider found himself shoulder to shoulder with 5,000 would-be passengers at the station. These cruceños had decided to forgo the traditional twenty-kilometer pilgrimage on foot and completely overwhelmed the system's capacity.

Across the Americas, from Ecuador to the American Midwest, the construction of railways collapsed time and space but rarely in uniform ways.[50] Their layout and scheduling produced uneven effects drawing distant locations together while pulling other seemingly proximate ones apart.[51] Santa

Cruz was no different. A new connection with "the growing commerce of the rich eastern provinces and Brazil" meant the introduction of new products and a new outlet for exports. As early as March 1955, *El Deber* marveled at the influx of Brazilian goods filling the plazas of the country. Bolivians were "avid in their demand for chosen and fine manufactured products." "We have connected ourselves," wrote the author "to the economic giant of Brazil, an example in development for the whole continent."[52] Another author celebrated the "vast and promising foreign market" for cruceño fruit. "Thanks to the railway," he reports, "products from cruceño storehouses are being sold in cities beyond our frontiers." It was with immense satisfaction that *El Deber* received a card from Ponta Pora on the Paraguayan-Brazilian border confirming that Bolivian oranges were being eaten in their town.[53]

That connection would be read in multiple ways: as a positive example (Brazil had begun its own March to the West two decades earlier), an economic opportunity, but also a threatening foreign influence. In the earlier account of pilgrimage to Cotoca, the true tragedy for the writer was not the increasingly predictable failure of the railway to deliver on its promise of efficient mobility but the degree to which the train had opened Santa Cruz to aggressive foreigners. In contrast with the fraternal embrace of Café Filho and Paz Estenssoro, on board, the pilgrim was jostled by "Brazilians, who act as if they own [the train]" and speak a "crude Spanish mixed with Portuguese." "Can the authorities really remain passive before these abuses" he asked, "is it not possible that a Bolivian could monitor this service and let the Brazilians know that we are not their servants [*pongos*]? Is it not possible that we could administer our railway for our countrymen, so they do not believe that we are some Brazilian colony?"[54] Highland newspapers also expressed reservations about a new Brazilian market. In the second year of the line's operation the La Paz newspaper *El Diario* published an article denouncing the "excessive contraband of food items to Brazil at Corumbá." Smugglers were selling UN food aid to Bolivia along with Santa Cruz's sugar and rice (subject to export limits) across the border in Brazil. The paper chastised unscrupulous cruceños whose "illicit commerce . . . makes useless the effort the government is realizing to increase production with a view to sending [Santa Cruz's] excess production to the interior of the country."[55]

These *El Deber* articles written during the first two years of the train's operation demonstrate the contested reception of newfound mobility. The performance of national, regional, and transnational unity that marked the railway's inauguration on January 5 masked this ambiguity. In the months that followed, commentators chastised the train's operators for prizing speed and volume over safety precautions. Yet those very complaints indicated the degree to which understandings of space and time were in flux. They were just as likely to denounce irregularities which caused delays, complaining of cars that "arrived late or never from Corumbá," stranding produce destined for the new Brazilian market.[56] Like the rail connections to Brazil and Argentina, cruceños eagerly awaited completion of the highway pictured in *A Little Bit of Economic Diversification*. It promised to be the first all-weather road to connect the lowland city and its undeveloped agricultural hinterland to the nation's heartland. That would also mean an immense new market for Santa Cruz's growing production. Yet here too, cruceños who had until recently denounced their abandonment by the nation, began to criticize the terms of their integration. While highland newspapers worried about Santa Cruz's illicit commerce with neighboring Brazil, cruceños rejected the idea that their sole purpose was to act as low-cost providers of necessities for the interior. One editorial in *El Deber* brought readers on a virtual tour of highland markets rife with discounted cruceño products. Meat, selling for 300 Bolivianos per kilo in Santa Cruz, sold for forty in Andean mining camps. Sugar and milk also "costs more in the zone of the producers," than in the highlands, he continued. Inverting an Andean-centric view of the nation, he argued that as a result of the new highway, the "epicenter [Santa Cruz], sees its food profit diluted to the periphery [the Andes] without collecting any benefit whatsoever for itself."[57] "Santa Cruz," he continued, "provides petrol but does not have water, wood but does not have public buildings, rice but does not have a hospital, sugar but does not have maternity centers, coffee but does not have a court, gas but does not have a refinery, cattle but does not have paved roads."[58]

Far more troubling for cruceños, the new highway also meant the arrival of thousands of highland migrants. One writer's nationalistic reaction to Brazilian elbows on the railway suggested a prizing of nationalism over unfettered trade. Yet commentators defined foreignness in regional as well as international terms. While denouncing their dependency on highland markets, cruceños were equally capable of treating their *brothers* from the interior as foreigners and racialized others. Highland officials were aware of this bias. Guevara Arze had been careful in portraying internal migration assuring readers of the *Plan inmediato* that future colonists would be fluent Spanish-speakers and not Aymara or Quechua monolinguists. Similarly, Jorge Ruiz ensured that his fictional Santos quickly left the urban space of Santa Cruz to settle on the frontier. From a national perspective, those narratives attempted to contain the worrying possibilities of interregional mobility. Nervous cruceños also used a range of discursive tools, from caricature to ethnography, to further

distance themselves from highland migrants cementing regional boundaries as impermeable racial barriers even as modern infrastructure annihilated
physical distance.

On December 14, 1955, *El Deber* reprinted a short story entitled "By the
New Highway," by Ramón Clouzet, a cruceño teacher and celebrated author
of hundreds of works of fiction.[59] The Santa Cruz-Cochabamba highway
had officially been inaugurated on September 24 of that year by a delegation
including the president, various ministers, and representatives from its U.S.
financers. Clouzet employed humorous regional tropes to define the meaning of this new form of mobility for his cruceño audience. The protagonist
of his story is Periquito. With a father from Santa Cruz and a mother from
Cochabamba, he is, like the fictional child of Pilar and Santos, the new mestizo.
Based in Cochabamba, he earns his living bridging those regional divides as a
truck driver along the old dangerous road to Santa Cruz. Periquito demonstrates a clear highland propensity for saving and hopes to own his own vehicle. After achieving that golden dream, he plans to marry his childhood
sweetheart, a highland woman suggestively named Robustiana, presumably
the *rubenesque* opposite of the slim cruceño ideal of feminine beauty. On his
frequent voyages on the old highway Periquito is accompanied by an assistant
Pascualito who is naturally "a true artist" at the *charango* (an Andean stringed
instrument).

Like other Bolivians, Clouzet reminds the reader, Periquito had been "anx-
iously waiting" for the completion of the new road. He is "tired of the bad
passes" particularly, "those places where the road bed became narrow . . .
where he would always make a prayer, entrusting his soul to God." On inauguration day all that promised to change. Periquito and Pascualito are in Cochabamba, preparing for their first trip to Santa Cruz on the new highway,
"a land that [Periquito] had great fondness for, being that of his father." In
their excitement the two down several glasses of *chicha kolla* (an alcoholic
corn beverage typically consumed in the Andes). Leaving Cochabamba with
a fully loaded trailer they stop in nearby Comarapa where the protagonist
makes a chance encounter with Robustiana. He proposes, and the lovers plan
to marry in the highlands upon his return. The assembled crowd celebrates
the engagement and, as Pascualito serenades them, the pair consumes several
more glasses of chicha before departing.

Back on the new highway, driver and helper make one final stop in Sa-
maipata, the last mountain town before reaching Santa Cruz. Residents are
celebrating the opening of the road and after several more drinks they head
out, disregarding the warnings of the townsfolk. An intoxicated Periquito

asks Pascualito to play the same song he had serenaded the new couple with in Comarapa. As "the notes cascaded, Periquito who considered himself the surest hand on the road, let go of the wheel and closed his eyes, visualizing himself back in Comarapa . . . the silhouette of his fiancée emerged, and he started to pull her towards him with such enthusiasm that he punched the accelerator and the truck went off the road and over the precipice." Arriving at the horrific scene, police find the corpses at the foot of the cliff and the charango still intact.

"By the New Highway" offered *El Deber*'s readers a cautionary tale, by turns playful and macabre, that diverges from the message of Ruiz's *A Little Bit of Economic Diversification*. In the latter, the protagonist's mobility results in social advancement and a stable nuclear family. Until its tragic ending, Periquito's story seems to promise the same. Along with geographical and social mobility, Clouzet suggests that mobility can also bring destruction, a result not unlike several news reports on the train service over the previous year. One explanation for such narratives, fictional and nonfictional, lies in the morbid fascination produced in the audience as dreams of rapid transit become nightmare scenarios. Yet while cruceños were quick to blame the train's administration for the death of one of their own, Periquito and Pascual's demise suggested something else. Fault lies not in the modern asphalted road but in the destructive drunkenness of highlanders, thus inverting a stereotype about lowlanders frequently employed by the ICB. As cruceños celebrated their belated connection with the nation they also worried about the crude enthusiasm of their siblings from the interior who were now free to descend on Santa Cruz whenever they wished.

As highland-lowland migration began in the 1950s, cruceño regionalism increasingly pivoted on similar ideas of difference and *El Deber* became a venue to fashion a racial identity distinct from, or even in direct opposition to, Andean Bolivia. This message, hardly subtle in Clouzet's "By the New Highway," is explicit in *El Deber*'s response to an official "Day of the Indian" first created by President Germán Busch in the 1930s and revived by the MNR to celebrate the promulgation of the Agrarian Reform Law on August 2, 1953. For the state, it is critical to note, celebrating *the Indian*, did not alter the degree to which the MNR imagined indigenous Bolivians as deficient rural subjects in need of transformation. Ruiz pursued both ends with early works like *Vuelva sebastiana, Donde nació un imperio*, and *Los urus* offering anthropological reflections on Bolivia's indigenous past, whereas later films like *Juanito sabe leer* (*Little Johnny Can Read*) attempt to demonstrate that through education, those same indigenous Andeans were being incorporated into the nation. As

Johnny demonstrates his new literacy to his village neighbors in the latter film, the narrator triumphantly proclaims, "Now you will not be a lost race in the shadow of Illampu [fourth highest peak in the Bolivian Andes]; light and knowledge will make you free."[60]

While *Little Johnny Can Read* imagined the paternalistic creation of new Indians to replace timeless rural existence in the Andes, one prominent cruceño rejected the idea that this vision of ethnic transformation spoke to his own frontier department. On August 2, 1956, the prolific cruceño sociologist Hernando Sanabria Fernández—faculty at Santa Cruz's Gabriel René Moreno University—commemorated the national Day of the Indian with an article in *El Deber*. Taking an anthropological approach to Guaraní, Mojeño, and Chiriguano life, Sanabría reminded fellow cruceños that "Our Indian" was distinct from the supposedly static Quechua and Aymara communities of the highlands.[61] Nomadic lowland indigenous communities were actually "open to all impressions of the outside world," could "easily change their lifestyle," and did not "root their existence with exclusivity in the piece of land where they were born," he continued.[62] Sanabría invoked an original moment of *mestizaje* in which this essential spirit had been fused with the "criollo cruceño colonizer and civilizer" of the eastern lowlands producing an intimate appreciation of the lowland environment and a "social mass that could populate the ancient jungle solitude." Like other elite strategic appropriations of a mythical indigenous past, Sanabría's valorization of lowland indigenous culture likely meant little in terms of cruceños' often exploitative relations with their indigenous neighbors many of whom were subject to coerced labor on lowland estates. Yet, as part of a discourse that emphasized Santa Cruz's unique regional identity, it could easily be marshaled in opposition to the nation-building project of the MNR and the Andean migrants that came with it. This would prove particularly valuable over the following two years.[63] From 1957 to 1959 Santa Cruz engaged in a prolonged conflict with the MNR over the allocation of royalties from its oil development. Over the course of what became known as the "civic struggles" or the "struggle for the eleven percent," residents of Santa Cruz engaged in strikes and demonstrations and the national government repeatedly sent militia and troops down the new highway to occupy the city. In the process, cruceños expressed increasingly contentious understandings of regional identity, development, and mobility challenging the happy resolutions to regional integration depicted in Jorge Ruiz's films.

In *Los Primeros*, Ruiz depicts the growth of the cruceño oil industry as a dynamic aspect of the revolution. The violent alteration of pristine nature unlocks its latent forces, generates wealth, and reorders a dysfunctional gender structure. In the process, Doña Ramona's delinquent son becomes a productive, and thus masculine, worker in the state oil company. The rosy picture suggests a universal sharing in the benefits of oil production from household to region to nation. Cruceños felt otherwise. In 1957, a regional advocacy group, the Committee Pro-Santa Cruz, began vigorously petitioning Bolivian President Hernán Siles Zuazo. Committee president Melchor Pinto Parada demanded that, as promised in a 1937 law, 11 percent oil profits would remain in the producer department providing funds for regional development.

As the conflict over oil royalties unfolded in the second half of 1957, contradictory images and narratives of Santa Cruz began to circulate at the national level. Some mirrored the stable visions of development put forward by the ICB in their films while others raised the alarm about a menacing new regionalism. On September 24, Bolivia's Congress rose to honor the region. The official holiday and obligatory homage to Santa Cruz, marking the day in 1810 when the city overthrew its local Spanish representative, had been signed into law by President Siles Zuazo only the previous day.[64] Members of Congress from highland departments that were sending large numbers of migrants to Santa Cruz spoke in glowing terms. A member from Oruro noted that in Santa Cruz, "lies the future . . . [it] will one day constitute the power of our patria, as this concept is one of the postulates of the National Revolution." Representatives from Chuquisaca saluted Santa Cruz and their own "campesino workers that daily, are working and laboring in the bush for the greatness of our patria."[65] Deputy Nuñez del Prado reiterated the themes of the *Plan inmediato* and pointed to the conclusion of paving on the Cochabamba-Santa Cruz highway through which cruceños "had escaped from their tropical enclosure to provide their harvests for the consumption of the habitants of Bolivia." In implicit dialogue with an *El Deber* editorial of the previous year complaining of the lack of infrastructure in Santa Cruz, Nuñez promised that, in exchange for cruceño rice, sugar, and beef the nation would "reciprocate" with "modern services . . . largely waited for with patience in the eastern capital, which will soon convert it in short order in one of the most populous and modern urban areas in our Patria." Nuñez also highlighted the importance of the two international railways that had made Santa Cruz the largest transport hub in the nation, a "network, whose arms will open in *a cross*, [will extend] towards all the confines of the soil of our patria [emphasis added]."[66]

While officials celebrated Santa Cruz as a future "node," *El Diario*, one of La Paz's leading newspapers, took a more sinister view of those interlocking transport routes. On the same day as the official homage, *El Deber* reprinted an *El Diario* article provocatively entitled, "Santa Cruz: crucified," rejecting the "cordiality of messages and official speeches." "Two railways converge like slow arrows on the heart of our Orient," he began, "one from Brazil, the other from Argentina." "Both will open this corner of Bolivia to the routes of economic and social change. Could it be that they will also initiate a political transformation?" He felt certain that creeping foreign influence would "balkanize" the nation. Without a corresponding railway to the highlands, "Santa Cruz will remain isolated from the country by two lines of parallel steel." The only option, the author concluded, was to "go to Santa Cruz, with machines, capital, men and a fundamentally Bolivian spirit, so that side by side with our brothers of the tropics we will make the country great."[67]

In his refusal to separate the economic from the political, the author broached a central problem in the March to the East, that would remain a lingering question in the region's future development. The image of a balkanized Bolivia, with Santa Cruz joining the neighboring nations of Brazil and Argentina, might have seemed farfetched to the readers of *El Diario*. Yet scarcely a month later, surprised cruceños learned from foreign news agencies and highland papers that the Bolivian government had declared martial law in response to their "revolutionary separatist activity." The activity in question was the return of exiled political leader Óscar Únzaga de la Vega. The founder of Bolivia's fascist-inspired Falange Socialista Boliviana (FSB) in 1937, Únzaga had run against the MNR in 1956 elections before fleeing the country. It was his return that led President Siles Zuazo to declare, on October 29, 1957, that "the policy of tolerance and respect of the government had served to feed conspiracy proposals, that had arrived at such extremes as to provoke reactions of a regionalist character, that threaten measures of order and preservation of national sovereignty."[68] The following two years would see attempted coups against the MNR that only ended with Únzaga's suspicious death in 1959.

The MNR was quick to link the seditious actions of Únzaga and the FSB with the demands of the Committee Pro-Santa Cruz for oil royalties. Cruceños attempted to challenge that characterization. In what *El Deber* described as a spontaneous and extraordinary reaction, an estimated 20,000 people gathered on Santa Cruz's central plaza on October 31 in flagrant violation of the decree. The crowd, "tired of infamies, of intrigues and calamities," listened to speeches by prominent cruceños including Hernando Sanabria, author of the regionalist "Our Indian" editorial of the previous year, before decreeing a twenty-four hour strike in protest of the martial law decree and the slandering of Santa Cruz's people as "separatists."[69] In a "Message to the Bolivian People," printed on the same page, the Committee Pro-Santa Cruz's president Pinto Parada offered his version of events for "our brother peoples" of the interior.

Far from fomenting rebellion, the Committee claimed to have sent a simple letter to President Siles Zuazo asking for the ratification of article 111 of the Petroleum code, guaranteeing 11 percent of oil profits for Santa Cruz, "proceeding pacifically and within juridical norms and without party sectarianism." Cartoons in *La Nación* indicating that Santa Cruz was requesting annexation from Brazil were baseless as was the claim that internal migrants to Santa Cruz had been "massacred by the [cruceño] people." Ours, he wrote, "is a civic movement without concomitant political conspiracy" supported by local members of the MNR, "together with the same sons of this land and residents of the interior."[70] Despite Pinto Parada's hope that the conflict would soon be resolved, the civic struggles of cruceños intensified over the next two years and included several violent altercations as the FSB launched attacks against the government and the army and militia occupied Santa Cruz, first in May of 1958 during a second period of martial law and again in June of 1959. Naturally, when MNR forces occupied Santa Cruz the second time, the minister of government who authorized the incursion was none other than Wálter Guevara Arze, the author of the *Plan inmediato* and intellectual architect of the March to the East. In both occupations, highland militia could rapidly descend on Santa Cruz, just like Clouzet's fictional Periquito and Ruiz's fictional Santos, along the newly completed asphalt highway.

Even as cruceños found themselves under supervision by a national occupying force and depicted as rebels, separatists, and anti-highlanders in the La Paz press, they continued to receive the comforting films of the ICB in their theaters. On April 25, 1958, *La Vertiente*, that interregional development romance featuring a paceña schoolteacher and a lowland crocodile hunter, screened in Santa Cruz. Over the preceding days, announcements had been placed all over town, outside the central post office, the university, and the police station. Radio Centenario celebrated the first full-length Bolivian film and critically, the first participation of a "cruceño actor" in national cinema. The showing took place in the Cine Teatro and drew a large crowd, even though the entrance fee of three bolivianos was equivalent to a meal in a decent restaurant. The ICB even made a newsreel short commemorating the screening and over the following six months, the film ran continuously—morning, matinee, and evening.[71]

An *El Deber* correspondent, Hugo Maldonado, attended the screening. Arriving with low expectations that *La Vertiente* would be little more than another ham-fisted ICB news bulletin, Maldonado was surprised to find a dramatic, plot-driven film. Aside from certain technical mistakes in the production, he found *La Vertiente* to be excellent "from the emotional point of view." Here the significance of watching a nationalist film in a regional setting became apparent. Though set in Rurrenabaque, an Amazonian town in the neighboring lowland department of Beni, Maldonado found great relevance for Santa Cruz, "and all people of the east who confront the same problems and have the same worries, who have truly been united since time immemorial for their traditions." Moreover, the male lead, Raúl Vaca Pereira, was from Santa Cruz. Thus, the ethnic romance of the film was in the reviewer's mind between a paceña migrant and "a countryman of ours."[72]

In the nationalist reading of *La Vertiente* that I offered previously, the teacher's outside intervention in the languid lowland setting fits easily within a highland developmentalist narrative that viewed lowlanders as deficient subjects in need of improvement. Viewed with Santa Cruz's civic struggles as backdrop, that meaning could change. Cruceños constructed the same links between gender, region, and race but inverted the outcome. Maldonado, for example, conceded that "a central role is portrayed by the *kolla* [teacher] as the initiator of the project," yet insisted that "this is completely absorbed by the gallantry [the role of the crocodile hunter] that according to the plot is a secondary role."[73] The final gendered lesson of *La Vertiente* is not that the east is improved by the west, but that the female is *absorbed* by the male. Vaca Pereira's wife, when reflecting on *La Vertiente* years later remembered Raúl, "winning over [conquistar] the teacher that led the movement."[74] Other Santa Cruz viewers suggest that the cruceño public saw "Raúl as gallant and bohemian, conquering the lady."[75] In an undercurrent that subverts the intended message, these viewers saw virile, eastern masculinity trumping development originating in the west. This may suggest that cruceños were more than passive viewers of government propaganda but also that the effectiveness of romances like *La Vertiente* lay in the ready availability of nationalist and regionalist readings.

It is worth moving from an analysis of the content of films like *La Vertiente*, to consider the setting in which they were viewed by cruceños. As one of the few healthy diversions in a small regional capital like Santa Cruz, cinema was still a novel form of spectacle in the 1950s. Films, from popular Mexican rancheros to Hollywood classics, could transport cruceños to modern or exotic locales, but the physical space of the city was equally transformed with the rapid construction of theaters in that decade. In 1951, the first modern cinema, Cine Santa Cruz, was inaugurated, quickly followed by Cine Metro, Cine San Isidro, and Cine Norte. Those attending the latter would depart from a screening "with a little fear because in those days it was still pampas [empty plains]," remembers Ruben Carvalho. Unlike Cine Norte which was located on the fringe where city rapidly became bush, the Cine Santa Cruz was intimately connected to ideas and experiences of urban modernity. The owners graded and graveled the street, the first of its kind in a city known for dirt roads that alternated between dust and mud. While most of Santa Cruz lacked electricity, large generators lit the streets in front of the cinema, powered the projectors, and offered air-conditioning to patrons seeking escape from the sweltering heat. Inside, viewers found decor modeled after the Gran Rex theater in cosmopolitan Buenos Aires.[76]

Cine Santa Cruz's 1,200 seats, already significant in a small city with a population of approximately 50,000 people, represented only a portion of the movie-going capacity of Santa Cruz in the 1950s. Many *rustic* theaters throughout the city and its hinterland, little more than "open sheds, with poor lighting and bad sound, with bleachers made of wood," catered to the popular classes. One entrepreneur even created a mobile cinema that toured the region.[77] In 1955, the small village of Puerto Pailas, on the western bank of the Río Grande, lacked enough houses for workers building the provisional railway bridge. Yet that same year municipal authorities were already passing safety laws, much to local chagrin, governing attendance at the village's overcrowded theater, which *El Deber* reminded readers was "the only distraction for residents."[78]

Given their centrality to urban and village forms of sociability, theaters had the potential to move beyond spectacle to become sites of conflict and political mobilization. Personal memories about the role of individuals in the civic struggles of the late 1950s often linked place and time to the movies. A young man named Jorge Roca was killed in a shootout with the Control Político, a security force loyal to the national government, on October 31, 1957. An angry mob took the perpetrators prisoner and, according to bystanders, planned to lynch them before the Cruceño Youth, the armed branch of the Committee Pro-Santa Cruz, brought them to be detained in the Palace Cinema. Orlando Mercado recounts that in 1959, when rumors arrived that a second highland militia was about to occupy Santa Cruz, Carlos Valverde, the leader of the Cruceño Youth, led 10,000 followers to the Cine Grigota to plan their defense.[79]

Given their popularity, sometimes the theater was simply the surest place to locate a wanted man. Leaving the Palace Cinema at 9:30 P.M. one evening in mid-1959, Raúl Vaca Pereira, the heroic crocodile hunting protagonist of *La Vertiente*, was approached by an agent of the police and instructed to appear at the Commissary to provide information. He was subsequently abducted by pro-government militia men and taken to Ñanderoga, a clandestine site where several fellow cruceños were detained. Although appearing as an agent of national integration in his role in *La Vertiente*, cruceños knew Vaca Pereira as a "brave civic fighter." Despite initially working for the government in the

Ministry of Campesino Affairs, where he first met Jorge Ruiz, Vaca and his wife both subsequently joined the Cruceño Youth and each spent part of 1959 in hiding.[80] Hernando García Vespa, who was only twenty when he saw *La Vertiente*, remembers Vaca Pereira as symbol of resistance or "the first national actor of cinema persecuted by a totalitarian regime [the MNR]."

As sites of spectacle and sedition, Santa Cruz's movie theaters were far more than those blank screens showing, by government decree, films of the nation's revolutionary project in action. Just as much as the images of felled bush, gleaming asphalt, and towering derricks that its screens displayed, the cinema was itself a contested space of development. There, as deputy Mendoza had suggested in his commendation of the ICB Cerruto, cruceños witnessed themselves "exhibited for the contemplation of the nation." But as with their responses to interregional road and international rail in *El Deber*, they drew their own meanings from those supposedly didactic representations of mobility, breaking from the scripted narratives of the March to the East. Occasionally they even broke from the scripted behavior of the movie theater itself as the cinema, the cruceño social milieu par excellence, became a rallying point in the "struggle for the eleven percent." Although promptly resolved on screen, the regional tensions engendered by eastern expansion would intensify over the following decades of migration.

Transposing the Tropics

It could mean something far different to watch a Ruiz film at the Cine Santa Cruz than to see that same film at the Tesla Theater in the heart of La Paz. In the former, one might swell with pride at the gallantry of a fellow cruceño. In the latter, a viewer could just as easily snicker at the backwardness of lowlanders. In the movement from highland capital to lowland frontier some understandings of development transitioned easily while others fell apart. Yet to speak of image and narrative in the March to the East as though they only drew and cast influence in a national or regional orbit is fundamentally misleading. Ruiz's itinerary took him across Bolivia, from arid *altiplano* to dense jungle, in those years, but it also led him to other Latin American nations with large indigenous populations that were engaged in projects of national consolidation and rural modernization. The repertoire that he and his production team cultivated in producing films like *La Vertiente*, proved as useful in a transnational context as it had in a national one. The same is true of his films which also circulated far beyond Bolivia. The U.S. Point Four Program had partially sponsored *La Vertiente* and officials found the finished product a powerful example of "community development promotion." Point Four's United States Information Service (USIS), representing the foreign branches of the United States Information Agency, was promoting development projects across Asia, Africa, and Latin America at the time, sending "Point Four media deep into Global South towns and countrysides."[81] In the late 1950s, mobile film units, crisscrossed Iran hosting impromptu public screenings in schools, parks, and plazas.[82] By the end of the decade, USIS claimed a global weekly audience of 150 million for its varied media. Amidst decolonization and Cold War realignment, incoming president John F. Kennedy sought to expand its presence in the "uncommitted world."[83] USIS produced more than 120 copies of *La Vertiente* for distribution while dubbing the film into multiple languages.[84]

Remarkably, copies of *La Vertiente* were also purchased by the USSR, Czechoslovakia, and the People's Republic of China. Eastern Bloc leaders were equally enthusiastic for images of rural communities engaged in heroic acts of self-driven modernization. At the outset of the Great Leap Forward in 1958 for instance, Mao Zedong had encouraged newly established agrarian collectives to produce collections of poems that, like Ruiz films, combined "realism with romanticism."[85] Beyond Ruiz's own claim that the film had a "massive diffusion," we have few sources to gauge how viewers in other regions reacted to *La Vertiente* or, whether the dubbed copies were actually shown in public squares or through mobile cinemas in places like Isfahan or Bombay, Krakow or Tianjin. Nevertheless, the money from global sales of *La Vertiente*, for simultaneous consumption by Eastern Bloc, Western Bloc, and nonaligned audiences, helped fund the ICB for the next several years.[86] What the example underscores is that *picturing* development, a potent tool for Ruiz in revolutionary Bolivia, was a highly mobile global practice that could transcend *and* profit from the ideological divisions of the era.

It is perhaps unsurprising then, that another Ruiz film served as backdrop when in August of 1961, leaders from across the Americas met at Punta del Este, Uruguay, in a defining moment of the emerging Cold War. They were on hand as part of the Inter-American Economic and Social Council to discuss the U.S.-led Alliance for Progress. The Kennedy administration claimed that Alliance funding (suggested at $20 billion over the following decade) would dramatically extend the reach (and systematize the distribution) of development assistance previously available through Point Four helping make Latin America "the greatest region in the world."[87] U.S. officials packaged economic development in apolitical terms, but Che Guevara, speaking on behalf of the Cuban delegation, famously challenged that assertion, claiming the vaunted Alliance was little more than a U.S. ploy to isolate Cuba. In what even U.S. diplomats acknowledged was a "masterful presentation of the Communist point of view," Guevara's speech moved from impassioned to humorous as he caricatured a U.S. vision of development and its expert practitioners whose obsessive focus on hygiene masked eugenics in the language of public health.[88] "I get the impression they are thinking of making the latrine the fundamental thing, that would improve the social conditions of the poor Indian, of the poor Black," chuckled Guevara. "Planning for the gentlemen experts is the planning of latrines. As for the rest, who knows how it will be done!"[89]

In his lengthy sardonic address Guevara turned to two issues central to Bolivia, the Alliance's agrarian and mass media components. The U.S. seemed to have accepted that the breakup of traditional landownership (latifundia) was inevitable and hoped to moderate this shift with the opening of new agricultural lands. "Agrarian reform," Guevara reminded his audience, "is carried out by eliminating the latifundia, not by sending people to colonize far-off places."[90] He also challenged the attempt to include an inter-American federation of press, radio, television and cinema within the program. This, he insisted "would allow the United States to direct the policy of all the organs of public opinion in Latin America." Media would be "managed, paid for, and domesticated . . . at the service of imperialism's propaganda plans."[91] These critiques could easily have been directed at Bolivia. There, the U.S. audiovisual center had been crucial to financing ICB films and a massive formal agrarian reform (like that instituted by Guatemalan President Jacobo Árbenz) existed alongside an ambitious program to send landless highlanders to colonize *far-off places* in Santa Cruz and the northern Amazon basin. Bolivia, the second largest per capita recipient of Alliance funding over the following decade dramatically extended this program and its publicity with U.S. financing.[92] Yet Bolivia was also the only country to support Cuba at Punta del Este leading Treasury Secretary Douglas Dillon to assert in a confidential Department of State memo that the Bolivian delegation had "followed a straight communist line throughout the conference, clearly taking guidance from the Cubans."[93]

Understandably Guevara did not publicly target his lone Bolivian allies, but he may have had a Jorge Ruiz film in mind when he critiqued colonization as an alternative to agrarian reform. U.S. officials had just premiered a Ruiz documentary called *Los Ximul* for delegates at the conference as an example of their vision for agrarian reform in Latin America. The film was set in Guatemala and documented a colonization program known as "La Máquina." As with Bolivia's March to the East, indigenous highlanders were being encouraged to settle frontier land in the nation's humid lowlands. Despite these strong parallels, the regime responsible for enacting La Máquina diverged sharply from the MNR. In Guatemala, colonization of the Pacific lowlands, as well as the Northern Petén jungle, was carried out by the right-wing military government of Miguel Ydígoras. Ydígoras had participated in the U.S.-supported overthrow of Guatemala's progressive president Jacobo Árbenz in 1954, ending a ten-year period of social reform known as the Guatemalan Spring and ushering in an era of widespread state repression. Promotion of Guatemala's post-Arbenz military regimes was a central aim of the U.S. media policy in Latin America in the late 1950s and early 1960s.[94]

How did Ruiz end up in Guatemala filming for a military junta that seemed to stand in stark opposition to the social goals of the Bolivian revolution? As with the broad circulation of *La Vertiente*, the example is revealing on several fronts. Ruiz's trajectory demonstrates that the U.S.-influenced cultural common market that Guevara feared was already well established. It consisted of a series of institutions, from embassies to research stations and audiovisual centers and beginning in the 1950s and continuing in the Alliance era, media practitioners like Ruiz from across Latin America trained and worked within this network. Ruiz's fluid movement between revolutionary and reactionary regimes also signals the near universal appeal of modernizing aesthetics for midcentury Latin American nations. This was particularly true when it came to picturing frontier development projects seemingly detached from the contentious politics of the nation's core.

Ruiz traces *Los Ximul*'s origins to serendipity and personal initiative. He first heard of La Máquina while reading a magazine in his orthodontist's waiting room in La Paz and left convinced that the theme "a history moved by the shocks of the agrarian drama in Latin America," would lend itself to film. At the U.S.-run Audiovisual Center in La Paz, Ruiz pitched the idea to his friend Loren McIntyre—a photojournalist who would work extensively in South America for National Geographic over the following decades. While waiting for a response, Ruiz took an advertising job in Chile—a common second occupation for his generation of Latin American filmmakers. He was still in Santiago when he was contacted by the U.S. embassy and informed that the project was approved. En route to Guatemala, Ruiz and cinematographer Augusto Roca made a brief stop in Costa Rica. They sensed that the subject at hand, a rollback of ten years of progressive land reform, was a "delicate" one that "would require the intervention of a brilliant scriptwriter."[95] At the headquarters of the Inter-American Institute for Cooperation on Agriculture (IICA) in Turrialba, Ruiz and Roca recruited Luis Ramiro Beltrán. The center of research and programming for agricultural services across the hemisphere, IICA's Turrialba office was founded in 1943 through Roosevelt's Good Neighbor Policy. In 1959, Beltrán, a journalist who had also worked with the Inter-American Agricultural Service in Bolivia was employed in IICA's communications center.

Beltrán's trajectory, much like Ruiz's, would eventually extend across the Americas. After completing a PhD at Michigan State University, he emerged as a key theoretician in the Latin American school of critical research in communications. Beltrán had provided the narration for Ruiz's acclaimed 1953 *indigenista* film *Vuelva Sebastiana*. The following year they collaborated on a film in Ecuador commissioned by the U.S.-run cooperative health service. The Bolivian team was tasked with providing a promotional film that would encourage the Ecuadorian Congress and public to continue to support anti-malaria campaigns in the countryside. They took a story from Bolivian scriptwriter Oscar Soria, another longtime Ruiz collaborator, set in Río Abajo outside of La Paz and transposed it wholesale to the valley of Huayllabamba outside Quito. Even the title (*Los que nunca fueron*) remained the same. The film was a success and the Ecuadorian government continued the public health campaign.[96]

Reunited in Guatemala, Ruiz, Roca and Beltrán spent three months in 1959 filming *Los Ximul*. Beltrán's narrative followed the trajectories of a father and two sons of the Ximul family. The younger son, Ramón leaves their village to work as a hired hand but quickly becomes disillusioned with exploitative nature of itinerant bracero labor. The older son Andrés moves with his family to the coastal colonization zone where he prospers after receiving land, housing, and technical assistance.[97] Their father, an ardent supporter of Jacobo Árbenz's land reform, remains on the small parcel of land given to him by Arbenz but that fateful day arrives when, as the narrator admonishes, "the piece of land of Don Santiago, given indiscriminately, without pay, without selection or advice, and distributed in a disorganized way among the first *campesinos* to arrive," becomes mired in legal problems. The conflict leads in short order to the destruction of Don Santiago's home, the loss of his land, and finally, his premature grief-stricken death. The two brothers return home for the funeral and the younger son Ramón, ultimately opts for "the road of productive work under the law" by following his older brother to the coastal colonies.

Ruiz's film, a dramatic endorsement of a reactionary regime, drew criticism at home and abroad. Beltrán's narration made clear that while documenting a single initiative in a small Central American nation, the U.S. supported La

Máquina as an example of its hemispheric Alliance for Progress vision for the "ninety million campesinos in Latin America [who] are born, live and die in the misery of monocultivation, ignorance and confusion."[98] Ruiz remembered critics suggesting that the film "was going against the grain of many [he] had filmed in Bolivia," particularly those like *A Little Bit of Economic Diversification* though he preferred to justify his work both pragmatically and artistically. "The colonization plan was becoming inevitable," he pointed out. More revealingly he felt that "these [negative] interpretations were contrary to my condition as an artist and entered in a detestable political field. I already affirmed that I never enjoyed party activism, I consider it an unpunished spoiler of the aesthetic."[99]

Although the regimes that supported them were ideologically opposed, the films in question and the projects they represented were remarkably similar. Bolivian officials were just as concerned as Guatemalan military leaders about the conflictive potential of landless indigenous highland communities. Both turned with relief to the frontier landscapes depicted in each film where strong productive families and modern farmers could flourish in a new space *over there*, one supposedly detached from the political. The two films could be both antithetical and identical because at heart, the medium of film and the concept of development remained as ambiguous as they were effective. Film's illusion of the real, of transparent and total representation, lent it an inescapable authenticity and immediacy that accounted for its didactic power. Similarly, development modernization's linkage to time, growth, and evolution also gave it the force of inevitability, making it difficult for Latin American leaders (from socialist to conservative, democratic or authoritarian) to represent their nation's futures in any other form. As Beltrán would later reminisce, we "began to utilize communication for development much before theories had been proposed for it and even before the term actually existed . . . in the first years of our professional work—the 1950s and a large part of the 1960s—we made gods of our means of mass communication as if they were capable of doing a lot of good for our people."[100] Beltrán echoes Brazilian dictator Getúlio Vargas who, already in the 1930s as he pioneered his country's March to the West, was describing film as "among the most useful educational agents available to the modern state . . . a book of luminous images through which our coastal and rural populations will learn to love Brazil."[101]

After Guatemala, Ruiz returned to Bolivia where in 1964, generals René Barrientos and Alfredo Ovando Candía staged their own successful military coup. The coup brought twelve years of MNR government to an end and Ruiz elected to briefly leave the country. He found ready work in neighboring

Peru where, ironically, liberal democratic president Fernando Belaúnde contracted Ruiz to make films rehabilitating the image of the military. He took to documenting an Alliance for Progress favorite (civic action)—the military's participation in road construction, rural development, and colonization which he had extensively filmed in Bolivia. Like Guatemala and Bolivia, Peru was also planning a dramatic infrastructure project that sought to link highland and tropical lowland and redistribute the nation's large indigenous population. In Bolivia, these hopes had been linked to the Santa Cruz-Cochabamba highway. In Peru, Belaúnde, an architect, sought to reimagine the nation through the construction of a "Marginal Highway of the Jungle." A 1963 agreement between Columbia, Peru, Bolivia, and Ecuador laid the foundation for this project which would extend into neighboring Amazonian nations throughout the continent. Before his overthrow by the very military whose reputation Ruiz helped restore, Belaúnde discussed with Ruiz the possibility of making a new film about this visionary, international project.[102] Temporarily derailed by the loss of his sponsor, Ruiz returned to Bolivia to become director of the news series *Hoy Bolivia* just in time to document construction on a Bolivian portion of the "Marginal." The road, subject of a 1968 *Hoy Bolivia* newsreel, linked colonization zones in the Chapare region of Cochabamba to the highlands and would eventually replace the old highway between Cochabamba and Santa Cruz built in the 1950s.

Following Ruiz's prodigious trajectory, it becomes less surprising than we might initially expect that his modernization narratives found favor in both revolutionary Bolivia and reactionary Guatemala as well as under liberal democratic regimes in Ecuador and Peru. Even though Guevara implicitly rejected the message of *Los Ximul* in his speech, Cuba embraced similar aesthetic sensibilities. Guevara returned to Cuba after Punta del Este lauding the support of the revolutionary Bolivian delegation whom he considered "Cuba's first cousins" and the gains of Bolivia's revolution which Ruiz had also depicted.[103] In the aesthetics of new roads, water towers, and harvested fields, Ruiz and his colleagues at the ICB emphasized that the transformation of latent landscapes into productive ones was the key to national development. This was not a significant departure from the modernist aesthetics of the Cuban revolution where a desire to visually represent revolutionary progress, from agrarian production to the construction of factories and per capita income growth, would fuel a booming cinema school.[104] Because of this crucial slippage, filmmakers like Ruiz and communicators like Beltrán were able to depict development across disparate regimes and national contexts in Latin America at midcentury.

Transposing narrative and aesthetic from one context to another proved both promising and compromising. For Ruiz, the temptation was often irresistible as when he saw Bolivia's Río Abajo in Ecuador's Huayllabamba valley; the Santa Cruz-Cochabamba highway in Belaúnde's "Marginal Highway of the Jungle"; and found in Guatemala's La Máquina a counterpart to Bolivia's March to the East, "taken to a head at the same time and with the same sense of opening new agricultural frontiers." Although he accepted that there were "specifics to both realties," in his opinion, Guatemala "suffered from a sociological conflict like that of Bolivia."[105] In his mind, narrative and aesthetic could draw together those two places where ideology and nationalism threatened to pull them apart. If Ruiz conceded that aesthetics could be politicized, he maintained that this was always something that happened *post*-production in a place and time such as Punta del Este in August of 1961.

Documenting a Human Transplant

Despite Cuba's objections, the Alliance for Progress was approved at Punta del Este launching Kennedy's decade of development for Latin America. Bolivia, socialist enough to be alarming to U.S. officials but moderate enough to be worth the effort, became a star member.[106] By 1964, it was the second largest per capita recipient of Alliance for Progress funding representing 20 percent of Bolivia's gross domestic product and 40 percent of its public expenditures. Moving beyond small-scale colonization programs, Bolivia, USAID, and the Inter-American Development Bank (IDB) began financing large-scale settlement across a broad swath of Bolivia's Amazon basin ushering in an era of migration that would permanently transform the landscape, economy, and demography of the Bolivian lowlands.

The beginning of the Alliance era coincided with the ten-year anniversary of the 1952 revolution and Ruiz was tasked with producing a film to celebrate the MNR's achievements. The result was the ICB's *The Mountains Never Change*. The film begins in a small Andean village and offers a hopeful narrative of highland modernization, but once again the eastern lowlands are the star location as Ruiz's camera swoops down mountain roads onto the expanding farming zones of Santa Cruz.[107] As much as a celebration of ten years of revolution, *The Mountains Never Change* was testament to a decade of Ruiz's work romanticizing frontier expansion in Bolivia and abroad. Yet it is legitimate to question whether the MNR was reaching potential colonists or simply carrying on a self-congratulatory conversation with itself. ICB films were costly to make and difficult to distribute in a country with few screening areas even if that number was steadily increasing. They were also prohibitively expensive for poor Bolivians, and media practitioners like Beltrán worried that messages were "restricted in large part to urban minorities of the upper strata."[108] In the 1950s, when colonization remained tentative, imagining the March to the East was likely as important as enacting it. The expansion of colonization in the 1960s called for new forms of visual media and the state increasingly spoke directly to the indigenous migrants it targeted for resettlement.

Flimsy illustrated pamphlets did not always possess the production quality of film but promised greater reach. Printed for mass distribution, they were written in simple language, relied on ample visual aids, and had titles like "How Will I Live and Work in My New Parcel?" or "We Will Form Our Cooperative." A simple sketch from the illustrated pamphlet, "What is the 10 Year Plan?" produced in the same year as *The Mountains Never Change*, and created by longtime Ruiz scriptwriter Oscar Soria, captures the demographic and geographic logic behind an ambitious new colonization program.[109]

Directed at potential settlers these pamphlets hoped to generate mobility. Accordingly, the most frequently repeated motif was the descent to the tropics itself. Illustrated maps sketched the route from La Paz to unfamiliar settlement zones in Santa Cruz and the lowland regions of Cochabamba and La Paz. The transition between ecosystems fostered an inevitable juxtaposition between arid altiplano and humid jungle."[110] In these images, the state reminded viewers that Bolivia was no longer limited to its Andean core. Other pamphlets sought empathy with migrants by capturing the anxious drama of colonization as a human endeavor. Images of Andean bodies in motion took several forms. Often this was done with cartoon-like characters with Andean clothing, a hat and poncho, moving from mountains to palm trees. In an image from "How to Live and Work My New Parcel," a prospective migrant leaves a hilltop with the eastern vista and rising sun in the background.[111] Photos of colonists tearfully embracing family members at departure points or trudging along jungle trails proliferated.

These representations of mobility, whether stirring or simplistic, were followed in each case by images that moved successively towards permanence. After all, at the point of departure, mobility was an essential ingredient in this state project but, once relocated, the continued mobility of colonists was often seen as a threat by officials worried about high rates of settler abandonment in new colonization zones. The question of how to manage migration's uncertainty while creating well-adjusted citizen farmers was thus a key one for the Bolivian state as it was for other Latin American nations that sought to

Image from the pamphlet "What is the ten year plan?" depicts the spatial logic of Bolivia's lowland colonization. Oscar Soria, Jorge Sanjinés, and Ricardo Rada, "Qué es el plan decenal?" (La Paz: Burillo, 1962).

move and then fix rural subjects in place.[112] The gender order of the family served as a tool for mediating regional tensions in Ruiz's films. In these crude pamphlets it also offered a means to promote the future stability of the colonist in a movement from migrant to settler.

In initial images of migration, the tropicality of land was frequently represented by a token palm tree, laughably distinct from the dense bush colonists would encounter. In subsequent frames, the management of this *primeval forest* was a frequent object of representation. While the bulldozer became as one historian has noted, "one of the grand stars of Bolivian cinema" in those years, the hatchet was a more familiar object for most colonists.[113] Bush clearing, was characterized by manual rather than mechanized labor and could easily lead new settlers from the relatively treeless Andes to lose heart. In these pamphlets environmental transformation was frequently pictured as the recovery of a masculine ideal for those who had abandoned their place of birth.

For Andeans to move from nervous, or confused departee to well-adjusted settler they would have to pass through a series of crucial steps in which they

"Cómo Viviré y Cómo trabajaré"
¿en Mi Nueva Parcela...?

Solo así te sentirás orgulloso y serás todo un hombre. Tu propiedad construida palmo, a palmo tumbando árbol por árbol, luchando firmemente contra la madre naturaleza,

Images from a pamphlet titled "How will I live and work on my new parcel?" depict the transition from confused would-be migrant to confident settler. "Como viviré y como trabajaré en mi nueva parcela" (La Paz: CBF, n.d.).

were frequently imagined not as individual actors with unique skills and goals but as empty vessels to be molded at the hands of *técnicos*. How pamphlets portrayed knowledge transfer was tricky in this regard as it involved a potentially emasculating, absolute surrender of authority to project officials (often an explicit condition of the settlement contract) followed by a dramatic resumption of independence at a point in the future when state support would be withdrawn. In the meantime, settlers were subject to a monodirectional barrage of inputs, medical tests, and vaccinations. They would receive instruction and technology packets from home economics agents, agronomists, and rural sociologists. In the process they would gradually be stripped of their former selves and inadequate knowledge and taught everything from planting unfamiliar crops to feeding their families and raising their children. "You wife will not feel *abandoned* either [emphasis added]," read the caption next to the image of a home economics agent scrutinizing a woman's use of a sewing machine.[114] As they passed through the stages of colonization depicted in these cartoons, their habits, clothing, and very bodies would be transformed. They would exchange alcoholic chicha for volleyball and soccer, don cotton shirts and straw hats in place of ponchos and llama wool ones, and transform stooped postures and thin waistlines into straightened backs and solid guts. The results of this invasive process that extended from field to household was inevitably pictured as the reconstitution of a stable, independent nuclear family with clear title to their land. Individual and familial success in the colonization zone fed vertically back up into the nation. "Assure the future of the worker's family" proclaims one caption next to a group about to embark for the lowlands. "The tropics await you, a true hope for [all] your sacrifice. The tropics are your future. Your work will make Bolivia great."[115]

Just prior to the 1964 coup that briefly sent him to Peru, Ruiz helped produce one of these pamphlets. "A Human Transplant" was ostensibly for distribution to prospective colonists but, curiously, was written in both English and Spanish. U.S. sponsors, it would appear, were hoping to see stable representations of their policies and dollars in action. Ruiz's pamphlet was replete with bodies and machines. Bulldozers perched on the edge of precipices and loaded trucks raced down winding mountain roads. Colonization, the accompanying captions stated, was the "perfect combination of man and machine."[116] Yet, in the images that followed, colonists had more in common with tropical nature than Caterpillars. People, like the bulldozed landscapes, were viewed as infinitely malleable. "They will soon be good friends," the narrator assured the reader below which the image of a young girl offering food to a parrot. While children got to know the local environment, "the few inhabitants of the chosen regions, have received with pleasure this contingent of *kolla blood*," the narrator assures the audience.[117] Other images showed happy settlers dutifully listening to technicians, playing volleyball and eating rations before heading out to clear the land. "A Human Transplant" invoked masculine honor while parading the strong families that colonization would produce. "Wellbeing for the mother of his children, this is the hope in the heart of every man," concluded the pamphlet with a shot of resettled indigenous women and children, "by whom and for whom man labors" able to "look with confidence towards the future." From technicians to family, Ruiz's pamphlet presented colonization as an immaculately scripted affair. Even the most candid and intimate scenes of settler life were ones in which the narrator reminds the audience "all details are foreseen."

Conclusion

In crude pamphlets and feature films Jorge Ruiz and other members of the ICB attempted to humanize eastward expansion. Through stories of individual migration, they reimagined economic and social policy as a family romance linking Bolivian development to assumptions about gender, region, and thus race. If their solutions to serious problems posed by colonization and regional integration seem pat, this diverse media still provides an excellent point of departure for a subsequent exploration of the conflicts that officials imagined would be so easily resolved. For all its simplicity, the vision of eastern expansion, which sought to represent national transformation by capturing bodies in motion, proved enduring and broadly appealing, claimed by planners, international financiers, and successive Bolivian governments for decades to come. The latter represent a spectrum of political ideologies and formulations from revolutionary nationalism, military-led development, and bureaucratic authoritarianism to neoliberal multiculturalism and pluriethnic socialism. It was also embraced by the hundreds of thousands of migrants, both foreign and national, who poured into the eastern lowlands over the following half century, radically reshaping the region's social and environmental history even if the latter's migrations looked little like the romances or meticulously planned movements pictured in films and pamphlets in which "all details are foreseen." Like the ever-resilient Turner thesis in American history, or the myth of racial democracy in Brazil, the March to the East was a vision that, once established, Bolivians and foreigners were forced to engage with even when, as I show in the ensuing chapters, they routinely and loudly spoke to its limits and brought their owning meanings to the unprecedented mobility it entailed.

As rail lines and asphalt drew residents of Santa Cruz into greater proximity with their highland brothers and foreign neighbors, cruceños produced their own images and narratives of eastern development at times distinct from those originating in the Andes. They shared with the MNR a fondness for images of the blessings of development and their department's natural and feminine fecundity. Yet their celebrations of arriving trains, newly paved roads, beauty queens, and bountiful harvests were interspersed with memories of the highland militia that had descended on the city in the late 1950s civic struggles. Guevara Arze might have been the intellectual author of the March to the East but for regionalist authors he was a criminal that had sent highland hordes to occupy Santa Cruz and would remain a reviled figure in the region for decades.[118] When cruceños watched *A Little Bit of Economic*

Diversification they saw drunken Periquito along with humble Santos coming down "the new highway." Viewing *Los Primeros* amid the "struggle for the 11%" they might identify as much with Doña Ramona, whose ability to earn a living selling oil had been appropriated by the state as with her son who dutifully served the national oil company. Screening *La Vertiente* at the Cine Santa Cruz, the take-home message for some was that the camba had won over the kolla and not that the former had been improved by the latter.

Beyond Bolivia, the narrative and aesthetic of the March to the East resonated, and at times directly intersected, with state-building projects throughout the region and across the globe. Neighboring nations with Amazonian frontiers also sought to broaden their national imaginaries to include these marginal frontier landscapes at midcentury. This much is evident in Belaúnde's desire to have Ruiz promote his "Marginal Highway of the Jungle" or in the attempts of Brazil's military leaders to advertise the colonization of the Amazon basin to potential settlers after the 1964 coup as "a land without people for a people without land."[119] As Ruiz collaborator Ramiro Beltrán explains, Latin American governmental embrace of mass media reflected a total view of its power, "almost to the point, of prompting the modernization of our nations in little time and practically on [its] own."[120] They were not alone in a postwar world characterized by ideological polarization and decolonization. In Africa, colonial authorities and postcolonial governments alike viewed film as a powerful "instrument of modernization."[121] Picturing the development of frontier landscapes proved broadly appealing to capitalist, Communist, and nonaligned regimes. Bolivia's new leadership for instance, embraced a revolutionary aesthetic that finds parallels with the Soviet Union, Mexico, and Cuba. Yet they effectively made use of Cold War fears to secure more U.S. funding between 1953 and 1964, in absolute terms, than any other Latin American nation.[122] While relying on U.S. financing for ICB films at home, Ruiz and Beltrán could also count on a network of U.S. support abroad and along with the colonists they frequently represented, they also became migrants. In doing so, they joined thousands of planners, academics, and missionaries that moved between state bureaucracies, regional development projects, research centers, and North American universities in the middle decades of the twentieth century. Tracing Ruiz's trajectory, threading the national, regional, and transnational, foregrounds the multiple and indispensable contexts in which image making and viewing took place. Like other mobile actors discussed in this book, his repertoire underwent crucial additions, translations, borrowing, and transformations at various stops along his endless pilgrimage. These transnational threads would in turn shape developments back home in Bolivia.

CHAPTER TWO

Military Bases and Rubber Tires

*Okinawans and Mennonites at the Margins of
Nation, Revolution, and Empire, 1952–1968*

Over three days of *Semana Santa* (Easter Week) in 1952 a heterogeneous group of miners, workers, students, police officers, and military officials joined members of the Nationalist Revolutionary Movement to take control of Bolivia's major cities and overthrow the country's ruling military junta. Jorge Ruiz was on hand to film the first interview with newly sworn-in President Victor Paz Estenssoro, who returned to Bolivia from Argentine exile on April 16. The interview was broadcast on NBC and emphasized the moderate nature of Bolivia's revolution to the United States and the world. The MNR hoped to mollify U.S. anxiety over its perceived leftist tendencies, but it also presided over a tenuous coalition demanding social and economic change. In October of that year, the MNR enacted one of the revolution's most radical measures by nationalizing the mining sector and expropriating the holdings of the country's fabulously wealthy tin barons. The following August, responding to peasant land seizures, the MNR promulgated an Agrarian Reform Law.[1] While ending forced labor and promoting the development of eastern agriculture, Decree 3464 also enshrined the link between citizenship and farming, defining the "social-economic function" of land as belonging to "those that work it."

Unexpected transnational undercurrents run parallel to these familiar narratives of revolutionary nationalism and agrarian citizenship. The same month as the mining decree, a small charter plane arrived in La Paz bearing six plainly dressed, low-German speaking farmers from Paraguay. The men— Mennonites from colonies in the Gran Chaco—were intent on investigating new settlement opportunities in the lowlands around Santa Cruz. A member of that delegation, Peter Regier, would later publish his first impressions in their home colony's newspaper, *Mennoblatt*.[2] Addressing a readership of a few thousand settlers, many of whom had fled Mennonite colonies in Russia during collectivization in the 1930s, Regier did not appear worried about the direction of Bolivia's social revolution or the MNR's policy toward foreign property holding—mine expropriations notwithstanding. He characterized the government as eager to promote agricultural production and willing to grant the special religious, social, and educational exemptions that his community of low-German speaking, pacifist, Anabaptists required. Regier also outlined a range of crops that flourished in the subtropical climate of the eastern lowlands and the new road and rail infrastructure nearing completion. "[In] the field of agriculture," he concluded, "sinfully little has hitherto been done, but the prospects are unexpectedly large."[3]

Mennonites were not the only foreign petitioners traversing the corridors of newly created revolutionary ministries that sought to present the MNR with an ethnic construction of agrarian citizenship. The year after Regier's visit and only a few months before the passage of the Agrarian Reform Law, the Ministry of Agriculture approved a resolution outlining a plan of colonization put forward by an organization known as the Uruma Society. Founded in 1950 and "by Japanese elements resident in Bolivia and some already nationalized," the Uruma Society's president was José Akamine, a longtime resident of Riberalta in the lowlands of northern Bolivia. Like other members of Bolivia's small Japanese community, he first came to South America as a railway worker in Peru and, facing horrendous conditions, had crossed into Bolivia to work in the Amazonian rubber industry before settling permanently in the region. Learning of the destruction and suffering in postwar Japan, Akamine advocated the same solution he had taken nearly half a century earlier: emigration. By the time of his 1953 petition the Uruma Society already possessed 50,000 hectares in Santa Cruz and had established a model farm on the site with seventy-three hectares of cotton and corn under cultivation as well as four houses where fifteen Japanese settlers lived. He now proposed to dramatically expand the scope of this test initiative whose "principal object was the development of agriculture [and colonization] in the region."[4] Over the following decade he planned to bring 3,000 Japanese, "or more correctly, Okinawan families," to Bolivia as lowland colonists.[5]

Bringing together these stories of Mennonites and Okinawans privileges mobility (both its practices and meanings) over more primordial concepts of identity.[6] Indeed for each of these diasporas, their identity as Okinawans and Mennonites was intimately and increasingly intertwined with their migratory experience. Okinawans' long history of emigration across the Pacific Rim was given new impetus by the establishment of U.S. military bases on the Ryukyuan islands after the Second World War. Okinawans were displaced and relocated by U.S. officials in the mid-1950s and these experiences informed the ways that they resituated themselves in Bolivia. Mennonites had been on the move for even longer before a combination of environmental, agricultural, and religious factors drove a large-scale emigration from Mexico in the 1960s that built on the small-scale colonies established by Regier and other Paraguayan Mennonites a decade earlier. Mennonite and Okinawan migrations overlapped in time but trended in opposite directions reflected in the structure of this chapter, which begins in postwar Okinawa before exploring the ways that several thousand Okinawan settlers and a handful of Paraguayan Mennonites established themselves in Santa Cruz. The final section of this chapter travels to northern Mexico to explore the factors that would contribute to the first large-scale migration of Mennonites in the late 1960s, just as Okinawan migration to Bolivia begin to wane. The subsequent history of Mexican Mennonites in Bolivia is treated in a separate chapter. Despite their distinct trajectories, some surprising parallels emerge when following the transnational networks that these two very different diasporas forged through the late nineteenth and early twentieth centuries. Okinawans and Mennonites each faced a tenuous claim to citizenship and belonging; were subjected to concerted attempts at assimilation by national governments; and engaged in migrations that were both forced, and enabled, by imperial and national expansion. Despite their largely rural backgrounds, Mennonites and Okinawans proved remarkably adept at crafting the sort of expansive global networks that one scholar has referred to as "transnationalism through parochialism."[7]

At the outset, the very terms *Mennonite* and *Okinawan* merit further scrutiny. After all, what was meant by the strange turn of phrase, "Japanese or more correctly, Okinawan," in the ministerial resolution endorsing Akamine's Uruma Society in 1953? From an administrative perspective, the term *Okinawan* designates residents of a Japanese prefecture that includes the island of Okinawa along with hundreds of others in the Ryukyuan chain, a physically remote archipelago extending far to the south of the Japan's four principal or *home* islands. Yet that neutral term masks a more complicated history.[8] Okinawa was an independent kingdom not formally colonized by Japan until the 1870s. Over the following half-century Japanese administrators subjected Okinawans to a concerted assimilation campaign. Targeting a range of cultural practices through national education programs, they treated the islands and their inhabitants as a culturally inferior colonial possession while frequently congratulating themselves on their own success by celebrating the new Okinawa as the "most Japanese of all places."[9] Beyond cultural imperialism, direct rule involved a transition to extractive plantation agriculture in Okinawa with profits largely returning to the Japanese metropole. Plantation agriculture displaced small farmers and in turn engendered emigration. Okinawans first traveled to the Japanese home islands and eventually followed sugarcane throughout the Pacific, transforming the island landscapes of Hawaii, the Philippines, Java, and New Caledonia in the process. After 1899, the Japanese government officially promoted emigration and large Okinawan (and Japanese) communities sprang up in Brazil, Peru, the United States, Canada, and elsewhere in the Americas. By 1938, 12 percent of the islands' population was living in diaspora, and the expansion of the Japanese empire produced a further migration of 50,000 Okinawans to work and settle in Japanese-controlled territory in Manchuria, Korea, and elsewhere.

After the deadly and protracted Battle of Okinawa and U.S. victory in the Second World War, Okinawa's status was transformed once again. Okinawans were repatriated from across the Pacific and from 1950 to 1972, the islands were occupied and administered directly by the U.S. military-run civil administration (USCAR). Okinawans were no longer citizens of an independent Japanese nation but stateless subjects of an occupying power. U.S. authorities sought to undo a half-century of Japanese assimilation. Their attempts to distance Okinawans, many of whom considered themselves to be fully Japanese, from Japan took many forms, but they were revealed most clearly in the U.S. insistence that Okinawans refer to themselves by the geographical neologism "Ryukyuan" rather than "Okinawan." Yet ironically, postwar population pressures soon led the United States to return to the prewar Japanese policy of encouraging emigration. As residents of a colonized periphery of Japan, Okinawans had engaged in what Symbol Lai describes as a form of "sub-imperialism . . . a colonial relationship emanating from one periphery to another," when migrating through Japan's expanding empire in the 1930s and 1940s.[10] Amid the escalating Cold War in the 1950s and 1960s they repeated this pattern, only now under U.S. auspices.

The term *Mennonite* describes a pacifist, Anabaptist faith that emerged during the Protestant Reformation yet encompasses a broad spectrum of religious denominations from Old Colony Mennonites similar in practice to their fellow Anabaptists the Amish, to evangelical, acculturated denominations that embrace modern technology. Bolivian Mennonites were overwhelmingly members of the former, an ethnically distinct, low-German speaking segment of the Mennonite faith that had historically pursued physical and cultural separation from surrounding society in endogamous, agricultural colonies whose church leaders attempted to enforce technological restrictions with shunning. Seeking religious and cultural exemptions, especially the right to conduct schooling in their low-German dialect and freedom from military service, they had migrated eastward in the centuries after the Protestant Reformation from Holland to Prussia to the Ukrainian steppe. Mennonites were initially welcomed into the Russian Empire in the 1780s as

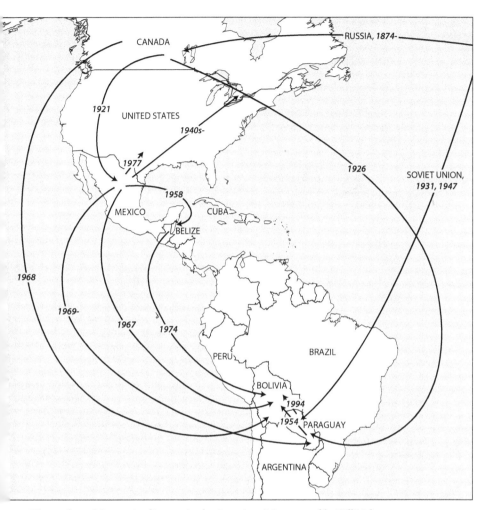

CANADA

RUSSIA, 1874-

1921 UNITED STATES

1940s-

1977

1958

MEXICO

CUBA

BELIZE

1926

SOVIET UNION,
1931, 1947

1968

1969-

1967 *1974*

PERU

BRAZIL

BOLIVIA

1994

1954 PARAGUAY

ARGENTINA

The evolving Mennonite diaspora in the Americas. Map created by Bill Nelson.

settlers in the newly conquered Ukraine by Catherine the Great. By the late nineteenth century, as Japanese assimilationist campaigns targeted Okinawan cultural, religious and linguistic practices, Russian Mennonites found themselves subject to a *Russification* policy on the part of Alexander III that threatened the military and educational exemptions they enjoyed in the Russian Empire. In the 1870s as Okinawans were settling the west coast of North America, the first Russian Mennonites arrived on the Canadian prairies in the newly created province of Manitoba. Although they were initially welcomed as frontier modernizers, the First World War led provincial Canadian governments to view low-German speaking Mennonites as threatening outsiders

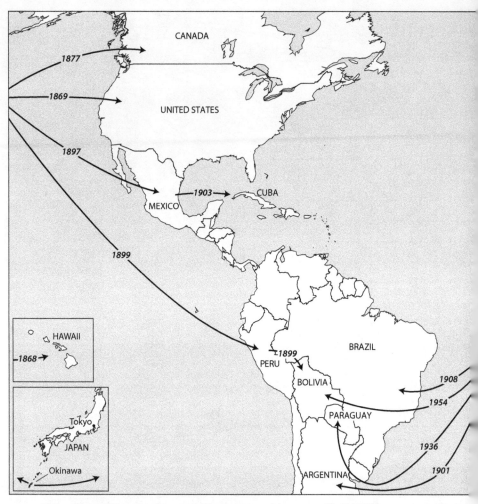

The evolving Okinawan diaspora in the Americas. Map created by Bill Nelson.

and, despite guarantees to the contrary, to subject them to public education.[11] As a result, Mennonites sent delegations to Latin America in the 1920s, where they were greeted as settlers of frontier landscapes just as they had been up north. By the 1950s, as plans for Bolivian colonization took shape, there were well-established Mennonite and Okinawan colonies in Mexico, Paraguay, Brazil, and elsewhere in Latin America. Over that decade, new geopolitical and environmental pressures would extend these diasporas to Bolivia. While presumably distinct from their Okinawan neighbors in Bolivia, Mennonites had thus also engaged in a historically recurring form of subimperialism as victims and agents of imperial and national expansion.

Agrarian and Militarized Landscapes in Postwar Okinawa

In a 1954 report, sociologist James Tigner wrote the following about Okinawa: "[They] have been a deficit area owing to a shortage of arable land, low-grade soils, virtually no exploitable natural resources, and severe population pressure for generations [offering] no more than a marginal, near-famine living standard."[12] The region presented Tigner, the scientific observer with "a corollary to overpopulation in a food deficit area [which] is the burden of continuing [U.S.] relief."[13] Curiously, it was just the sort of pitiable technocratic assessment that the MNR often directed at the Bolivian altiplano in the post-revolutionary period. Not only did Tigner's concerns about demography and food security resonate with the language of Wálter Guevara Arze's *Plan Inmediato*, his report, "The Okinawans in Latin America: Investigations of Okinawan Communities in Latin America, with Exploration of Settlement Possibilities," was published the same year. Andean and Okinawan landscapes posed distinct problems for Bolivia's revolutionary government and Okinawa's U.S. occupiers respectively in 1954, but both Tigner and Guevara Arze would identify colonization in Santa Cruz as an ideal solution.

Tigner's report was the product of his travels between Okinawa and Bolivia in the 1950s.[14] In contrast to the private initiative shown by Akamine and the handful of Okinawan-Bolivians that formed the Uruma Society in 1950, Tigner benefited from the powerful institutional support of the U.S. Armed Forces. With the San Francisco Peace Treaty of 1951, the U.S. officially claimed sovereignty over the former Japanese prefecture of Okinawa including the island of Okinawa and hundreds more in the 1,000 kilometer long Ryukyuan archipelago. Those islands faced a serious demographic crisis owing to wartime destruction and the postwar repatriation of 176,000 members of the Okinawan diaspora. In the early 1950s, the U.S. military expropriated a full quarter of the islands' arable lands for base construction, producing a new round of displacement. In response, the military tasked Tigner—a member of the Stanford-based Hoover Institute—with traveling through the constellation of Okinawan communities scattered across Brazil, Argentina, Bolivia, and Peru to find a suitable location for this now *surplus* population. In Bolivia, Tigner met Akamine and he advocated for expansion of the Uruma Society's fledgling colonization plan in his final report. Over the following decade, U.S. military support resulted in the emigration of 500 Okinawan families to Bolivia, a total of 3,200 people whose individual migrations linked U.S. geopolitical strategy in the Pacific with Bolivian's revolutionary state-building project in the eastern lowlands.

Tigner claimed that Okinawa's problems "were inherited by the United States."[15] Like Guevara Arze's discussions of overpopulation in the altiplano, this disingenuous comment naturalized a political landscape.[16] U.S. land expropriation was driven by a vision of global security that included permanent military bases in the Far East. Within the context of the Korean War, the islands offered a staging ground for U.S. intervention and they remained central to U.S. conceptions of global security in the Eisenhower and Vietnam eras. In claiming Okinawan sovereignty, the U.S. also took official responsibility for more than half a million Okinawan "wards."[17] USCAR would manage the Ryukyus in consultation with a local Government of the Ryukyuan Islands (GRI). The U.S. attempt to reconcile the conflicting obligations of hegemon and guardian in the emerging Cold War climate of East Asia defined postwar Okinawa. This inherent paradox of U.S. rule emerged most clearly over access to land. Tigner's 1953 report coincided with a new round of military land acquisitions. In the contest between militarized and agricultural landscapes in the 1950s, Okinawans employed a range of strategies from petitions and physical confrontations with soldiers to interviews with local and foreign newspapers in which they emphasized the tragic circumstances of their removal. Displacees also held large public rallies, wrote protest songs, and plastered military bases with leaflets. These tactics have been explored in the extensive historiography on U.S. occupation, but two elements of this repertoire, one ideological and the other agrarian, merit further exploration.[18] Although both were unsuccessful in their principal aim, the former pushed the U.S. to support Bolivian colonization, and Okinawans would employ aspects of the latter in Santa Cruz.

Nervous USCAR officials dutifully gathered and transcribed a mass of materials related to the land conflict. Their obsessive cataloging of even the most mundane details were inseparable from the paranoid ideological climate of East Asia in the 1950s, and Okinawans often framed removal in terms of that geopolitical tension. Some Okinawan petitioners, for instance, spoke to the anticommunist and prodemocracy rhetoric of the United States. One newspaper article on removal that appears in USCAR archives includes a quote from an elderly woman who complains that, "the muzzles that should be directed at the Communists are directed at the islanders."[19] A petition posted at a U.S. military base in Naha, Okinawa's largest city, on Veteran's Day expressed outrage and disbelief that "such conducts were done by the United States of America which is leading the world with the spirit of freedom and equality and philanthropy."[20] In addition to highlighting U.S. hypocrisy, Okinawans increasingly turned to radical politicians who vigorously protested removal.

In his 1953 report, Tigner had been frank in this regard pointing out that "traditionally farming and ownership of land is one of their most cherished desires in life." Evincing a deep-seated U.S. fear, he argued, "restiveness and dissatisfaction will inevitably accompany the waning prospects" and presented the youth of Okinawa as "a potentially vulnerable element" to Communist infiltration.[21] For anxious U.S. officials, such fears appeared to become reality in 1953 when Kamejiro Senaga, leader of the left-leaning Okinawan People's Party, led two separate protests of military removal. Senaga failed to block removal but the unsuccessful protests were closely monitored by USCAR and further encouraged U.S. officials to support the emigration program.[22]

While often engaging with Cold War discourse directly, Okinawans' resistance to land occupation also drew on a compelling script of agrarian citizenship in which they appeared as model farmers, a genre of appeal that Okinawans (and Mennonites) would later employ in Bolivia. In a petition addressed to USCAR's Deputy Governor, residents of Maja hamlet on Ie-Jima—an island to the west of Okinawa Island—contrasted the agricultural and militarized landscapes of their home which had become the site of a U.S. gunnery range. Ahagon Shōkō who led hunger strikes and became a key organizer for Maja farmers, boasted an impressively transnational résumé. Born at the turn of the century he had spent a decade and a half laboring in Cuba and Peru—two of the larger migration destinations of the Okinawan diaspora—before returning to Okinawa in the 1930s and purchasing a farm on the island.[23] With Shōkō's leadership, Maja residents framed the tragedy of their eviction in relation to their history of successful farming and highlighted their achievements as winners of the region's annual agricultural and forestry contest over the last four years.[24] Prior to eviction, theirs had been a "peaceful and model village that had previously produced no offenders," yet it currently faced an epidemic of theft. "Crimes have been committed one after another as if to entirely change the aspect of the village," they continued, juxtaposing the model farmer with the delinquent and dangerous evictee.[25] Other petitions from that year also drove home the image of productive Okinawans engaged in farming and thus leading a peaceful life, in opposition to images of displaced Okinawans at the mercy of U.S. forces.[26] The contrast between agrarian and militarized landscape was performed quite literally in a protest supported by Kamejiro Senaga of the Okinawa People's Party (OPP). Villagers, who had been farming land slated for expropriation, refused to harvest their crops prior to removal. U.S. authorities seeking to avoid the image of crops destroyed by construction teams called in soldiers to serve as ad hoc field hands. Senaga and his fellow protesters rejected the conciliatory gesture and threw their harvested cabbages under the treads of advancing U.S. bulldozers.[27]

Okinawans succeeded in attracting enough attention to the land conflict to provoke a U.S. congressional hearing in 1955, but they were surely disappointed with the results. Congress's 1956 report simply reiterated the policy of cash payments for land acquisitions, at a slightly increased rate. But the land struggles of 1955 are important for two reasons. First, as they would do in Bolivia, Okinawans demonstrated strategies for negotiating over land with a state that did not consider them to be citizens. Second, because of the hearings, additional funds were made available for resettlement (within the islands) and emigration (off the islands). The former initiative, which sought to "increase the scope of arable land and agricultural productivity in the Ryukyus," offered a minor mitigation of Okinawa's population woes.[28] Indeed, Japan had been promoting Okinawan resettlement on the remote Yaeyama islands for more than half a century to little effect. In contrast, the large-scale settlement projects in Latin America proposed by Tigner increasingly appealed to the United States. From a financial standpoint it was significantly cheaper to send Okinawans around the globe than to refurbish land on the volcanic Ryukyuan islands. Congress had allocated $5.7 million for resettlement within Okinawa. Between 1955 and 1959, $2.1 million was used to relocate only 650 families within the Ryukyuan archipelago. In that same period over 7,500 individual Okinawans left for South America, principally to Brazil. The majority emigrated without any official sponsorship as call-in migrants financed by members of the Okinawan diaspora in Latin America. Even the 1,173 people directly funded by USCAR to settle in Santa Cruz in that period did so at lower cost. The average expense per emigrant to Bolivia was $421, whereas resettlement within the Ryukyuan islands—in areas not requiring costly land reclamation of up to $10,000 per hectare—cost from $466 to $766. "This is one of the most, if not the most austere, efficient and effective of all economic programs pertaining to the Ryukyus," noted one USCAR official. "Even though it might be considered as exterior to the Ryukyuan economy, its contribution to the fundamental economic problem of the Ryukyus is considerable."[29]

From a political perspective, the U.S. interest in managing dissent by physically displacing dissident elements on to empty lands finds a parallel in Bolivia where a similar spatial politics was employed by the MNR, which sent cooperatives of ex-miners to isolated regions in the lowlands with the hope that their revolutionary message would find a limited audience. Long relying on emigration in the face of challenging circumstances at home, Okinawans may have been fully aware of the U.S. motives for resettlement. In oral histories conducted by sociologist Kozy Amemiya, Okinawan settlers in Santa Cruz, some of whom had taken part in leftist politics, remembered the U.S. emigration scheme as a "thinning policy" for disposing of an undesirable element.[30] Yet cynical or otherwise, many Okinawans were quick to avail themselves of the opportunity provided by Latin American colonization. According to officials at the Department of Social Welfare in Naha (administrative center of Okinawa) in 1960, a staggering 42,000 families had expressed interest in settling in South America. Projects varied. An agreement had been reached to settle Okinawans around Brazil's new capital, Brasília, on the country's tropical savannah frontier. The GRI's most fervent praise was reserved for Santa Cruz. Officials noted that the "settlement site in Santa Cruz is so fertile as to grow crops without fertilizing for twenty years."[31] Infrastructure linked the region to every place in the country and the "native population offers friendly and positive cooperation to the Okinawan immigrants." The Department of Social Welfare felt certain that for those reasons it was "obvious that the emigrants settling into Bolivia and Brazil will be successful and prosperous."[32]

Japan also looked to emigration in the postwar era, resettling 50,000 individuals in Latin America.[33] Although the majority came to Brazil, more than 1,000 settled in the colony San Juan Yapacani in Santa Cruz. In the Bolivian context, Japanese and Okinawan migrants were frequently conflated and adopted similar responses to local xenophobia. However, in transit Okinawans faced a series of logistical problems related to their status as U.S. wards, distinct from those of their fellow migrants from Japan. By 1961 USCAR had already been sending Okinawan emigrants to Bolivia for six years. Yoshihide Higa chaperoned the eleventh group to travel by boat through the South China Sea, the Indian Ocean, around the Cape of Good Hope, and across the Atlantic to the Brazilian port of Santos, where they would board a train for the Brazilian-Bolivian frontier before switching onto the Santa-Cruz-Corumbá line constructed the previous decade.[34] Their ship left the port of Naha on February 18 and included nearly two hundred Japanese migrants that, like their fellow Okinawan passengers, were bound for Bolivia and other destinations in South America. A month into the voyage, they made port at Lourenço Marques (the present-day city of Maputo in Mozambique) with everyone eager for some freedom from the confined space of the ship. If Japanese and Okinawan emigrants had felt common cause on the voyage thus far, difference was here reinscribed. Japanese passengers, passports in hand, easily obtained entry permits to disembark and tour the city. Okinawans faced an unexpected problem. Portuguese colonial officials did not recognize the curious "Certificate of Identity" issued by U.S. authorities, which was clearly not an official entry document. After a long delay and some intervention from the American consulate, Okinawans were eventually able to obtain permits.

The awkward moment could be dismissed as a typical bureaucratic bungle, one that consular officials had successfully resolved. After all, how were officials in Portuguese East Africa to know about Okinawa's special status? Yet the incident highlights a central aspect of postwar Okinawan life that had still not been resolved after six years of U.S.-supported emigration to Bolivia and would continue to affect Okinawans in Santa Cruz. Okinawans remained stateless, no longer Japanese yet certainly not American. This was not the first humiliation that Okinawans had experienced. USCAR officials noted that another case from May of 1958 was receiving wide circulation in Okinawa with newspapers reporting on the struggles of an Okinawan stranded in Northern Italy after attempting to enter Switzerland with a U.S. Certificate of Identity. The hapless emigrant wandered between the U.S. and Japanese embassies as each delegation attempted to claim he was the responsibility of the other.[35]

Throughout the early years of resettlement USCAR officials repeatedly confronted the issue of Ryukyuan nationality and residual Japanese sovereignty. In January of 1956, General Lyman Lemnitzer, Commander of U.S. Forces in Japan, noted that some Ryukyuans took U.S. unwillingness to grant them American citizenship as evidence that "they remain full-fledged Japanese citizens." In contrast, "most Ryukyuans" he noted, "simply are confused and regard themselves almost as stateless persons with no official allegiance."[36] Deputy Governor General James E. Moore, expressed frustration. Efforts to create a designation "citizens of the Ryukyu Islands" had been rejected by foreign governments, and he urged the United States to continue to work toward a "citizenship status for Ryukyuans." If the result, "would be to create a mess of dual nationals . . . it should not have any more deleterious effect than the present situation where there is doubt they have any citizenship."[37] A 1958 USCAR memo revealed that even the Bolivian government referred to Ryukyuans as "neither fish nor fowl" because some colonists turned to the U.S. State Department for assistance while others appealed to Japanese diplomatic officials.[38]

In 1960, the year before Higa's voyage, USCAR had held an internal conference to discuss ways of gaining wider recognition for the Certificate of Identity. At the outset, participants conceded the obvious: The "travel document issued by USCAR is not a passport" and that "while it may be advantageous to call it a passport, this could not be done because it does not identify the nationality of the bearer."[39] "What was the nationality of these individuals," one participant wondered, proposing that USCAR could "refer to Ryukyuans as Japanese nationals." Another responded, "the U.S. can only attest to the nationality of its own citizens." Ultimately, attendees agreed to push forward with a campaign to "familiarize" governments with the document while altering its format to include, "issued in lieu of a passport immediately following and under the words 'Certificate of Identity.'" They broached the idea of employing the term "Okinawan" rather than "Ryukyuan" in the document "inasmuch as some people recall Okinawa from World War II days whereas the term 'Ryukyu Islands' is a lesser known geographical term," but feared it was too closely linked to the prior existence of Okinawa Prefecture and thus a history of Japanese governance.[40] With the easing of Cold War tensions unlikely, U.S. officials continued the deliberate process of detaching Okinawan sovereignty from Japan that began with the postwar repatriations of Okinawans from the Japanese home islands. Their absolute insistence on the neologism "Ryukyuan" over "Okinawan" implied a perpetual stateless status for Okinawans in this new geopolitical order.

Settling Okinawans in Santa Cruz

As USCAR supported the emigration of several thousand Okinawans to the plains of Santa Cruz it imagined an indefinite period of U.S. rule on the islands necessitating the future migration of many more. With tens of thousands of Okinawan families asking to emigrate, new settlement sites across the tropical regions of Bolivia offered Okinawan colonists—and their U.S. sponsors— nearly limitless possibilities for expansion. Those migrants, according to one report, would eventually exceed the capacity of the initial colonies simply named Okinawa 1, 2, and 3 and later form Colonies 4, 5, 6, and 7 in the late 1960s and through the 1970s.[41] U.S. geopolitical concerns in East Asia had led to the conversion of island farmland into military bases and this exclusionary landscape, like the earlier nineteenth century transition to Japanese-controlled plantation agriculture, engendered the displacement of thousands of Okinawans. Meanwhile, U.S. fears of Communist influence in Bolivia resulted in massive support for the MNR and its plans to convert its lowland frontier into productive farmlands. In short, U.S. Cold War policy created the conditions for Okinawan displacement *and* resettlement, securing a receptive host government disposed to receive U.S.-sanctioned displacees.

Back in 1954, when U.S. support for the Uruma initiative was first being considered, Victor Andrade, the Bolivian ambassador in Washington, told USCAR officials that his country would welcome an unrestricted number of Okinawans because Bolivia "admires their courage, diligence, willingness to work, adaptability and many other fine qualities."[42] However, the official agreement between the Uruma Society and the Bolivian government was terminated in 1955 after a disastrous early colonization attempt. Several colonists died from a mysterious disease and rumor spread within the Bolivian government and among cruceño leaders that "Uruma disease" (as it came to be known) was endemic to Okinawans and could spread to the rest of the population.[43] With the relocation of the colony and the abatement of the epidemic such fears faded but Bolivian officials insisted on admitting future Okinawan settlers on a case-by-case basis. As a result, USCAR officials attempted to carefully assess national and local attitudes as they prepared future migrants to enter the country.

In 1957 the Bolivian Development Corporation (CBF), the state institution in charge of settlement, let the U.S. embassy know that it still held Okinawans in "high regard."[44] Despite the tragic epidemic, Okinawans had "maintained a high morale and have steadily pushed forward to gain a foothold on the virgin soils to which they have been assigned." Consular officials happily noted that the industrious characteristics of the settlers had won them a core group of government supporters. However, USCAR was aware that cruceños did not always share the confident outlook of highland officials and, in 1959, embassy officials attempted to assess local racial dynamics in the wake of the polarizing "civic struggles" conflict over oil royalties. While the "lower classes . . . show few signs of emotional opposition to the Orientals," the embassy reported that, the "creole middle and upper class" controlling Santa Cruz were ambivalent about Okinawans and favored European immigrants.[45] On the one hand, individuals questioned were, like national authorities, impressed with the work ethic of Okinawans and their economic contribution. On the other hand, they harbored a more deep-seated racism in opposition to any intermarriage they felt "would weaken Cruceñan racial stock."[46] According to embassy officials, local elites also tended to define foreigners in regional and racial rather than national terms. Development assistance given to outside colonists was frowned upon but for cruceños, *outsiders* could just as easily reference "altiplano Indians or kollas" as Okinawans. As both groups staked their claim to the frontier over the following decades, this regional definition of *belonging* would prove fateful.

Conversations on Okinawan settlement were not limited to confidential memos. Over the following years Bolivians, Europeans, and Okinawans (the latter frequently conflated with Japanese) were brought into explicit dialogue in the pages of Santa Cruz's leading newspaper *El Deber*, illuminating the interplay between local ideas of foreignness and race. While the U.S. military went to great lengths to distinguish their Okinawan wards from the Japanese, who were also migrating to Santa Cruz in those same years, Bolivian elites rarely made such a distinction. These commentators leveled their critiques at an undifferentiated group of real and imagined *japonés* migrants. In early May of 1955, the newspaper *El Deber* reported on a supposed plan to settle 20,000 Japanese families in an article entitled, "Immigration to Bolivia should be carefully selected."[47] The meaning of *selection* was not left to the reader's imagination. "We demand refined races," the author stated, particularly those that had a common language and "telluric influence," listing Germans, Spanish, Portuguese, and Italians and citing the tenth Inter-American Conference held in Caracas the previous year in which attendees had "pronounced in favor of white immigration." The author also pointed to the exodus of Bolivians who left highland frontier departments for Argentine sugar fields. "Before they think of mass immigration, they should return [braceros] to the patria" instead of "abandoning them and resorting to a foreign element to intensify agriculture."

In advocating the return of Bolivian braceros in place of Japanese immigrants, the author slipped from a long-standing, if increasingly untenable, Latin American racial discourse of "whitening" (through the promotion of European migrants) to a more stable xenophobic nationalism based on economics.[48] Despite the noble idea of bringing conationals back home, some cruceños had nothing more than rhetorical interest in the fate of Bolivia's braceros, most of whom were highlanders and thus equally foreign in their eyes. In fact, in that same issue, *El Deber* described the first government efforts at internal migration of the "altiplano population to the east."[49] "What end does this program have[?]" the author asked, criticizing the haphazard "decongestion of the mining centers" with scarce attention to the climate, environment, or race of the receiving regions and insisting that there were "plenty of empty spaces in the highlands where ex-miners might settle." He maintained that the people of Santa Cruz and the Beni "are special races," and surmised that they "do not need these injections of another class. Their own campesinos are white, literate, and capable."[50]

Regionalism, nationalism, and eurocentrism comingled in *El Deber*. Its authors enthused over Italians, who were migrating en masse to Canada, Australia, and Argentina in the postwar period and who had established a small colony in Santa Cruz. A 1955 article linked racial desirability to positive environmental change.[51] The author, waxing poetic, hoped that "in place of these 20,000 [Japanese] families, Italians will come in an exodus from Libya, that formed of its sand dunes, the productive magnitude that international justice stole from the patrimony, the German in the south of Brazil, . . . that created the grand industrial emporium of that country, or the Arabs of São Paulo, the strongest and most prosperous industrial power, the Spanish in Mexico . . . these are, without a doubt the migrations that are good for all countries, others no."

This overt racism did not go unchallenged. In June, *El Deber* published a critique by an author identified as AJF who denounced the press and reminded readers that "men of races of supposed racial superiority" had utterly failed in colonization schemes and could be found frequenting Santa Cruz's hotels and bars.[52] In contrast, he saw the Japanese as "admirable and easy in their complete adaptation to the new patria," citing long-standing Japanese communities like Riberalta, home to Uruma Society founder José Akamine. He challenged the "careful euphemism" of race, which "in these times is an absurd prejudice." Yet, the majority of articles in *El Deber* continued to champion European migrants and reject "this strange invasion of the sons of the Chrysanthemum Empire."[53] In late 1955 the paper greeted the arrival of engineer Felipe Bonoli, who promised to build an Italian colony in Santa Cruz and claimed that Italians, undergo "easy assimilation in our environment, [due to] their industriousness, religious beliefs and the similarity of their customs."[54] The following year, a poem by Luis Simón García appeared opposite *El Deber*'s editorial page. In "My Salute to Santa Cruz," the poet expressed his sincerest wish to see Santa Cruz, "populated, not only with the beautiful Andalusian type but also with the Italian and the American who are [men] of business, energy and light."[55]

Even as it attacked Japanese immigrants, *El Deber* also became increasingly hostile to Andeans, the other visible outsiders who were settling in the region. In August 1956, the paper sardonically referred to the "latest novelty: colonizers that decolonize" in reference to a new colony of highland migrants established along the Yapacani River across from the Japanese colony of San Juan.[56] The paper claimed that the settlers were destroying the once flourishing estates and private property of prominent landowners "whose rights come from family tradition."[57] Ultimately the paper's writers constructed both Andeans and Japanese (the former brought from the highlands, the latter, "from another continent") as unassimilable outsiders in Santa Cruz by blending discourses of race, public health, and environmental fitness. *El Deber* portrayed Japanese migrants as a people "that does not improve our race and are foreign to our environment," while insisting for instance that the poor health of Andean miners-turned-migrants made it impossible for them to carry out productive work in the tropics.[58]

Accusations of racial and environmental incompatibility conveniently ignored the fact that many Andeans had been engaged in productive labor in the semitropical sugarcane enclaves of Northern Argentina and that Okinawan and Japanese migrants had an even longer history of transnational labor in the tropical sugar fields of Hawaii, the rubber zone of Bolivia's Amazon Basin, and the farming regions of neighboring Brazil. In the insular but rapidly modernizing climate of Santa Cruz in the mid-1950s, such willful ignorance was far from surprising. When Yoshihide Higa arrived with a group of Okinawan colonists in 1961, he likened the city to the set of a Hollywood western even though it had more than tripled in size in the past seven years.[59] Furthermore, race-baiting and more aggressive forms of discrimination characterized the Japanese and Okinawan diasporic experience since the early days of their migration to Latin America in the nineteenth century. Yet race prejudice held a paradoxical relationship to the spread and status of *undesirable* migrants with the depth of such antagonism in the press and elsewhere just as likely to indicate the success of a migrant community.[60]

Long accustomed to xenophobia, Japanese and Okinawan migrants effectively negotiated nativism in Santa Cruz through displays of agrarian citizenship. Given their distinct diplomatic representation (Japanese and U.S. governments) and the cultural divide between Okinawans and Japanese, they rarely did so in concert. The Japanese government, wishing to avoid a conflict of overlapping jurisdiction with U.S. authorities in Bolivia, also stayed out of Okinawan-Bolivian affairs.[61] Physical distance played a role with one hundred kilometers of rough road separating Colonia Okinawa and San Juan Yapacani. In the early 1960s one settler in Okinawa Colony remembered that it would often take two to three days to travel forty kilometers to nearby Montero in the rainy season.[62] Even after regional roads improved and the Japanese government took over control of Okinawa Colony in 1968, this separation persisted. The two colonies would not create a joint organization until the 1990s. One possible explanation for this missing solidarity is the degree to which Japanese settlers in Latin America, at least in the prewar era, disparaged Okinawans, speakers of distinct languages (two principal *dialects*, as they are often termed) as "the other Japanese," replicating a colonial relationship from across the Pacific.[63]

Despite their distinct paths, Okinawan and Japanese settlers engaged in surprisingly similar strategies of agrarian citizenship. A common recourse was to emphasize agricultural contributions to region and nation just as José Akamine had done in his initial plans for the Uruma Society. Unlike Akamine, such performances could also rely on the support of powerful backers in the

Japanese, U.S., and Bolivian governments. An official letter from the Ministry of Agriculture to the mayor of Santa Cruz preceded the arrival of a Japanese settlement commission led by Minuro Takata. The ministry requested city hall's collaboration and noted the tight relationship between the Japanese mission and the revolutionary government's policy of economic diversification.[64] When Japanese families began arriving in Santa Cruz in 1955, the colony of San Juan had already established an agricultural cooperative and its president Toshimichi Nishikawa wrote to the mayor reminding him of the official agreement with Japan, and "asking and not doubting for one moment of the help and support of your distinguished person" while reiterating the colony's aim to mechanize and intensify Santa Cruz's "agricultural potential."[65]

At times enlisting national and local officials to quietly ensure the cooperation of local officials, Japanese and Okinawan settlers were attentive to public perception and occasionally confronted attacks in the press head-on. Through 1957 and 1958, *El Deber* continually repeated the tired tropes of racial fitness to attack Japanese colonists. Journalists became amateur historians drawing examples from other failed migrant initiatives to drive home the point that "only the Bolivian" (by which they truly meant, "only the cruceño") could "resolve the problem of agrarian production."[66] Japanese migration was most commonly linked to Jewish migration, which, *El Deber* claimed, had been premised on agrarian production but resulted in a flood of "rich merchants" and absentee landowners whose Bolivian peons worked the land. The invasion of the false farmers has begun, they concluded.[67]

Shortly after that last article appeared, the *El Deber* office received a personal visit from Yoei Arakaki, President of the Committee of Reception of Japanese Immigrants for Okinawa Colony. Citing the inflammatory article, Arakaki invited the newspaper's staff to visit the colony and ascertain for themselves whether Okinawans were true farmers. *El Deber* accepted the challenge declaring that "after [the visit] we will emit the judgment that this development action deserves."[68] A few days later the paper sheepishly published a brief note that, while less than a full retraction of their prior invectives, conceded that "the impression received is generally satisfactory."[69] The next day staff followed up with a lengthy assessment of Okinawa Colony's progress. "We have confirmed," wrote the editors, "that there are buildings, dispensaries of good rice, people in the fields harvesting, machines opening roads and people disposed to work, happy to have arrived at a country that is hospitable and of prodigious land."[70] Although it remains unclear whether the initiative came from the newspaper or from the colony, *El Deber* made a similar visit to the Japanese colony of San Juan that year. Again, humbled reporters noted productive farms, cleared land, and new construction and concluded favorably, "in our opinion . . . the Japanese Colony is fulfilling its work commitments and we wish them success."[71]

The performance of good agricultural citizenship had not been successful at halting U.S. military bulldozers on Okinawa, but it did mollify nativist opposition in Santa Cruz. With agricultural production so closely linked to ideas of development and state sovereignty in the minds of both the MNR and elite cruceños, ostensible outsiders could claim the former to ingratiate themselves at the national and local levels. For immigrants across the Americas this grudging acceptance, unlike legal citizenship, required continual performance and was highly susceptible to economic change.[72] Okinawans engaged in repeated bargaining over their belonging in Santa Cruz through the late 1950s by broadcasting their agricultural prowess to cruceño elite. In December 1959, amid continued press attacks against Japanese colonists, the mayor of Santa Cruz received a Christmas card from the head of the Okinawa Colony, Sieryo Nagamine. The note included a résumé of annual production and infrastructure. "The 1500 residents of the growing colony persist," wrote Nagamine, "in their mentality of making Colonia Okinawa a rural center and a model colony in Bolivia."[73] He concluded with the hope that the colony could continue to rely on "the cooperation of the cruceño people and its distinguished members." As a concrete manifestation of the colony's promising future, Nagamine included a bag of the colony's rice claiming that "though small in quantity, [it was] donated sincerely as a homage to our contribution of labor for the greatness and development of the people of Bolivia."

Acceptance of these performances of agrarian citizenship was profoundly spatial—contingent upon positioning migrants in nonthreatening places. In 1959, the U.S. embassy recommended establishing Okinawans in remote regions, admitting that while the creation of colonies along the central Santa-Cruz-Montero agricultural corridor would be preferable, an "excessive concentration of Orientals in the Santa Cruz service community . . . may result in strong opposition."[74] Accordingly, Okinawa I, II, and III were established along the Río Grande, marking the eastern limit of Santa Cruz's "Integrated North." Good agrarian citizenship also meant producing in specific ways. Two years after Nagamine's symbolic gift, the Okinawan colonies had risen to national prominence for their rice production, which Yoshihide Higa proudly explained to USCAR officials, set the standard and price of rice in the nation.[75] In 1962, Okinawans and their rice were even briefly celebrated in Jorge Ruiz's *The Mountains Never Change*, perhaps best demonstrating their ability to incorporate their "ancestral race" (as the film's narrator described them) with national and regional ideas of the March to the East.[76] Production of one of the essential ten commodities outlined by Guevara Arze in the *Plan Inmediato* was acceptable, even commendable. But locals found it far more threatening when Okinawans moved into urban markets already monopolized by cruceños. Okinawans faced strong resistance when they attempted to market native fish from the nearby Río Grande in Santa Cruz. Higa reported that Okinawans were bringing a truckload of fish to the city's market when they had been met by a group of the city butchers and fishmongers who complained that the colonists, "were supposed to work on the farmlands not to fish." The colony immediately halted shipments, but the accusation that Okinawans were fishermen, and not farmers, was added to the arsenal of antimigrant writers. Little is known about what happened to the fish.

Although Japanese and Okinawans could circumvent local resistance on the ground, the indirect economic appeal of their immigration also became particularly attractive to the Bolivian state. With guidance from the International Monetary Fund, President Hernán Siles Zuazo initiated a broad austerity program to combat inflation in 1956. Internal budget shortfalls were in part ameliorated by an increase in U.S. foreign aid, meaning U.S. pressure to accept Okinawan immigrants carried significant weight.[77] As the Japanese postwar economy recovered, several Latin American countries began to court direct Japanese investment. Led by Ambassador Victor Andrade, Bolivian embassy officials in Japan worked to attract the interest of wealthy businessman.[78] In 1959, a Japanese businessman, Heizo Tsukui, the head of a Tokyo-based company named "Bolivia Industries," expressed interest in developing a sugar refinery and colonization project and contacted Adolfo Linares, the head of the CBF (the state agency responsible for colonization and road-building projects). Linares provided a series of details for the "gentleman investor" before learning from ambassador Andrade that the supposed corporation did not exist and that Tsukui was a notorious confidence man considered *persona non grata* by the Japanese Ministry of Foreign Relations.[79]

The following year Andrade came to Linares with a sounder opportunity, informing him that Japanese officials were interested in investing in a large-scale irrigation and colonization project near the town of Villamontes.[80] This CBF project had stalled for lack of funds after the introduction of Siles Zuazo's austerity measures, and in early 1959 a commission of Japanese technicians visited the region. Based on their positive report, the Japanese government was willing to invest $3.5 million into the Villamontes scheme on the condition that 800 Japanese families be settled in the project area. Bursting with excitement, Andrade pressed Linares for details to provide the donors.

Two months passed without a response and in frustration he wrote Linares from Tokyo in June advising him that with competing investment opportunities in Paraguay and Peru, any delay, "would mean the failure of the initiative."[81]

The reason for Linares's silence emerges clearly in his personal correspondence. On May 4, 1960, technical supervisor Ricardo Urquidi had already provided him with an assessment of the Japanese proposal.[82] Urquidi acknowledged that Japanese investment in the stalled project would stimulate production and allow Bolivia to cease importing a million dollars of cotton annually but warned that the appreciable economic benefits needed to be weighed in relation to "other very serious aspects for the future of our country." He feared Japanese *penetration* in Villamontes—as he framed it—would lead to a dramatic, and in his view undesirable, shift in the region. Urquidi assumed the 800 families would total a population of 4,000 settlers which, based on 1950 census data for the population of the southeast (30,000 inhabitants) would constitute fully one-quarter of the regional population. Shoddy arithmetic aside, Urquidi stressed that the "tradition of our country is clearly western" and that the Japanese would bring "social problems" typical of "other countries of marked racial conglomerates."[83]

The CBF continued to discuss the Japanese proposal throughout 1960. In October, Augusto Valdivia circulated a memo that also noted the potential increase in cotton production in the zone. At present, imported Argentine cotton hurt Bolivia's balance of payments and produced "a continuous current of depopulation," that, by his count, drew 20,000 Bolivian harvesters (conveniently the exact estimate being circulated in the press about proposed Japanese migration) to Argentina every year from the Villamontes region alone.[84] Like other authors, Valdivia drew on the plight of the Bolivian bracero in Argentina to frame his opposition to Japanese settlers. The latter he claimed would never assimilate, constituting "racial islands" while the army would surely reject any "foreign nuclei" along a historically contested frontier.[85]

Ultimately the Japanese settlement in Villamontes never took place and Okinawan-Japanese colonization in Bolivia was limited to the San Juan colony to the north of Santa Cruz and Okinawa 1, 2, and 3 along the Río Grande. Despite the ability of colonists to negotiate nativism, popular pressure could place limits on the MNR's willingness to allow large-scale immigration. These local and national considerations intersected with major economic changes across the Pacific to ensure that the Japanese and Okinawan presence in Santa Cruz remained relatively small. Nineteen-sixty was a transitional year. As ambassador Andrade explained at the time, "one of the greatest preoccupations of the government of Japan is the constant growth of its population."[86]

USCAR officials were similarly looking to mass emigration as a solution to population pressures in the Ryukyuan archipelago. This window closed definitively with the Japanese economic miracle already underway that same year. The mass migration of 20,000 Japanese families that cruceños had railed against in the 1950s simply did not take place. The once *surplus* population quickly found employment in Japan's rapidly expanding industrial sector. In the early 1960s, U.S.-supported Okinawan migration continued at the steady but slow rate of a few hundred settlers a year permitted by the Bolivian government. But Okinawans were also increasingly likely to migrate to burgeoning Japanese industrial cities like Toyota after the United States returned administrative control over Okinawa to Japan between 1968 and 1972. The confident projection of USCAR officials that Okinawa Colony would expand from its initial three settlements never materialized. By 1977, when colony Okinawa 8 should have been under construction, Okinawan immigration had stalled for more than a decade.[87]

Tentative Mennonite Migrations to Santa Cruz

In the late 1950s, as government officials and cruceños expressed private unease and public outrage over a potential large-scale Japanese immigration, they were oblivious to another foreign nucleus in their department. When Augusto Valdivia of the CBF weighed the merits of Japanese participation in the Villamontes venture, he drew a brief comparison with cotton production in neighboring Paraguay among members of three large Mennonite colonies established in 1926, 1931, and 1946. What Valdivia failed to mention was that Mennonites were not only producing cotton across the border. By 1960, they had already established two small settlements in Santa Cruz a few kilometers from Okinawa Colony. The contrast might have proved revealing. Mennonites did not simply raise the specter of racial islands. Rather, their settlement was driven by that premise and repeatedly enshrined in law by Bolivian governments. Initially small-scale, Mennonite migration would vastly outpace Japanese and Okinawan migration. As the latter waned in the late 1960s, Mennonites began to arrive in Santa Cruz by the thousands from new destinations including the northern Mexican state of Chihuahua. Before turning to the causes of that large-scale emigration from Mexico, I briefly explore the scarcely noticed Mennonite migrations (from Paraguay, and later Canada) that preceded and enabled it.

In 1954, the Santa Cruz municipal government filed a settlement agreement between the Algodonera Boliviana and the "September 24th Cooperative" in the public registry.[88] The cooperative was composed of ten foreign families from the Mennonite colony of Fernheim in Paraguay. Their leader, David Wiens, was a member of the six-man delegation that had toured the country in the months following the 1952 National Revolution. On February 10, 1954, in a document notarized by the cruceño notary Emilio Porras (who had written similar contracts for Italian and Okinawan migrants), Wiens was given special powers to negotiate a sharecropping agreement.[89] Mennonite colonists would provide the Algodonera Boliviana with a portion of their production in exchange for the rent-free lease of ten lots of sixty hectares each. Each family was to devote twenty hectares to cotton, but colonists were also encouraged to raise dairy cattle, hogs, and tobacco. The Algodonera promised to provide financing of five million bolivianos. The stated aim (like the Uruma Society), was to develop "a work of beneficial colonization for the country and to intensify the agricultural production of Santa Cruz."[90]

Both parties agreed that the contract was conditional on approval by the national government. The following year the Mennonite settlers received official sanction when President Paz Estenssoro signed a supreme decree affirming that the "Mennonite collectives which establish themselves in any zone of the republic to dedicate themselves to agricultural labors will benefit from broad guarantees on the part of the state."[91] This *Privilegium* (a set of privileges or special exemptions dating back to the Roman Empire) was essentially a facsimile of the one Canadian Mennonites had negotiated with the Paraguayan and Mexican governments in the 1920s. Its central provisions included the right to affirm in court with a simple "yes" or "no" in place of swearing an oath; freedom from military service; the right to administer orphans and widows funds along with fire insurance; and most importantly, the permission to found and conduct schools and churches in their low-German dialect.[92] Supreme Decree 4192 stipulated that such schools would receive lessons in civics, geography, and history from national instructors but, given their absence in the lowlands, this provision was essentially meaningless.

Even as Mennonites sought freedom from outside interference, in naming their cooperative "September 24th," the date Santa Cruz celebrated its participation in Bolivian independence, they made a small claim to place their initiative within the regional milieu in which they were settling. They also succeeded in framing their unusual migration in the nationalist terms of the March to the East. In the text of Supreme Decree 4192, the stated rationale for Mennonite privileges lay in that "one of the propositions of the National Revolution is to populate agricultural zones susceptible to development, for which it is necessary to encourage the immigration of family groups that will dedicate themselves to exploiting this agricultural wealth."[93] Because of their reputation for doing precisely that, the Mennonites, the decree concluded, should be extended guarantees despite their "peculiar customs and habits."

The extraordinary provisions granted to Mennonite settlers in 1955 attracted no notice from *El Deber*, whose writers were otherwise preoccupied denouncing Japanese migration and pining for Italians. There is suggestion that the issue was debated within the government, but no mention appears in transcripts from the Chamber of Deputies or the Senate who devoted entire sessions weighing the merits of bringing another dangerous migrant, "Indian" Zebu cattle from Brazil to mix with local herds in Santa Cruz. Perhaps the migration of ten families who arrived across the Chaco frontier rather than through the airport or train station in La Paz was simply too small to register among local pundits or national legislators or, as ostensible Germans, even those of peculiar practices, Mennonites were construed as racially desirable but unremarkable in a region with a small, well-established German merchant community.[94] Mennonites also made no use of diplomatic representation in negotiations, preferring to work with local counsel and this strategy likely left fewer traces than Okinawan or Italian migration.

While faring better than the Okinawan Uruma colony (whose first settlement was an unmitigated disaster) the "September 24th" initiative also failed. Lands on the "Hitapaqui" estate near Cotoca were not suited for cotton and Mennonites found their contract with the Algodonera difficult to fulfill. They soon moved to a new settlement site named Tres Palmas. Despite the inauspicious start, a second group of 138 Paraguayan Mennonites, led by Peter Fehr and Abram Doerksen, arrived in Bolivia in April of 1957. Both men were in their forties and had been born in Canada. Their children were Paraguayan citizens. Had the group included any elderly members—whose admittance was permitted under Decree 4192—there certainly would have been a few Russian citizens (born in Mennonite colonies in the Ukraine) among them. Just as Okinawans traced their migration histories through Hawaii, Peru, and the outposts of Japan's wartime East Asian empire, Mennonites could recite a generational or even personal trajectory that included stops in frontier zones across the Americas and further afield.

Like their predecessors, the 1957 migrants employed local actors to insert their migration within the agricultural vision of the Bolivian state. Given authority to act on behalf of the colony, lawyer Felix Pérez Baldivieso traveled to La Paz where he met with the President, members of Congress, and officials in the Ministry of Agriculture. His letter of introduction mixed a nineteenth century language of settlement with the technological discourse of mid-twentieth-century development assuring the MNR that Mennonites had come to Bolivia to "populate the land and especially to develop agriculture under scientific methods."[95] Although frustration with a rigid cooperative system in Paraguay was a principal reason for the emigration of these colonists, Pérez played to the MNR's interest in developing modern cooperatives in the lowlands giving assurances that Mennonite settlers would form an "agricultural cooperative, with new methods while working with specialized machinery." The accompanying documents repeated these concepts to the point of mantra. Pérez and the former landowner, Ricardo Hurtado Medina, promised that Mennonites would improve "the methods of production with imported seeds from Canada and other places according to the climatological conditions and the quality of the terrain, with the aim of increasing in a grand manner agricultural production." Like the Uruma Society, they also intended to "establish training schools for the instruction of specialized agriculture and ranching and other skills [for surrounding farmers]." In short, colonists would "improve in every order the form of production and the quality of products, to advance and contribute in this most efficient and effective manner, to the economy of the country."[96]

The *Canadiense* settlement, as it would become known, significantly expanded on the scope of the earlier Mennonite migration. Fehr and Doerksen had negotiated with a group of landowners led by Hurtado for the purchase of a parcel of 4,000 hectares. Hurtado affirmed that the property already contained several buildings, coffee processing equipment, cattle, and fruit trees and vouched for the credentials of "these scientific gentlemen farmers" who would enter the lowlands not as humble farmers—like many internal migrants from the Andes—but as "technicians in agriculture." Working with "modern methods," substantial capital, and machinery, imported seeds, and improved livestock, they would eventually farm 10,000 hectares and incorporate 200 members.[97]

The reality of settlement rarely matched such confident projections. In Santa Cruz, as was the case across the tropical and semitropical world, outsiders often incorrectly assumed that the lush vegetation and ample rainfall in a region were indicative of a uniform fertility. Regier and the other Mennonites that arrived in 1954 were soon forced to abandon their initial settlement because it was not suitable for farming. The 1957 migrants arrived with scarce capital and were hardly the transnational agroindustrial pioneers described in the settlement petition. When North American Mennonite J. W. Fretz visited the Bolivian Mennonite colonies in 1960, he was alarmed, to find the scientific gentlemen farmers that Pérez had successfully promoted barely scraping by, their children "so undernourished as to show outward signs of malnutrition."[98] Peter Fehr's son and David Neufeldt, eight and eighteen at the time, recall the extreme poverty that drove their departure from Paraguay and continued in Bolivia.[99]

Fretz published the story of the struggling Bolivian Mennonite colonies in the monthly North American periodical *Mennonite Life*. The English-language magazine had a limited readership among low-German speaking Mennonites, but news of the possibilities of settlement in Bolivia spread through other publications like the Manitoba-based *Mennonitische Post*. As with the evolving Okinawan diaspora, "ideas, as well as letters and international money orders" tied together far-flung Mennonite communities throughout the Americas transmitting information about new opportunities.[100] Fretz's article may read as a dismal prognosis for the future of Mennonite settlement in Bolivia, but the tone is fundamentally misleading. Mennonites from across the diaspora flocked to the region over the following years often explicitly seeking the sort of challenging conditions that he identified. The idea, as Regier had claimed in his 1952 report in the Paraguayan Mennonite newspaper *Mennoblatt*, that "sinfully little had hitherto been done," on the plains of Santa Cruz was paradoxically appealing. A rugged region of rough roads and scarce (if nascent) agricultural production, was an especially welcome proposition for those colony Mennonites that sought the maintenance of traditional ways of farming through a qualified relationship to modernity and mechanization at the margins of the secular nation-state. In 1963 twenty families from Canada joined some Paraguayan Mennonites in founding a third colony (Bergthal) in Santa Cruz. Before settling, they decided to confirm that the 1955 privileges remained in effect. In 1962, they obtained a new decree from Paz Estenssoro—serving a second term as President—confirming that the *Privilegium* would apply to future Mennonite colonists. The only addition to the 1955 decree was a provision freeing Mennonites from import duties and visa fees.[101] For impoverished Paraguayan Mennonites who entered the country with little capital in the 1950s such an exemption would have been insignificant. However, over the following decades, thousands of Mennonites left colonies from across the Americas, bringing millions of dollars of duty-free machinery with them to establish large colonies in Santa Cruz.

Drought and Diaspora in Northern Mexico

The largest immigration of Mennonites to Bolivia originated in the northern Mexican border state of Chihuahua. Much like the six Mennonites whose visit to La Paz in the wake of Bolivia's 1952 Revolution opened this chapter, an earlier six-man Mennonite delegation arrived in Mexico City in February 1921 at the tail end of the Mexican Revolution. Johan Loeppky and his fellow delegates sought freedom from imposed English-language education in Canada and negotiated a *Privilegium* with President Álvaro Obregón that included their right to conduct their own low-German school system. As they would in revolutionary Bolivia, Mennonites managed the nationalism of the Mexican Revolution through an effective performance of agrarian citizenship.[102] In doing so they were able to mitigate the obvious paradox that expropriated hacienda land in Mexico's most violent revolutionary state was being sold to conspicuous foreigners.[103] As Russo-Canadians, they served (in the minds of politicians) as a bulwark against U.S. investment-based encroachment along the frontier. As European immigrants, they escaped the xenophobia directed at Chinese immigrants, the other major foreign community in northern Mexico.[104] Although Mennonites encountered resistance to their settlement, they made effective use of relationships with national leaders to mitigate hostility on the ground.[105]

Good agricultural citizenship was predicated on the transformation of landscape. Mennonites settled in high, semiarid valleys sandwiched between the rugged Sierra Madre Occidental to the west and the Chihuahua Desert to the east. They faced severe adjustments in the forms of farming they had grown accustomed to on the Canadian prairies but soon abandoned wheat for oats, corn, and beans and wooden structures for adobe brick.[106] In under a decade, the Mennonite colonies of Chihuahua were hailed as a veritable *emporium* in the national press. "Wastelands Turned into Orchards: Brilliant Success of the Mennonites," began one author, citing local opinion that "[they] will manage to convert a vast region of Chihuahua into a true granary."[107] When a group of Mennonites had difficulty entering the country in 1931, the Chamber of Commerce of Ciudad Juárez intervened. "We are dealing with expert technicians in agricultural work," they began, noting with a sweeping temporal flourish, that while the fields of Chihuahua had not been improved since the time of the conquest, the lands of the Mennonites "have been converted into true gardens."[108] Even those who recognized that the premise of Mennonite migration was cultural isolation and "centuries could pass and the Mennonites would maintain themselves in their colonies" were liable to rationalize the state-settler bargain by citing the economic benefit of the colonies and their value as an "example of collective organization and frugality."[109] By the 1940s, *El Nacional* counted 106 Mennonite villages in Chihuahua with a population of 12,000 producing "milk products such as butter and cheese of excellent quality," and generating two million pesos for the regional economy.[110]

Unbeknownst to these authors, signs of trouble were already apparent in the Mexican Mennonite *granary* in Chihuahua in those years. In 1944, J. W. Fretz—the same U.S.-Mennonite sociologist who would visit struggling Bolivian Mennonites in 1960—traveled to Mexico on behalf of the philanthropic organization the Mennonite Central Committee (MCC). His visit was provoked by the arrival of an increasing number of poor Mexican Mennonites to Canada, and Fretz discovered a situation that contrasted with the accolades in the Mexican press. Like Okinawa, where James Tigner had claimed that farming constituted, "one of their most cherished desires," colony Mennonites insisted that the individual family farmer was the only acceptable model for Mennonite youth. Combined with an extremely high birth rate, this produced a cyclical landless problem as colonies aged. Mennonites had similar experiences in Canada and Russia as frontiers became crowded with new settlements but in Mexico, with a large native population pushing for land, the issue was more pronounced, forcing Mennonites to rent land on surrounding *ejidos* (communal landholdings created by the agrarian reform).[111]

As in Okinawa, the enduring land issue in Mexican Mennonite colonies was exacerbated by an external force. In the former case, a geopolitical factor—postwar repatriation and U.S. land expropriations—strained the agrarian aspirations of Okinawan society. In the latter, an environmental crisis—a prolonged drought unprecedented in the modern history of Mexico—compounded landlessness. Both situations extended from the late 1940s through the 1960s and became especially acute in the 1950s. In Mexico, as in Okinawa, increasing local tensions and transnational migration resulted.

The effects of northern Mexico's midcentury drought crept into the intermittent news coverage of the Mennonite colonies, influenced their periodic negotiations with the Mexican state, and led to the creation of seasonal and permanent migrations. Like other natural disasters from hurricanes and earthquakes to landslides, this *mini-dust bowl* offers a unique vantage point to explore latent tensions and social, cultural, and political cleavages at a moment when the unspoken relationship between humans and the natural world was laid bare.[112] The long drought of the postwar era is also remarkable as it extended through an era that is typically imagined as a golden age of prosperity and modernization in Mexican history.[113] Northern states faced environmental challenges during this period that challenged prior economic structures and engendered migration. The Mennonites were no exception, particularly given the imbricated nature of their agrarian practice, community structure, and religious belief.

Droughtlike conditions began in earnest in 1948 and worsened in the 1950s revealing cracks in the agricultural emporium that Mennonites had constructed in northern Mexico. In May of 1953, the leaders of the Mennonite colony of Los Jagueyes in Chihuahua came to Mexico City to meet the head of the Department of Migration and requested that fees for processing of their migratory documents be waived.[114] Though the Los Jagueyes leaders did not state the reason, their home state had been under groundwater use prohibitions (*vedas*) for the past three years as it suffered through the worst of the drought.[115] In August of 1954, with these restrictions still in place, Daniel Salas López, a lawyer who frequently represented Mennonites, wrote to the Department of Population with the same request. Though insisting that he was not refusing in the name of his clients, he hoped the Department would consider, that this was not "an isolated colonist," but "hundreds who in their grand majority have paid [processing fees] punctually." He asked on behalf of a colony "that for lack of rains has lived with positive sacrifice to the degree that it has taken out bank loans to meet necessities."[116]

In 1955, the heart of the drought intersected with a different aspect of Mennonite-state negotiation. Mennonite colonists had learned of the passage of a law of mandatory participation in Mexican Social Security. Feeling that this violated a provision of their 1921 *Privilegium* allowing colony-based fire, orphan, and widow's insurance, some Mexican Mennonites protested to the government, often engaging sympathetic locals, while others began to investigate settlement opportunities in newly independent Belize (British Honduras). Drought and migration formed a critical subtext to these discussions. In 1957, Gabino Aguilar, a teacher from Chavarría station on the Chihuahua-Pacific railway wrote to President Ruiz Cortines. He requested that Mennonites be exempted from Social Security noting that they maintained distinct institutions "that attend to their necessities."[117] He emphasized the "unequaled importance" of the economic production of Mennonite farmers "with whom many in this state have lived alongside." Unlike those campesinos, "they have never asked for help . . . in these enormous droughts," continued Gabino incorrectly, "and none of them have asked to go as braceros [to the U.S.] and abandon their lands."[118] Yet he assured the president that "the Mennonite colonies have also suffered this collapse." He would repeat this contrast in a 1962 petition to President Gustavo Díaz Ordaz complaining that a local campesino leader was inciting unrest by demanding the expropriation of Mennonite lands. He pointed out that the city of Cuauhtémoc was booming thanks to colonists and added that Mennonites were "an example of morality in their style of living and working" who unlike campesinos, "never go work as braceros."[119]

Writing from Durango, a sympathetic Manuel Schmill invoked similar comparisons, claiming that Mennonite colonists remained environmental stewards while those "poor and unfortunate" campesinos, stood "ready to abandon it at whatever moment to go and look for fortune abroad."[120] He framed traditional horse-and-buggy Mennonites as rooted to the land in the face of an environmental crisis and in opposition to their Mexican neighbors who abandoned their country as braceros. While the juxtaposition was useful to those that wished to defend Mennonites or criticize the Bracero Program (which brought over four million contracted Mexican laborers to the U.S. over roughly the same years as the drought), it was hardly true. In a 1955 letter to Ruiz Cortines, the lawyer Daniel Salas López painted a very different picture of colony villages in which middle-aged Mennonites worked failing farms with the help of only their elder daughters and youngest children. Young Mennonite men, "though all born in this country have ceased contributing because they have had to emigrate."[121]

These Mennonites may have joined their campesino neighbors in migrating north to look for work but, because of their unique migratory history, they had access to an alternative labor market.[122] As the Canadian government eased border restrictions in the postwar period many Mexican Mennonites—who had access to Canadian citizenship—renewed expired passports, or applied for citizenship, and were able to temporarily or permanently return home. Arriving in southern Manitoba as early as 1947 where the sugar beet industry was booming, they were viewed as easily manageable, cheap labor by growers, many of them Mennonites who had remained in Canada in the 1920s. New opportunities also emerged in the tomato fields around the Heinz ketchup plant in Leamington—immortalized in the 1969 #1 hit, "The Ketchup Song" by folk singer Stompin' Tom Connors.[123] By the 1960s this well-established but unregulated recruitment network attracted the attention of the Canadian Department of Manpower and Labor. A special task force reported that Mennonites enlisted in the Mexican colonies, boarded car caravans, and traveled the 2,000 kilometers from Chihuahua without rest. Authorities stopped one camper packed with twenty Mennonites—some suffering from dysentery.[124] Despite such hardship, the drought conditions in Chihuahua and Durango continued to push desperate colonists north. Like other Mexicans, Mennonites returned home from their annual pilgrimages where they invested their earnings in new land or struggling farms.

Mexican newspapers, prone to superficial and infrequent coverage emphasizing timeless or humble traditions rooted in the soil, entirely missed the highly mobile nature of Mennonite life in the drought years. However, migration officials bore witness to this growing transnationalism. In June of 1954 Heinrich Zacharias, from a Durango Mennonite colony, applied for a visa. He informed Mexican authorities that his ultimate goal was to obtain his passport in order to "move to Canada in virtue of the fact that currently this place is experiencing a frightening economic crisis caused by the drought that has ruined this region."[125] Some migrants provided few motives though the timing of their movements—corresponding to the Canadian harvest season—is telling.[126] Others followed dizzying itineraries in the 1950s and 1960s as periodic migration became a way of life. By the end of the latter decade Frank Petkau had filled every page in his original travel document. Born in Manitoba in 1943, he grew up in Mexico but returned to Canada in 1957, 1958, and 1959, where he likely worked alongside his family in the fields. In 1962 he visited Mennonite colonies in Belize before returning to Manitoba in 1963. He also made numerous forays through El Paso to Kansas, Oklahoma, and Arkansas in the 1960s (likely joining enterprising Mennonites who purchased used farm equipment for resale in Mexico) and was back in Canada in 1968 and 1971.[127]

Some Mexican border officials gained an intimate familiarity with the colonies as they encountered border-crossing Mennonites traveling for work, to escape debt or abuse, and to reconnect families. In 1962, Jorge Domínguez reported his exhaustive intervention on behalf of Aganetta Klassen.[128] Accompanied by six young children, she had been stopped by U.S. officials while attempting to cross into El Paso and sent back to his office in Ciudad Juárez. Interviewed by Domínguez, she claimed she was en route to Ontario to join her husband but had insufficient funds to complete her trip. Domínguez not only took the entire family 400 kilometers back to Ciudad Cuauhtémoc but, once there, headed to a nearby Mennonite village to see Jacob Peters, a man he knew to be colony leader (*vorsteher*), hoping to convince him to intervene on Klassen's behalf. Peters was absent and Domínguez spent an hour "pleading with [Peters's neighbor] to help their unfortunate compatriot."[129] Undeterred by their refusal, he returned to Cuauhtémoc and located David Redecop [*sic*] "a prominent resident of Mennonite origin," who helped him locate one of Klassen's brothers in a nearby village. It was finally decided that Klassen would leave the youngest children in his care, traveling with the others to Canada, and would send her husband back to collect them. This left the issue of Klassen's passage which was resolved by generous "Mennonite elements" in Cuauhtémoc who paid her fare and gave forty dollars for expenses.[130]

Mennonite unwillingness to help Klassen, albeit anecdotal, is suggestive of a broader conflict. Cross-border migrant labor has often been linked to ideas of modernization in the eyes of sending societies and the migrants themselves, particularly when returning braceros brought new consumer goods and cultural practices home with them.[131] Mennonite braceros were no different and their frequent migrations—although an effective response to the drought—were a significant source of worry for Mexican Mennonite elders (*Ältestern*) attempting to impose restrictions on modern conveniences, from the car to the rubber-tired tractor, whose use was relatively limited in Chihuahua but widespread in Canada. These leaders also noted changes in farming technology brought on by the drought, and in their minds a dependence upon rubber tires, aquifer-based irrigation, or assistance from North American Mennonites, who had become aware of the dire situation in the colonies through interaction with Mennonite braceros, were of overlapping concern. While occasionally petitioning for tax relief, both elders and colony vorsteher were understandably reticent about the internal difficulties they faced during the drought given that their relationship with the Mexican state was premised on the image of a harmonious and productive agricultural enclave. In fact, Mexican Mennonites had historically performed the role of administering, rather than requesting, relief through public donations—in the form of food and animal fodder—to Mexican communities affected by natural disasters.

The earlier correspondence of Daniel Salas López sheds light on their undisclosed plight. In addition to his officially contracted legal work with the colonies, Salas was a self-proclaimed booster for Mennonite migration; in drought-ridden northern Mexico, he actively petitioned the government, often without colony leaders' knowledge. "My friends have not put me in charge of asking for anything," he wrote to President Ruiz Cortines in 1954. Nevertheless, he decided to act on their behalf, insisting, in a tone of exaggerated patriotism, that it was his "obligation, as a good Mexican."[132] The drought had forced colonists to depend entirely on small-scale milk and cheese production, rather than cash cropping, and Salas requested a program to create hardy and more affordable pasture to support dairy cattle which "during this period has been the salvation of the colonies." Presciently, forecasting a new environmental order of irrigated agriculture in Chihuahua, Salas also asked that state-commissioned geologists locate aquifers, drill wells, and establish dams in the region as they had in the Laguna district of neighboring Coahuila. Although the colonies had typically depended on rainfall, he explained that "all the

Mennonites are now inclined towards irrigation." A decade later state forester Andrés Ortega Estrada, claimed that Mennonites were leaving the country because "their lands no longer give them sustenance."[133] For those that remained, a transition to agriculture based on pumping water from deep aquifers—with an attendant need for electrification—was underway.

Along with irrigation, another seemingly innocuous technological innovation produced a serious conflict in Mennonite colonies during the drought years. In the 1940s, most tractor manufacturers ceased production of the cumbersome steel wheel, equipping new models with rubber. Mennonite elders in Mexico, concerned that increased speed was turning the farming tractor into a surrogate rural automobile (for trips into town and surrounding communities) issued a prohibition on the use of rubber tires. Colonists were limited to horse-and-buggy for village transport and steel-wheeled tractors (the wheels often manufactured in colony workshops) on the farm. Initially these restrictions were not difficult to enforce given the paucity of motorized transport in rural northern Mexico. But as sojourning Mennonites began to embrace the modern transport they encountered while working up north, the question of rubber—on cars or tractors—became a particularly contentious issue. Land shortages compounded the problem. Returning Mennonites often bought or rented land beyond the limits of overcrowded colonies demanding the greater mobility offered by rubber-tired tractors and personal automobiles. Other braceros initially purchased vehicles to facilitate their annual trips to Canada before employing them at home in Mexico. Modernization in Chihuahua also played a significant role as former outposts like Ciudad Cuauhtémoc began to protest the unsanitary presence of Mennonite wagon teams and regional road networks expanded. By the mid-1960s, some Mexican Mennonite colonies were in a state of open rebellion over the rubber tire.

The midcentury drought also opened Mennonite colonists up to the spiritual influence of a new group of U.S. Mennonites who established themselves in northern Mexico as the Mennonite Service Committee (MSC)—initially a branch of the Mennonite Central Committee—in the mid-1950s. While arriving with the stated purpose of assisting drought-stricken colonists, these young MSC volunteers were from U.S. and Canadian churches of a decidedly evangelical outlook and often viewed colony Mennonites as a spiritually stagnant branch of the faith mired in *backwards* tradition and in need of *rebirth*.[134] Some Mexican commentators shared this perspective, acknowledging Mennonite agricultural contributions, "so progressive in some aspects and retrograde in others," while hoping it might be possible to "adjust them to a general social change without hurting the interests of the inhabitants."[135]

In drought-stricken Mennonite colonies in Durango, which lost five consecutive harvests in the 1950s, the American Mennonites opened schools and clinics and gave money to struggling colonists to purchase fodder for their dairy cattle.[136] Poor Mennonites accepted MSC help at the height of the drought, but colony leaders became increasingly hostile to this influence by the 1960s. Landlessness and rubber tires would be remembered by a future generation of Bolivian Mennonites as the incentive for their parents' migrations to Santa Cruz, but the evangelical conflict in the colony also played a key role. In 1963, aware that many MSC members were entering the country on tourist visas while actively proselytizing among Mennonite colonists, colony leaders sent repeated petitions to Mexican migration authorities. Not unlike Okinawans who employed a discourse separating destructive military landscapes from harmonious, productive agrarian ones, leader Isaac Harms juxtaposed the work of MSC members who were "attempting to form a division and disrupt the order" of the colonies from colonists that simply wanted to "continue working the fields," without the intervention of people "*foreign* from ourselves [emphasis added]."[137] Leader Daniel Loewen noted that the North American Mennonites were "invading the colonies . . . sowing division and discontent among all the colonists that will come to prejudice the agricultural work that they do." In another letter, colony leaders reminded the government that colonists numbered 20,000 and "are respectful of the authorities [and] solely dedicated to work and the production of basic food articles," in contrast with evangelicals who "do not produce anything."[138]

Colony leader Aaron Redekop Dyck even claimed a more sinister intent on the part of the MSC. He sent authorities a translated letter (although the original was never produced) allegedly written by Andrew Shelley head of the General Conference Board of Missions in Newton, Kansas. In the letter Shelley admitted to membership that the mission board faced a severe budget shortfall and encouraged all missionaries to strenuously solicit as many donations as possible from overseas mission areas.[139] Colony leaders pointed out that colony funds—threatened by Shelley's scheme—were used for maintenance of rural roads and routinely offered to support Mexican disaster relief, citing the recent aid they had given during a flood affecting the city of Tampico.[140] Noting the growth of missionary schools in the area, colonists who had long resisted government schools argued, without a trace of irony, that "it is the government that should tell us what to do and not a group of foreigners."[141]

Unaware of the profound differences separating colony and evangelical Mennonites, officials may have initially been bewildered by these accusations. Nevertheless Redekop, Harms, and Loewen were successful in having an investigation launched into the activities of the MSC in Chihuahua including an official report prepared for Secretary of the Interior, and future president, Luis Echeverría.[142] As a result several of the young evangelicals were deported for operating outside the scope of their entry visas. The conflict revealed a colony leadership well-versed in negotiating with the state and brought Mennonite ideas of agrarian citizenship and national belonging to the fore. Their success was temporary. Those same MSC volunteers soon reentered the country with proper documentation and continued their divisionist labor in the colonies. The invasion of rubber-tired tractors, personal automobiles, and evangelicals combined with ongoing labor migration and landlessness convinced Mennonite leaders to act. From 1966 to 1969, the first wave of 1,300 colonists left behind their fellow Mennonites in Chihuahua and a contentious debate about modernization. Over their decades of settlement and expansion in Santa Cruz, the subject of chapter 5, Mennonites would continue to engage in transnational migration, resist evangelical overtures, and face internal struggles over technological innovation and environmental change. As in Mexico, they would also refashion their ethnoreligious version of agrarian citizenship in dialogue with an ideology of revolutionary frontier settlement.

Conclusion

In the 1950s and 1960s, the routes of two global diasporas converged on the forested lowlands of Santa Cruz. Despite their distinct transnational trajectories, Okinawans and Mennonites negotiated similar pressures including attempts at cultural assimilation by the state, landlessness, and overcrowding. Each struggled to maintain farming communities when faced with external factors (land expropriation and environmental change, respectively) and turned to labor migration and resettlement in response. In electing to migrate, Okinawans and Mennonites employed strategies that had been cultivated over generations of emigration to a range of interconnected settlement sites. Indeed, their migrations to Bolivia in the 1950s and 1960s were enabled by the earlier arrival of members of their respective diasporas like Akamine and Regier. In June of 1968, Bolivia-bound Mennonite colonist Martin Dueck succinctly expressed this history of mobility to a Mexican reporter.[143] Although silent on the intense theological disputes over technology and agrarian politics that had motivated their departure from Mexico, Dueck conjured an image of a perpetually evolving diaspora for the curious reporter. Many postwar Okinawan migrants to Bolivia could recite a prodigious prewar history of migration from outposts of the Japanese empire to communities across the Pacific Rim and throughout Latin America, a period when according to one report, nearly a third of the Okinawan population lived outside the Ryukyuan islands.[144] Hiro Chibana, who served as a Ryukyuan government representative for Okinawans in Santa Cruz in 1961, had been born on Okinawa in 1905, trained as an agronomist, and served as a leader of colonization initiatives in Japanese-controlled Manchuria. After postwar repatriation to Okinawa he began to work with USCAR on land issues before joining the Bolivian resettlement initiative.[145]

In their respective migrations between nations and regions Mennonites and Okinawans troubled state-centric concepts of identity—a practice that caused trouble for them as well. In the 1950s the Mexican press referred to Mennonites as "men without a country," but colonists made every effort to renew or maintain their Canadian citizenship when it helped them find work in Manitoba or Ontario while vigorously defending themselves as Mexican when confronted with an onslaught of foreign Mennonite missionaries from Canada and the United States. Okinawans faced greater documentary ambivalence as they attempted to cross international borders with flimsy U.S. identity certificates. The document's insufficiency highlighted the tenuousness of their new subject positions as Ryukyuan wards of an informal U.S. empire. In the conclusion to an article on U.S.-administered Okinawa, Earle Reynolds, a visiting American anthropologist critical of U.S. rule, questioned an Okinawan who wrote "Ryukyuan" on an official travel document and asserted provocatively, "there is no Ryukyuan nationality." He noted with satisfaction when the man, after acknowledging the artificiality of the term, snapped, "I quit. I am Japanese. I will put myself down as such."[146] In unfamiliar Bolivia, Okinawans were also increasingly desperate to normalize their Japanese status trying, in the words of a sympathetic U.S. official, to "maintain a link between the place they left and the new land they chose."[147]

Facing challenging contexts in the Ryukyuan islands and the northern Mexican desert, these farming peoples developed forms of negotiation that would prove useful in the new context of Santa Cruz. Mexican Mennonites established identities as model agrarian citizens in the four decades after the Mexican Revolution, and this served as a crucial model for resituating themselves in revolutionary Bolivia. Although Okinawans were not always successful in challenging military land acquisitions, their intransigence forced the United States to actively pursue Bolivian colonization in the 1950s and they would once again perform the role of model farmers amid a nativist reaction they encountered in Santa Cruz. As a result, it was not local opposition but external factors, the dramatic economic recovery of Japan in the early 1960s, that ultimately limited Okinawan migration.

As Okinawans and Mennonites arrived in Bolivia from neighboring Paraguay, northern Mexico, and East Asia they joined a much larger stream of Bolivian migrants descending from the Andean altiplano into Greater Amazonia with whom they also shared some unexpected similarities. Members of this Andean diaspora had also experienced harsh environmental conditions (including drought) and a shortage of arable land in their home communities. Like Mennonite seasonal migrants to Canada, many Andeans had become transnational braceros at midcentury, traveling in the hundreds of thousands to neighboring Argentina. Like Okinawans, Andeans would rely in part on U.S. financing to assist their establishment in lowland colonies. Andeans, Okinawans, and Mennonites had all been subject to past policies of cultural assimilation and were still seen as partial or noncitizens by the state and Bolivian elites even as they each laid claim to a form of agrarian citizenship based on their new roles as frontier farmers in an unfamiliar landscape. Foreigners often arrived in Bolivia with the external funding, personal capital, or a reputation for farming marginal lands and thus passed as technicians in agriculture. Such a designation was more difficult for impoverished Bolivian farmers whose extensive agricultural knowledge was not valued by Bolivian officials. Yet as the subsequent chapter will show, Bolivians negotiated those assumptions of cultural deficiency to make explicit demands on the state. Even more forcefully than Okinawan or Mennonite settlers, they were able to link agriculture and citizenship through the revolutionary legacy of the MNR.

CHAPTER THREE

Abandonment Issues

Speaking to the State from the Andes and Amazonia, 1952–1968

The power of nationalism is as evident in the gesture of the *Niño héroe* [cadet] who wraps himself in the flag and dies for his country as it is in the gesture of the peasant who invokes his citizenship when petitioning for land, or the small-town notable who claims that his villagers and himself descend from Aztec ancestors when he petitions for a school.

—Claudio Lomnitz, *Deep Mexico, Silent Mexico*

We need specialized technicians, agronomists and engineers, we need seeds for pasture and crops. We want to develop ourselves and not remain beaten down by poverty. If [it is true] that all Bolivians should live as equals and not be marginalized . . . we should unite and work together one helping the other so that one day our children will be good citizens and good Bolivians in order to defend our patria and escape from misery. We hope that our words will not go with the current or be carried away by the wind.

—*Petition from Acción Rural Agrícola, Corque, Oruro to Ministry of Agriculture,* December 10, 1968

Throughout the 1950s and 1960s, thousands of letters flowed from established highland communities and new lowland colonies into the government ministries of La Paz and the office of Bolivia's president. These petitions, still preserved in state archives, provide an invaluable source for exploring how Andeans engaged with the logic of the March to the East. Would-be migrants and new colonists took up the language of settlement and civilization, of agricultural production, and good citizenship to legitimate their claims for land and resources along the nation's frontiers. In appealing to the scientific, technical, and agricultural imperatives of the revolutionary state, these Andeans adopted similar strategies to Mennonite and Okinawan immigrants to lay claim to the project of settling the Bolivian lowlands in the years after the 1952 revolution. In their evocative appeals they treated agrarian nationalism as "a currency, that allows a local community or subject to interpellate a state office in order to make claims based on rights of citizenship."[1] By offering to cultivate the nation's lowland frontiers, Andean petitioners simultaneously sought to cultivate a relationship with the revolutionary state.

The MNR's plan to transform a fugitive frontier landscape into a site of intense agricultural production is easily recognizable as a form of midcentury high modernist engineering in which planners sought to construct legibility and reorder unruly space into interchangeable, and thus marketable, units.[2] Individual architects and planners like Le Corbusier and Oscar Niemeyer, along with leaders from Josef Stalin to Julius Nyerere often stand in, as synecdoche, for these modernization initiatives. Yet such projects depended on the active participation of thousands, and at times millions, of individuals whose trajectories often then escaped the intentions of planners. In Bolivia, a range of actors laid claim to the meanings and trajectories of midcentury modernization campaigns. As Sarah Hines argues, postrevolutionary dam-building initiatives in Cochabamba reveal "a more inclusive vernacular vision of modernity, albeit a messy and incomplete one."[3] In terms of the MNR's March to the East, there was substantial overlap, as well as divergence, in the aims of state and settler. Turning to the perspectives of the latter inverts a standard approach to high modernism, by highlighting forms of *speaking to*, rather than, *seeing like* a state.

Such subaltern practices have a long tradition in Latin America. Colonial historians identify the myriad ways in which indigenous subjects made use of the Spanish legal system, as well as extrajudicial appeals written directly to the sovereign, in their struggles over land and labor.[4] In their oft desperate pleas to the crown, indigenous peoples extensively employed *naïve monarchism* (the stated belief in the benevolent nature of the sovereign) to challenge misbehaving local officials. This tactic continued in the regional rebellions of the Middle Period (1750–1850) and was one which independence leaders often viewed with bewilderment or scorn.[5] They understood such appeals to the monarchy to represent the illogical, reactionary, or conservative nature of indigenous society. Yet direct appeal to rulers offered subalterns a useful and effective way of framing their refusal to cooperate with authorities at the local level.[6] These tactics had a particular importance in the Andes, where large autochthonous communities existed at the margins of nascent national independence movements.

In the modern era, rural populations continued to employ the personal petition effectively addressing emerging caudillos that depended on their support. In mid-nineteenth century Bolivia, the charismatic President General Manuel Isidoro Belzu gained a following among the popular classes in La Paz by denouncing "the insensible mob of aristocrats [that] has become the arbiter of your riches and your destinies [whose] monstrous fortunes have accumulated with your sweat and blood."[7] Although anathema to urbane liberal politicians in Bolivia and elsewhere, this strategy reemerged in the turn to populism in the 1930s. Subalterns could be found petitioning populist leaders, most famously Getúlio Vargas in Brazil and Juan and Eva Perón in Argentina. Their desperate voices emerge in stacks of letters often preserved in presidential archives.[8]

Populism may have been muted in Bolivia in the first half of the twentieth century but governments, especially in the post-Chaco War era, began to address the "Indian question" through tentative reformist programs in public health and education. Across the Andes, such programs were steeped in condescending ideas of uplift but they also provided indigenous petitioners with a framework for making demands on the state that exceeded their original limited scope. As Brooke Larson argues, elites imagined early twentieth-century indigenous education as "mapping the political and cultural parameters within which Indian men and women would be allowed to enter the future nation," yet this unapologetic paternalism was met by "an extraordinary escalation in petitions for the rights to schools" from indigenous communities.[9] Investigating communal landholding, Laura Gotkowitz arrives at a similar conclusion, that state illusions of control did not stop rural Andeans from also addressing officials and articulating forms of agrarian radicalism that "did not simply parrot the state's civilizing discourse."[10]

In the 1950s and 1960s, the MNR's sweeping campaigns in education, public health, and voting encouraged a vibrant state-citizen discourse that resulted in a "peaceful deluge of petitions."[11] The MNR encouraged such requests to be directed through newly created, state-sanctioned, agrarian unions, often as a way to undermine the strength of local leaders. These novel organizations soon claimed hundreds of thousands of members. But as Silvia Rivera and others have pointed out, the encounter between state paternalism and local aims varied widely from complimentary, to mediated, to antagonistic in the valleys of Cochabamba, the altiplano, and the north of Potosi, respectively. Moreover, petitioners often exceeded the aims, progressive in some respects and cynical in others, of the MNR. They used agrarian unions as a springboard for more radical demands. Others circumvented the official process altogether.[12] During the presidencies of Paz Estenssoro, Siles Zuazo, and Barrientos, thousands of unsolicited letters were written by individuals and groups in rural communities and sent directly to the president or to state institutions. Emboldened and encouraged by the rhetoric of a revolutionary state, petitioners like the Acción Rural group from Corque, demanded medical services, land grants, schools, and technical training for their struggling communities. As Carmen Soliz explains, postrevolutionary agrarian petitioners were not only landless tenants (*colonos*) on large estates broken up by the agrarian reform demanding "land for those who work it." They also included members of independent indigenous communities (*comunarios*) who, in referencing previous decades of dispossession, sought the return of "land to the original owners."[13] In addition to this vibrant discourse, other petitioners neither asked for property they had previously owned, nor land they worked without title. Instead, they wrote to the Ministry of Agriculture, its director of colonization, or to the CBF, staking their claim to land in new colonization zones opened by the MNR.[14]

This chapter begins with letters written by would-be colonists who were often former miners or small farmers from highland departments like Potosí, La Paz, and Oruro. In their petitions, these Andeans pushed the state forward and reminded the MNR of its obligations to the rural populations it sought to "incorporate into national life."[15] They employed the language of empty, abandoned lands, a frigid altiplano, rational scientific production, territorial integrity, food sovereignty, and the state's obligations to its own revolutionary legacy. While the Bolivian state sought to rationally manage its population through colonization, these petitioners offered—and at times insisted on their right—to settle the nation's frontiers. Such initial petitions are full of hope for the prospects of colonization. As Andean migrants entered the lowlands, these lofty projections encountered a daunting range of barriers including a tropical lowland environment that was radically different than (but just as challenging as) the arid altiplano they were leaving and a state that lacked funds to fully realize the extensive promises it had made in pamphlets, radio ads, and films. The second half of the chapter shifts from demands to be included in colonization to the denunciations, complaints, and accusations of state abandonment that emerged as Andean migrants became colonists and found state support utterly lacking. These subaltern voices sit in uneasy tension with those of national politicians, who deplored the exodus of highlanders to Argentina, and governmental officials and foreign experts, whose explanations for the failure of colonization initiatives often rested on assumptions about Andean cultural inferiority.

As with *Okinawan* or *Mennonite*, the term *Andean*—along with similar terms such as *highlander* or *kolla*—merits scrutiny. Andean simply references a mountain range—one intimately associated with Bolivia by historians and the public despite occupying less than half of its national territory. It encompasses both the altiplano—the high arid plateau extending through the departments of La Paz, Oruro, and Potosí—and the eastern valleys located

predominantly in Chuquisaca and Cochabamba. In practice, the vast majority of "Andeans" were indigenous peoples who spoke Quechua or Aymara;
indeed the 1950 Bolivian census defined the category of "Indian" primarily
through language though in practice it was also often linked to rural dwelling,
or, in the case of urban indigenous populations, to dress.[16] In the postrevolutionary period a shift in terminology took place that betrayed the MNR's
underlying hope that colonization and other reforms would transform Indians into citizens. Officials eschewed the pejorative term *indio* and used the
nonethnic marker *campesino* to refer to all rural Bolivians regardless of ethnic
background.[17] The shift from indio to campesino was common across twentieth-
century Latin America in nations where revolutionary regimes confronted
large indigenous populations. For their part, Andean petitioners rarely described themselves as indigenous—a fact that was likely self-evident to writer
and recipient given their origins in rural communities where the overwhelming
majority were Aymara or Quechua—and often wrote as campesinos or simply
vecinos (residents).

If the MNR hoped to erase indigeneity with a change in terminology, in
the lowlands that ethnic identity reemerged with force amid a strong nativist
reaction. A heterogeneous mixture of Andeans became kollas, defined in opposition to the lowland's equally diverse camba population. Those two terms,
kolla and camba, were ostensibly broad geographical references but in the
eyes of the cruceño elite, they were racial or ethnic markers separating Bolivians into indigenous highlanders and mestizo lowlanders. Furthermore, as
Sanabria's discussion of "our Indian" in the first chapter makes clear, Santa
Cruz's mestizo elite symbolically embraced a Guaraní indigenous past rather
than an Aymara or a Quechua one. They also identified an alternative historical
and racial genealogy, considering themselves to be descendants of Spanish
conquistadors that had come up the Río de la Plata to settle Argentina and
Paraguay, a distinction that they claimed made them whiter than Bolivia's
Andean population.[18]

Imagining the Lowlands from the
Highlands of Revolutionary Bolivia

The petitions in the archives of the Institute of Colonization range in format
from simple letters rife with spelling and grammatical errors to elaborate
project proposals with budgets, work plans, and rolls detailing migrant names,
ages, and former occupations. This diversity is indicative of the variety of places
that petitioners wrote from but also in the range of occupations that they

held. Some petitions came from elderly veterans of the Chaco War and others from young men who had just completed their year of obligatory military service. Although many letter writers were small farmers and herders from rural hamlets, the most vociferous claimants were often former miners from some of Bolivia's largest and most conflictive mining camps. Along with petitions that came from individuals or families, Andeans formed youth brigades, school groups, and farming cooperatives and their appointed leaders wrote, and signed, on behalf of their members. Petitioners also enlisted the help of authorities to serve as intermediaries from departmental prefects to *corregidores*—a designation dating back to the colonial period equivalent to a mayor of a small town—and the MNR's local representatives.

Andeans conjured a powerful landscape imaginary in their petitions. In constructing their home communities as deficient, harsh, and uninviting spaces and lamenting their miserable or desperate circumstances they hoped to push the government to support their cause. While facing environmental challenges distinct from those encountered by Okinawans on the Ryukyuan archipelago or Mennonites in the high desert of Chihuahua, Andean petitioners also spoke of overcrowding, drought, landlessness, and poor soils not suitable for farming. The most evocative petitions typically came from Potosí and Oruro—two departments dominated by Bolivia's high, arid plateau. In 1954, the Corregidor of Santiago de Andamarca in Oruro wrote to the director of colonization. Residents of the small hamlet sitting at close to 4,000 meters above sea level had gathered at a public meeting to discuss joining a colonization program. "Our lands are *arenales* (shifting sands)," wrote the corregidor, "without any benefit to human life."[19] He elaborated by noting that their meager harvest did not even produce sufficient fodder for their herds and that many desperate residents, who "are the sons of soldiers who fell in the Chaco War," were leaving the community in search of work.[20] The following year, a group of would-be migrants asked the government for the right to settle in the semitropical south Yungas, a transitional zone which descends from the eastern flank of the Andes at an elevation of more than 4,000 meters down to the tropical Amazon basin in less than one hundred kilometers.[21] The group described their home communities on the meager grasslands of the altiplano where they had "led a life of suffering" in a region "extremely unfavorable" to agriculture where even herding was not profitable.[22]

In January of 1955 petitioners from the municipality of Llica, the tiny capital of Daniel Campos province in Potosí, wrote to the ministry asking to be sent to the lowlands. At well over 3,000 meters, the area receives less than 200 millimeters of rain a year and sits on the edge of Bolivia's Uyuni salt desert, a challenging environment for small-scale farming among its majority Aymara population. A cyclical history of mobility characterized the region dating to at least the nineteenth century.[23] Residents would typically work their fields from December to May, bring produce to regional markets and then, along with landless community members, travel to labor in the nitrate fields along the Chilean coast in June and July. The authors of the petition, Moises Colque and Casimiro Flores explained that despite their best intentions to modernize local agriculture and ranching, the people of the zone could not "prosper because of the unproductive conditions and climate of the region."[24]

The same year as the aforementioned petition, a group from Mojinete, in southern Potosí along Bolivia's Argentine frontier, appealed to the ministry. The petitioners from the remote, rural, predominantly Quechua region linked their desire to migrate with the MNR's revolutionary project, noting that on the anniversary of the 1952 revolution, community members had gathered in a public assembly to discuss colonization. They enlisted the help of the head of the local school district to draft their petition, which contained a census of willing migrants including twenty-seven heads of family between the ages of forty-four and nineteen. In a letter signed by the mayor and the local representative of the agrarian reform, they explained that their current lands in the department of Potosí did not provide for even 10 percent of their annual consumption. They characterized their highland environment, consisting of jagged mountains and saline lakes, as "a frigid and narrow place surrounded by suffering on all sides."[25]

In 1957, Toribio Tarqui, head of the "Sora" agroindustrial cooperative, wrote to the Ministry. The MNR hoped to transform highland agriculture through mechanization and other modern methods. Tarqui explained that his associates shared this goal but that their attempts to modernize agriculture in their highland department of Oruro had failed. The cooperative purchased tractors but despite the "efforts and sacrifices" of its members who "worked with tenacity and pride" their efforts had been "in vain" overcome by an inhospitable climate.[26] Tarqui was likely referencing a devastating drought that had struck the Central Andes during the 1956 growing season. According to reports from an experimental farming project operated by UNESCO, the drought had resulted in crop losses ranging up to 94 percent. Campesinos were unable to collect seed for the following year's planting and could not prepare the dried, cracked earth for the following season.[27] For the mechanized Sora co-op, as with many small-scale farmers, labor migration to Argentina was the only remaining option.

Letters like these continued to flow into the Ministry of Agriculture throughout the 1950s. As a genre of petition, they sought to marshal misery to generate intervention on the part of the state. Their authors emphasized their humble, hardworking nature in the face of insurmountable environmental conditions with a tone of intimate familiarity. This lent an air of realism to their desperate portraits. For state officials who had little firsthand knowledge of everyday life in Llica or Mojinete it was difficult to question the veracity of such heartwrenching depictions. Conditions in those regions were likely as trying as the petitioners made out. Even today Mojinete is a remote region of few roads where nearly all residents are without electricity, while Llica faces extreme poverty. Some petitions include a resolution, but it is unclear if the majority were successful. Yet the fact that Andeans employed these tactics, presupposing their purchase, indicates a dramatic shift, perhaps less in the conditions of daily life on the altiplano and more in the relationship between state and subject in revolutionary Bolivia.

In contrast to their quotidian experience of the highlands, Andeans knew little about the areas that the Bolivian state was opening for colonization. Settlement sites ran throughout the half-moon of the lowlands, beginning in the tropical Yungas to the north of La Paz and extending in a broad arc along the northern and eastern flank of the Andes through the Beni, the Chapare region of Cochabamba to the forested plains of Santa Cruz. Unfamiliarity with these new lands fed rather than limited expectations. Petitioners readily constructed the lowlands as an obvious solution to the considerable problems found in their home communities. Letters typically imagined the region in intertwined economic, agricultural, or territorial terms as rich, fertile, and abandoned. The former two designations carried evident promise for future migrants and state. Although *abandoned* may seem an unlikely description for migrants to embrace, it functioned in two important ways. First, abandoned lands were presumably unoccupied and thus open to potential settlers. Here indigenous migrants, as much as the state, were silent about the presence of lowland indigenous communities. Abandonment also invoked a question of state sovereignty—that unworked state lands, in remote frontier areas were liable to be usurped by neighboring countries. This was particularly resonant in a nation that had lost sizeable portions of its territory to Brazil, Chile, and Paraguay. At times all three designations came together in a single petition. In 1955, a local representative of the agrarian reform wrote to the Ministry of Agriculture on behalf of a group of three hundred campesino families, asking that they be sent to the "fertile and rich lands of Caupolicán [province]" in the Northern Amazon basin.[28] In addition to "populating an abandoned region," he assured the ministry that this "will also be a method of securing our international border with Peru."

The nature of petitioner requests varied in accordance with their diverse backgrounds. Individuals often wrote to the ministry asking for a single plot of a few hectares, while larger organizations proposed resettlement schemes occupying thousands of hectares and promised large-scale investments in infrastructure and machinery. Some petitioners sent advance delegates to inspect and identify suitable locations in Santa Cruz and elsewhere. For others, the entirety of the lowlands (from the Amazon near La Paz to the Santa Cruz plains) appeared as a homogenous whole and they hopefully asked for free transportation to open lands anywhere in eastern Bolivia. At the same time, migrants began to speculate about the challenges, as the well as the possibilities, posed by the tropical nature.

In the previously mentioned petitions from Mojinete and Llica, letter writers put no prerequisite on where they ended up. The former would-be migrants were willing to go to "Santa Cruz or another zone,"—although they politely asked the Ministry of Agriculture to inform them "to which department and province they would migrate and in what month" once a decision had been made. They also hoped it would "not be too far from roads or railways [and not in] zones of epidemics."[29] The latter group noted that they had resolved to move to eastern Bolivia because those "fruitful regions offer great prospects" for those "with will and strength (voluntad y fuerza)."[30] Evorista Mayorga, the leader of a different group from Potosí, speculated about the fundamental differences a migrant might encounter moving between highland and lowland. He reminded state officials, "to take into account that we are from a frigid climate and that Villa Tunari [in the lowlands of Cochabamba] is tropical" and thus that medicine for malaria and other diseases should be made available to colonists.[31] Arriving from the treeless altiplano he also suggested, with a considerable degree of understatement, the need to be provided with "hatchets and machetes because perhaps there will be vegetation and trees that we need to clear."[32]

Petitioners drew portraits of difficult conditions in their home communities and tentatively promising ones in lowland colonization zones reflecting both lived and imagined environments. As farmers, ex-miners, and seasonal laborers imagined themselves moving to Santa Cruz, the Alto Beni, or the Chapare, they also framed their mobility in ways that resonated with the immediate aims of the Bolivian revolution. Although there is no direct reference to Wálter Guevara Arze's *Plan inmediato* in the following petitions, the terms and concepts employed by letter writers appear to directly paraphrase that policy document. We can imagine the many ways that economic policy may have been diffused in revolutionary Bolivia as Ruiz's films and pamphlets circulated while Guevara Arze's foundational paper was read and re-read among government officials, engineers, and extension workers and picked up by local teachers, mayors, representatives of the Agrarian Reform. Those intermediaries helped interpellate the individual desires of migrants into revolutionary policy.

In February of 1954, Felix Oroza wrote to the Ministry of Agriculture as a representative of a three-member agriculture and ranching society, El Yeso. He pointed out that a short distance from Santa Cruz there were lands suitable for agriculture "totally unpopulated and without close neighbors," and asked that five hundred hectares be given to the society as such an act would be "in harmony with the Supreme Government's plans for agroindustrial diversification."[33] In December of 1955, Julio Menses presented a petition through the Labor Federation of Rural Workers which was eventually passed on to the General Director of Colonization. Menses represented a small cooperative of seven young individuals from Oruro between the ages of nineteen and twenty-nine. They sought land in Masicuri in the Vallegrande region of Santa Cruz in "accord with the social, political, and economic reality" of the nation and "taking into account the plan of economic diversification put in place by the government." The group, Menses continued, felt it was both "a right and obligation of all good Bolivians to transform the rhythm of life, [especially in] the most remote places of our patria."[34]

These Bolivian nationals employed similar strategies to Mennonites and Okinawans who also framed their mobility with deliberate references to the goals of the March to the East. Like those immigrants, the aforementioned societies and cooperatives outlined work plans and listed capital that they could bring with them to the lowlands. This was not the case for most impoverished highland migrants. Yet even with scarce resources to move independently, Andeans who offered little more than their grain of sand could also employ a patriotic discourse to demand inclusion in the March to the East. At times they did so in direct opposition to foreigners. Menses, representing the least wealthy of these petitioners, argued that highlanders had finally awoken to the reality that for too long they, "had believed ourselves inferior in all aspects in relation to foreigners, who had pursued their own interests [in Bolivia] served by our inexhaustible sources of wealth."[35] Other Andean petitioners drew similar links to criticize foreign participation in the March to the East to the exclusion of Bolivians. In December of 1957, a group of five men petitioned the ministry for fifteen hectares each in the colonization zone of Caranavi in La Paz department. With no response forthcoming they wrote again the following year. Their leader, Angel Cossio, implied that the ministry's inaction was hypocritical and unpatriotic. He bitterly recounted that in early 1958 he had read a newspaper article detailing the "fabulous sum" spent on Japanese colonization in Santa Cruz. "We, the Bolivians," Cossio continued, "do not doubt that we will be received with preference," and assured the minister—with a persistence that bordered on the sinister—that he would be waiting "outside the door of your office" for a resolution.[36] No response to Cossio's letter is included in the archives, but the internal correspondence of the CBF related to Japanese participation in the Villamontes project, discussed in the previous chapter, suggests that such petitions were effective in pushing the Bolivian government to give preference to internal migrants.

Historian Fernando Coronil identifies the bodily metaphors that infused discussions of nature and oil-driven nationalism in Venezuela. In revolutionary Bolivia, Andeans also articulated personal and national interests through the corporeal, employing an emotive strategy largely unavailable to foreign colonists.[37] National bodies under threat as well as patriotic agrarian landscapes (the "Bolivian soil") were evoked in numerous Andean petitions. Menses referred to foreign exploitation of "our beloved land." Petitioners also brought their own bodies into play to solidify their claims. A group of agriculturalists from Cochabamba provided the government with a rational, scientific explanation for their request for land. They pointed to their agricultural expertise and a "historical reality" that the state surely knew and could not afford to ignore. However, their claim was an embodied logic as much as an intellectual observation. "We are Bolivians" they informed the government, "and we feel in our flesh the present reality."[38]

The most radical and frequent petitions sent to the Ministry of Agriculture in the 1950s came from miners and ex-miners who made explicit claims on the MNR's revolutionary legacy. That legacy could be interpreted to mean something similar to that articulated by Menses or Oroza, the need to diversify the economy through food production. A group of miners from the closed La Chojilla mine were looking for land in Caranavi, "so that we can produce for our country" in accordance with the government's plans.[39] But petitions for land in the wake of the national revolution oscillated between revolutionary promise and revolutionary threat. Miners were conscious of their political weight and often emphasized their radical proclivities. In the mid-1950s, representatives from Siglo XX, the largest tin mine in Bolivia, petitioned the national government. These miners from Potosí had played a central role in the MNR's 1952 revolution and the mine (formerly owned by tin baron Simón Patiño) was expropriated by the state mining company in the aftermath of the revolution. Ironically, the nationalization decree, which had been formally signed in a ceremony held at Siglo XX mine in 1952, resulted in a dramatic reduction of the number of miners on the state payroll. In the face of high unemployment, the miners formed numerous agricultural, ranching, and industrial cooperatives in the hope of moving former workers to new colonization zones and regularly reminded forgetful authorities of their promises to set aside land for ex-miners.[40] One group of ex-miners formed a cooperative named after mining labor leader Mario Torres and, in 1955, expressed frustration with the bureaucratic hurdles and stalling of colonization authorities. They reminded the Ministry of Agriculture of their "high revolutionary spirit," and that the mine had "sacrificed the most during the oligarchy." Some referred to historical, economic, and social realities to justify their request, but this group of ex-miners brought politics into play, warning the government that their "revolutionary conscience told them that they would not be defrauded in their call for [state] collaboration."[41]

No direct response to the petition exists but the state accommodated subsequent petitions from the large mine. The following year, Victor Hugo Zelada, another representative from Siglo XX, asked the director of colonization for free transport for three members of his cooperative who would travel the Cochabamba-Santa Cruz highway in March to select lands that were suitable for colonization. In a rhetoric standard to the colonization petition but bearing additional weight given the political power of the Siglo XX cooperatives, Zelada concluded by stating "we do not doubt [in the fact] that we are deserving of attention."[42] He also requested and received an official letter from the Ministry of Agriculture certifying the group's mission and requesting cooperation from local officials—not unlike that received by visiting Japanese settlement delegation leader Minuro Takata two years earlier.

In the late 1950s, residents of Santa Cruz accused arriving Okinawans of being fishermen and not farmers with the aim of blocking their future immigration. Okinawans responded with a performance of good agricultural citizenship in which they invited locals to tour the colonies and presented officials with symbolic gifts of their agricultural products. Ex-miners also felt obliged to justify their credentials as suitable farmers in lowland colonization zones. Wálter Guevara Arze provided a rationalization in the *Plan inmediato*, arguing that most miners had spent childhoods laboring on small family farms and large haciendas and typically returned to these occupations during harvest months.[43] One group of miners made a similar claim in their petition to the state. In 1958, Segundio Gómez wrote to the Ministry of Agriculture on behalf of miners from Milluni and Viloco mines in La Paz department who had voluntarily resigned due to a shortage of work in the wake of Siles Zuazo's austerity measures. The miners wanted lots in Caranavi and insisted that they were specialists in agriculture, as they had "been born and raised in the countryside." They promised to bring their families with them so that "with our labors, in a day not too distant, we will see these [lowland] regions prosperous through the force of good Bolivian farmers."[44]

The miner-turned-colonist was a script that was easily embraced by both the Bolivian state and Andeans. It fit within a national development narrative in which excess populations would be redistributed and latent landscapes reordered as Bolivia shifted from an extractive to a cultivated state. In those same years, Andeans also drew on another discourse, one with an impressively transnational yet patriotic dimension. As Mexican and Mennonite migrants traveled north to work as field hands in the United States and Canada in the 1950s, tens, and later hundreds of thousands of Bolivians also became braceros as they migrated southward to work in the Argentine harvest. Bolivians had engaged in transborder migration since at least the early twentieth century. However, these migratory patterns increased dramatically due to demographic and social changes in rural Argentina, where a large-scale exodus to urban centers was well underway in the 1950s and remaining agricultural laborers, emboldened by Juan Perón's progressive social legislation, began to unionize and demand better working conditions. Both factors further drove northern growers to look for undocumented, cheap labor from Bolivia, producing an impressive network that relied on local recruiters to canvass remote highland communities. By the early 1960s, nonunionized Bolivian labor dominated the sugar harvest in the northwestern provinces of Salta, Jujuy, and Tucuman that had expanded because of the U.S. embargo on Cuban sugar. Over time, Bolivians also took on new roles in rural and urban environments throughout the nation. When the sugar season ended they traveled to Mendoza's vineyards and still further south to pick apples in Patagonia. Between harvests, an increasing number of Bolivian braceros settled in Argentine cities from Jujuy to Buenos Aires. In 1960 there was a reported 89,600 Bolivians living in Argentina. By 1969 that number had grown to an estimated 450,000 (some suggested it was nearly double that) representing nearly one-tenth of Bolivia's population.[45]

As the current of Bolivian braceros first began to increase in the early 1950s, their undocumented presence became a subject of congressional debate. MNR representatives had first encountered their conationals while living in exile before the 1952 revolution. According to one, the bracero problem represented, "a shameful current of depopulation."[46] As previously indicated, the bracero issue was also present in intragovernmental memos and newspaper reports on immigration with some commentators, disingenuous or otherwise, advocating a return of fellow nationals to colonize the lowlands before the entry of undesirable foreigners. Few evoked the bracero issue more evocatively than author Fernando Antezana who had signed bracero legislation as a member of the MNR in the 1950s. In an exposé entitled *Bolivian Braceros: Human Drama and National Bleeding*, Antezana invoked a corporeal strategy of a national body under threat, while cataloging a series of abuses in a manner reminiscent of Mexican-American labor activist Ernesto Galarza whose investigative reporting on the U.S.-Mexico Bracero Program "Merchants of Labor" had appeared only two years earlier.[47]

While working in the Bolivian frontier city of Tupiza throughout the 1950s and early 1960s, Antezana became accustomed to the annual appearance of thousands of migrant workers that passed through on their way to the sugar harvest. In 1966 he decided to publicize the plight of his conationals who "abandon their patria, their home, their people and their field of labor," to seek a profit abroad.[48] He claimed that a total of 784,000 Bolivians were "nomads without patria and without home." This included nearly 200,000 who made the annual pilgrimage to Jujuy and Salta. Antezana followed braceros from their homes in remote highland-sending communities in Potosí and Oruro, where many walked to the closest rail line that would take them south across the border and into the sugar zones of Argentina. Utilizing personal testimonies and songs from the migrants, Antezana detailed the grueling conditions of the harvest and the difficulties braceros faced in bringing even a portion of their meager earnings back home. Like earlier commentators, he placed the tragedy of bracero emigration in explicit dialogue with Bolivia's colonization zones "that are waiting for the fecund labor of the sons of this prodigious land" and international organizations, "disposed to promote the development of the community, the colonization and the exploitation of the riches of Bolivia."[49]

In an article published in the widely circulated La Paz newspaper *Presencia*, Wálter Guevara Arze addressed the bracero issue in terms of lost labor power.[50] He compared the annual emigration of braceros to "the evacuation of a major European city during war."[51] Like Antezana he also pointed out the difficult conditions the braceros faced and the "great detriment to the future life of the country, . . . especially in terms of the labor that our agriculture needs." However, he was forced to admit that economic incentives would continue to drive Bolivians south where, even though "illusions of wealth" were rarely realized, cane cutters could earn twice as much as they did for identical work in the expanding sugar harvest in Santa Cruz. Around the time of his article, the Ministry of Agriculture was actively attempting to recruit settlers from among Bolivian braceros. In 1966, Roberto Lemaitre, the director of the newly formed National Institute of Colonization (INC), received a letter from Minister of Agriculture Rogelio Miranda. Citing the same statistics as Antezana, Miranda informed Lemaitre that the question of Bolivian braceros, "has created interest in all areas of the country and abroad."[52] He advised the INC to "consider with preference the possibility of settling braceros and their families in the zones of colonization," as those migrants were currently living "in subhuman material and social conditions in a foreign country."[53]

This discourse was not the exclusive province of the revolutionary government. If we return to the petitions discussed earlier, it becomes clear that the figure of the bracero was often invoked—alongside a harsh highland environment—when Andeans wrote the government asking to take part in lowland colonization. This massive and embarrassing loss of valuable bodies to Argentina with its emotional hold on national officials, gave Andeans another key element in their strategic repertoire. When the aforementioned group from Mojinete in Potosí petitioned the state in 1955, it might have been enough for the head of the school district to invoke a desperate regional context in which residents were eager to "move to the east" to push state officials into action. But he also reminded the government that, by becoming colonists, Potosinos could avoid making the annual migration to the sugar fields of Argentina. He claimed that this loss of labor power was an affront to national sovereignty given the shortage of *brazos* in the lowlands. While they honorably sought land to "cultivate and harvest goods in our Bolivian soil," they were instead impelled to "waste our forces in the neighboring nation of Argentina."[54] Their evocative petition was successful and the director of colonization assured them that though they might need to continue to sell their labor abroad in the short term, the ministry would soon provide funds to resettle them including technical assistance and medical support. In 1958, another group of Andeans writing to the state insisted they "were virtually dying of hunger" and needed to find a place "that offers us better living conditions." They had considered following the lead of their neighbors from Potosí who migrated to Chile or Argentina, "where they need brazos [labor] but for reasons of patriotism we prefer to remain within our patria."[55] Still they informed the state that if no help was forthcoming they would be forced emigrate. In framing this unfortunate abandonment as their only remaining option if they were not given land in Santa Cruz, the Alto Beni or the Chapare petitioners effectively drew together real and imagined mobility across Andean, Amazonian, and Argentine landscapes.

Some Andeans may have merely threatened to emigrate to shame the state into offering material support. However, there is some evidence from migrant petitions that others followed this transnational trajectory from Bolivia's highlands through Argentine sugar zones before settling in colonization zones in the Bolivian lowlands. The archives of the Ministry of Campesino Affairs contain a note related to settling two hundred families of campesinos "displaced from the sugar zones of Argentina," in an area on the edge of Chaco formerly managed by the state oil company.[56] In May of 1962, a man wrote to the Ministry of Agriculture claiming he had been living in Argentina since 1937 and working a variety of jobs in sugar mills and ranches. He had left Bolivia to escape "the chaos of the oligarchy," but he had returned home in early 1962 with the idea of taking part "in the magnificent plan of colonization" run by the CBF and the Alliance for Progress in the Alto Beni.[57] Ironically, U.S. policy in Latin America pulled Andeans in both directions in the 1960s. An expanded Argentine sugar quota after the 1960 Cuban embargo drove the growth of sugar cane in northwestern Argentine and demanded a steady stream of Bolivian braceros. Meanwhile, U.S. Alliance for Progress funding—ostensibly another means of isolating Cuba—bolstered colonization programs that attempted to recruit these same migrant laborers as national colonists for lowland colonies.

Oral histories reinforce these connections between the Andes, Argentina, and Santa Cruz. Francisco Condori, a settler in Santa Cruz's Yapacaní colony, was born in 1943 in the province of Nor Chicas in Potosí, an area that produced a significant number of braceros. He remembers that he was only fourteen when "because of poverty I had to abandon my home and go to Argentina."[58] Prohibited from traveling as an officially contracted bracero because of his age, he made most of the journey on foot with several uncles who had established a prior labor relationship with an Argentine *patrón*.[59] Whereas Antezana wrote evocatively about the exploitation of poor highlanders in those years, Condori remembers his time in Argentina as a positive and formative experience—one in which he learned about new production technologies and "gained a vision of how to work" that would help him as a settler in Santa Cruz. Like other Bolivian braceros, Condori and his family likely experienced elements of racism in Argentina—a nation that defined itself as European and nonindigenous in opposition to its northern neighbors. Argentine sociologists cast indigenous braceros as a public health concern while documenting their growing presence in slums (*villas miseria*) in cities like Mendoza.[60] Although those discussions of immigration and public health were rarely far removed from ideas of race, the Argentine press rallied against Bolivian braceros in more explicitly racist terms. The Buenos Aires newspaper *El Mundo* published an article in 1965 that caught the attention of Bolivian diplomats. The article insisted that Argentina had every right to control "this massive entrance that is causing a deterioration of its social and cultural levels."[61] Condori's silence regarding the everyday manifestations of this racial discourse likely has more to do with its similarity to his subsequent experience in Santa Cruz where highlanders were also pathologized as a public health threat and constructed as racial outsiders by a lowland elite that defined itself as whiter and more European than highland Bolivia.

Even as members of Condori's home community were traveling south, the Bolivian government was attempting to convince them to head east instead. In 1958, members of the United Nations World Food Program arrived in Potosí. In cooperation with the Bolivian government, the relief program offered free transport, credit, and supplies for volunteers joining a new colonization program near the town of Cotoca on the outskirts of Santa Cruz. Condori's uncles enlisted in the Cotoca endeavor, but he continued working the annual harvest in Argentina until 1962 before he too ventured to Santa Cruz. There, his path intersected with those of other global migrants. He and an older brother first worked as laborers on Japanese and Okinawan farms and eventually settled in Yapacaní (a short distance from the Japanese colony of San Juan and the Mennonite colony of Las Piedras). Condori's travels linked the home communities of the highlands to the sites of the Andean diaspora, first in Argentina and later in the Bolivian lowlands. In Yapacaní, he homesteaded alongside dozens of other Condoris as well as settlers with other typical Quechua and Aymara last names like Quispe and Mamani. In that sense Andean colonists were not unlike two other groups—Mexican Mennonites who headed north to work in Canadian harvests before pioneering settlement ventures in Bolivia and Okinawans who had migrated through the Japanese diaspora and Japan's empire in East Asia before joining USCAR's resettlement program in the 1950s—that they settled alongside in Santa Cruz. Andeans constituted a diaspora of their own, both in Argentina where they were obvious foreigners and in the Bolivian lowlands where they were also seen as outsiders by local elites. One cooperative of ex-miners from Potosí appropriately named their colonization cooperative "Exodus" reflecting the idea that struggling highland communities were undergoing a form of dispersal—a central element in the diasporic experience.[62] Whether Andeans were threatening to emigrate as braceros or actively doing so like Condori, developments in the northern Argentine sugar zone shaped Bolivian colonization and were central to the *discourse* of Andean petitioners who drew on this transnational flow in their evocative letters to the state.

Abandonment Issues: Letters from Bolivia's Settler Frontier

In a stream of letters flowing into state offices, Andeans cultivated the state by placing their own personal migrations—as small farmers, miners, and transborder braceros—in conversation with the 1952 revolution. That epistolary tradition did not cease as Andeans moved into the colonization zones they had so readily petitioned for. In their earlier letters, petitioners recited lines about patriotism, national progress, food production, and strong families. Once established in colonies like Cotoca, Yapacaní, and Caranavi, those seamless narratives often jarred with the reality of life in remote rural areas. Initially presented as promising future homes, they now experienced settlement as bitter lived reality. A core element of their critique centered on the idea of *abandonment*, a ubiquitous and versatile term employed by regional elites, state authorities, and settlers. It could refer to the status of land, the action of colonists, or the inaction of the state. For example, when cruceños demanded infrastructure and other aid from the Bolivian state in the early twentieth century, they frequently referred to the idea that their department had been abandoned by the nation. Over time, that discourse shifted as they found themselves the central object of the March to the East and began to denounce abandonment's opposite, a so-called invasion of highlanders that were displacing cruceños, the *true pioneers*.[63] The Bolivian state also invoked abandonment imagining the lowlands as a sleeping, dormant, or latent place abandoned by prior oligarchic regimes. Colonization would help Bolivia achieve true sovereignty by reestablishing the territorial integrity of a disarticulated nation. Yet, as colonization progressed in the late 1950s and early 1960s, the MNR became increasingly concerned with another form of abandonment—worryingly high rates of settler attrition in new colonization zones that often exceeded 80 percent. While accused of invading the lowlands by hostile regionalist writers and of abandoning their new colonies by the Bolivian state, Andeans also made use of the idea abandonment. Writing from the Andes, they had often cast the lowland landscape as abandoned and thus open to settlement. In correspondence coming out of the colonization zones in the 1960s, colonists charged the state with abandonment and cited multiple failures on the part of state institutions to provide for the necessities guaranteed in films and pamphlets. Complicating this issue further, colonists were also likely to accuse one another of abandonment of land particularly when such an accusation could open a supposedly empty plot for acquisition by a newcomer.

The Cotoca colony, which members of the United Nations Andean Mission enlisted Condori's uncles to join during a promotional tour through Potosí in the late 1950s, was one of several early directed settlement initiatives in Santa Cruz; Others included the military-administered colonies of Cuatro Ojitos, Huaytú, and Aroma. Those first attempts were intended as something of a social and biological experiment to see whether highlanders could adapt to the rigors of tropical life. As the first generation of colonies in postrevolutionary Bolivia, they reveal an emerging discourse about abandonment that planners would repeat over the following decade, a period during which the Bolivian government opened new colonization zones in partnership with USAID across the tropical lowlands of La Paz, Cochabamba, and Santa Cruz. In those latter colonies, state officials and foreign experts obsessively tracked abandonment rates, held conferences, and prepared studies on the theme, invoked the lessons of Cotoca and sought to engineer new colonization ventures (to be discussed in chapter 4) in response.[64]

Various explanations circulated among national and international actors involved in colonization as to why colonists so often failed to remain on the land. Examining the archives of the Bolivian Development Corporation the explanation of settler abandonment appears clear cut. When deciding to leave the zone, settlers submitted signed letters to colonization authorities explicitly stating the reasons for their withdrawal and promising to pay back the costs of their transport and stay. In September of 1956, Martín Quispe signed off on a document confirming his abandonment of the colony of Huaytú. The military administered this colony where young Bolivian conscripts opened roads and cleared fields for future colonists and were given the option of remaining as settlers themselves when their year of obligatory service was done. Quispe attested to the fact that his departure was voluntary and that he simply "could not perform the [required] work because of the climate and owing to the insistence on the part of his family [to return to the highlands]."[65] Despite spending a mere sixteen days in the colony before reaching this decision, Quispe insisted that he had "received good treatment on the part of the military authorities," including "tools for my work, clothing, and food" for himself and his family. In February of 1957, the colonist Eustaquio Ayaviri Villca signed a similar letter. First brought to Cotoca by the UN's Andean Mission, he had left that colony and presented himself as a colonist before military authorities at Huaytú at the end of January. Only two weeks later he decided to leave once again, "considering myself inept for agricultural work in the environment of the zone."[66] Like Quispe he thanked the administration and promised that in "returning to my land of birth I will not distort the truth," about anything pertaining to his departure for would-be colonists eager for news. The cause had been the inability of his wife and children to "become accustomed to the climate."[67] Another abandonee, José Alba López, also referenced his inability to acclimatize to the zone and confessed that as a mason by trade he was poorly suited to agricultural labor. In these signed official letters, carefully absolving the military and other colonization authorities of any wrongdoing, colonists structured their abandonment within a logic of personal shortcomings and environmental incompatibility. They expressed gratitude for the role of the state and the actions of its representatives. Such letters, written in the first person but produced on official typewriters, tell more about the concerns on the part of colonization authorities to appear to have "provided every assistance" than they reveal as to the experience of their signatories. The power differential between military and settler precluded a frank discussion of any shortcomings by the former.

In the lowlands, foreign technicians often acted as proxies for the Bolivian state. The UN's Andean mission brought transplanted international development workers in direct contact with transplanted migrants. In 1955, the Haitian social worker, Jeanne Sylvain, began working for the UN at Cotoca. From 1948–50 she had collaborated with Alfred Métraux in a study of the Marbial valley in Haiti where UNESCO was initiating a community development project.[68] As a body of transnational expert practitioners drawn to lowland development, these foreigners provided an alternative rationale for settler abandonment. Sylvain, for instance, leaned toward political and cultural, rather than environmental, determinism. During her extensive recruitment trips through the highland departments of Potosí and Oruro, she frequently encountered government officials who placed obstacles in her way. In September of 1956, after an exhausting but successful tour of highland mining camps, Sylvain found that COMIBOL (the state-run mining organization) was prohibiting her new colonists from leaving for Cotoca. She initially dismissed these and similar frustrations with national agencies "confident that there could be nothing more than a slight misunderstanding between the COMIBOL and the [United Nation's] Andean Mission."[69] However, over the course of her two-year tenure she would increasingly complain that "international technicians were unfairly painted with the mistrust caused by national technicians and due to political issues."

In addition to bureaucratic infighting Sylvain identified several other culprits for settler abandonment from local antipathy to poor administration. However, she ultimately privileged sociocultural explanations or as she would bluntly state, the "fundamental distrust of the altiplano Indian to anything and anybody alien."[70] These were evident in her accounts of the repeated instances in which she and other project officials attempted to dissuade colonists from leaving Cotoca. In August of 1956, new colonists had "declared themselves to be happy," but only a month later one colonist left without notice. As the sole colonist from Cochabamba, Sylvain imagined that the man "probably could not feel well integrated among his companions [from Potosí]."[71] The explanation, suggested that cultural incompatibility might occur not simply between highlanders and lowlanders but also within the former group, which included colonists from across a diverse Andean cultural geography. Worse still, this single abandonment appeared to spread to other colonists with the aggressiveness of an infectious disease. That same evening another group of colonists announced that they would be returning to their home community of Calcha for All Saints Day and to plant crops on their highland plots. "These are Quechua-speaking people," acknowledged Sylvain, "for whom the religious significance of All Saints Day and fear of retaliation from the neglected dead are vivid."[72] She suggested that the Andean Mission would "assuage possible terrors" by recreating a version of the ceremony in the colony. By focusing on their spiritual motivation for leaving the colony Sylvain ignored the clear agroeconomic incentive that colonists had in maintaining both highland and lowland fields.

The following year the situation at Cotoca deteriorated further.[73] Whereas in November of 1956 there were just over 200 colonists at the site, by January of 1957 a quarter had left. Sylvain's assessment of this low retention rate and the poor adaptation of new colonists again drifted toward the cultural—particularly its gendered dimensions. According to Sylvain, single men were prone to homesickness but were ultimately more malleable than women or those with families. This was apparent to project officials who were engaged in far more than transplanting farmers. They sought to encourage a series of social transformations ranging from attempts to transform the recreation, housing, alcohol use, literacy, hygiene, and sanitation of colonists. Sylvain reported significant initial progress in limiting the consumption of typical highland products like chicha, a fermented corn beverage, and coca leaves, a ubiquitous stimulant with a strong spiritual and cultural significance in the Andes. Officials saw these native practices, along with the continued use of Quechua and Aymara, the indigenous languages spoken by the majority of migrants, as deficient cultural traditions that needed to be reformed. However, Sylvain lamented that "with the arrival of their wives," men that had once shown signs of transformation quickly began to regress.[74] She counted a total of "seven women in the last contingent of settlers [that] regularly engage in chewing coca and in two cases husbands who have abandoned this custom," had returned to it. Furthermore, these women made no progress with Spanish, insisting on speaking their native Quechua instead.

In addition to a lingering Andean culture that prevented migrants from adapting to life in the lowlands, Sylvain blamed abandonment on the culture of colonization being cultivated at new settlement sites. In her reports, Sylvain acknowledged that shortcomings on the part of project technicians, uncertainty over future ownership of land, and colonists' unwillingness to engage in forced cooperative labor hampered colonization. Abandonment was a response to all these insoluble issues but, at heart, it was also "a manifestation of the simple dependence without responsibility [of the colonists] because of the satisfaction of basic needs for the families for too long."[75] Her terminology—linking abandonment with dependency—merits scrutiny. Dependency was an omnipresent term in midcentury Latin America, often used to describe the underdevelopment of peripheral nations by economists and theoreticians and to justify protective measures and subsidies for emerging industries. As a theory it emerged from the work of Raúl Prebisch as a member of the UN Economic Commission for Latin America and the Caribbean and was later popularized by radical socialist theorists like Andre Gunder Frank.[76] Here *dependency* took on a different meaning when employed as a critique of excessive state intervention on colonist's behalf. In fact, by suggesting that pampered colonists should be cut loose to fend for themselves, Sylvain was arguing for the exact opposite of the policy favored by most dependency theorists.

The rationale was both dubious and convenient given the frequent financial shortfalls of the Bolivian state. Yet this alternative dependency theory spread among the crowd of national and international development workers that took part in lowland colonization and would be cemented as the definitive lesson of the Cotoca project—that directed colonization was a failure. In subsequent colonization reports from a range of officials and observers, a simple mention of "Cotoca" came to be synonymous with excessive state paternalism.[77] Other more concrete problems from secure land title to effective marketing of agricultural production, which Sylvain had mentioned, and colonists had explicitly stated were the "principal cause of their discouragement," would be swept aside. In practice the reaction to a crisis of colonist dependency emerged in several ways. Officials decided to decommission the cafeteria that served colonists at Cotoca and replace it with a straightforward food allowance intended to force colonists to independently manage their own consumption.[78] Replacing a cafeteria with ration cards was, after all, a simpler solution than reforming a convoluted titling process or guaranteeing reliable and fair prices for colonists' produce.

The aforementioned sources on Cotoca drawn from military authorities and foreign technicians provide several explanations—from the environmental to the cultural—that ultimately attribute abandonment to shortcomings on the part of Andeans. Settler petitions in the first decade of state-sponsored colonization turned this discourse on its head and specifically implicated the state and its agents. Settlers complained of three interrelated issues—titles, technical advice, and transport—that had been raised, and purportedly resolved, by the state. In state propaganda, Andeans dutifully followed the directions of técnicos, promptly received title to their lands, and immediately saw their produce efficiently whisked to regional and national market on new roads, but in letters from within the colonization zones, that vision appears woefully incomplete. Colonists were acutely aware of this and invoked the MNR's vision of the March to the East even as they pinpointed its shortcomings. In doing so, they proved themselves to be more than passive recipients of state intervention—or lack thereof. Yet as several of the following examples indicate, state shortcomings and scare resources also pitted some colonists (many of whom arrived in the lowlands spontaneously) against those who had come as part of directed settlement ventures. That combative settler–state and settler–settler discourse jarred with the image of a harmonious and meticulously planned human experiment in the tropics.

A letter from colonists living at Cotoca colony nearly a decade after Sylvain's tenure is particularly revealing. By the mid-1960s, abandonment had taken its toll on Cotoca and the adjoining settlement of Campanero. The few remaining colonists petitioned President René Barrientos asking that the government help solve the problems that "hold back the social development of the colony."[79] They recognized their role to "populate the Orient" and in the "experiment to acclimatize the man of the altiplano on the [eastern] plains." But after ten years and with five thousand dollars invested per colonist, they found the results of the Andean program to be "very sad." In that, they aligned with development authorities that considered the program to have moved from a potential model colony to a conspicuous failure. However, they differed with Sylvain on the causes of this failure—particularly on her assertion that state largesse had cultivated settler dependency and hindered their development. Instead settlers insisted that the "help they had received bears no relation," to Cotoca's extravagant budget. They specifically targeted people like Sylvain, the "distinguished experts and authorities," who had devalued their knowledge claiming the "situation would have been different" had authorities listened to their views which, "came directly from experience [things] which only they [the colonists] had felt," but ultimately they had been led by the criteria of others and now "suffered the disastrous consequences."[80]

In addition to this attack on a dubious development expertise, colonists made a litany of denunciations against corrupt and incompetent personnel administering the base. They charged that the head of the center, Armando Salinas had "made a fraudulent sale" of a colony vehicle that was intended to transport the sugar cane of colonists.[81] A bulldozer belonging to the station was frequently rented but the money earned was not invested in its maintenance and the machine was now broken. Four tractor tires had been placed as a guarantee for a colony business transaction in Santa Cruz but were never returned. More sinister were the frequent late-night thefts. A sewing machine had been stolen at 3 A.M. Medical service was nonexistent because the very beds had been removed from the health center. The base's chief agronomist "does things unrelated to agronomy," they continued while characterizing the head accountant, Eduardo Cajías as "a degenerate drunk who committed abuses against the colonizers, demonstrates immorality and refuses to work saying 'I will not attend to these Indians.'"[82]

No response was made to the Cotoca petition—which is filed alongside thousands of others in state archives—but the litany of complaints presented by the remaining colonizers and directed at colony personnel provide us with an alternative understanding of both abandonment and technical expertise. Here it is not settlers who abandon their plots because their deficient Andean culture prohibits them from acclimatizing to a new environment. Rather, they are the ones who have been abandoned and mistreated by an absent state and its corrupt and racist representatives. This explanation had emerged in a restrained way in one of Sylvain's reports from March of 1957 which hinted at "all kinds of failures of technicians to direct the settlers" and even confided that "the fact that the superior technicians of the center were themselves abandoning the center was pretty demoralizing for the colonists which tended to confirm the fallibility of the project."[83] Yet for Sylvain these deficiencies were bizarrely listed as accidental rather than essential causes of settler abandonment, perhaps indicating the unwillingness of an international expert on short-term assignment to engage in a more robust critique of the state.

In its denunciation of the arrogance of foreign experts and the overt racism of national authorities, the letter from Cotoca's colonists makes for a compelling read. But simply cataloging an inverted state–settler discourse—where the former is accused of abandonment by the latter—does not capture the complexity of colonization in the Bolivian lowlands. It falsely homogenizes a diverse group of colonists who were as likely to denounce one another as to challenge the state. A few months after the remaining colonists of Cotoca wrote their petition, Barrientos received another letter that also made specific reference to the failing endeavor. On March 1, 1966, a group of colonists from Jorochito (another new settlement along the Cochabamba-Santa Cruz highway to the west of Santa Cruz), led by Luis Fernández Prado and including representatives of the local cooperative and agrarian union, wrote to the president. They were aware of Cotoca the "so-called base of development" they sneered, "which for more than a decade has received all sorts of assistance unknown by other campesino communities."[84] Fernández felt that the base, with a scant eighty colonists, should be dismantled and transferred to their own community where its rusting machines could be put to productive use by the "thousands of campesinos that also came from distant regions of the country."[85] Was the colony headquarters at Cotoca simply a "display case," he wondered, or was it there to serve "the humble farming people?" Even if the former were true, Fernández insisted, "it would be better to move it somewhere more appropriate for its exhibition."[86] He boasted that the "majority of cruceños," were entirely unaware of the Cotoca center and only visited the region for the annual pilgrimage to a local shrine; in contrast, "none denied the existence of Jorochito" which produced an impressive quantity of goods for the regional and national market.

Fernández concluded his letter with a demand that in addition to moving the languishing agricultural center the state should send true "técnicos and not tourists" to advise the settlers of the region. What was meant by the disparaging alliteration? For their part the Cotoca colonists had denounced an "agronomist who does things unrelated to agronomy." While the Jorochito petitioners criticized their fellow colonists in Cotoca as belligerent because they "never permit the machinery to leave to lend help to other colonists," they shared the perspective that the frequently publicized technical assistance of the government had been without true substance—little more than tourism on the part of the visiting authorities. The response from government officials confirmed this dismal assessment with the Ministry of Campesino Affairs apologizing that due to a shortage of funds and personnel it was impossible to move the center or extend coverage to all colonists of the region though they "hoped that one day it will be extended."[87]

In the highland sending communities of Oruro, Potosí, and La Paz, rural Bolivians expressed similar frustration with the vaunted technical advice of authorities. They often did so in letters addressed directly to the president. For example, residents of the town of Humacha on the eastern shore Lake Titicaca wrote a petition to René Barrientos in 1966 after a survey commission had passed through. They noted that they had lacked irrigation systems, "since time immemorial in this forgotten corner of the patria."[88] They also requested extension services and tools, "to escape from these systems that are completely antiquated." Addressing the president directly, they explained that "the indigenous race . . . [seeks] the level of civilization and progress that all other people aspire to." Like the petitioners from Jorochito they were wary of assuming true change would arrive with a technical commission. They nevertheless hoped that the resulting study would lead to "an effective and tenacious commitment, . . . and not be a mere *visit* which has occurred so many times before, cheating our aspirations [emphasis added]."[89]

From across the colonization zones settlers sent complaints, ranging from absent technicians to missing tools, to state authorities. Representatives from Santa Fe colony criticized the Institute of Colonization in October of 1956 explaining that new colonists arriving from the altiplano were "very sick because we are entirely lacking in medicine." A month later they were still protesting that sanitary services were deficient.[90] Members of the Colonia Juan Lechín Oquendo—named after the leader of Bolivian Worker's Congress and then vice-president of Bolivia—wrote to Paz Estenssoro in late 1962, furious that they had waited for four months for the arrival of forty machetes and hatchets destined for the colony. The minister they claimed, "does not demonstrate a single bit of interest in helping us and . . . is sabotaging the labor of men of the party [MNR] who gave up all personal comfort in the city and decided to work in agriculture for the self-sufficiency of the country."[91] A few years later colonists settled in the lowlands to the northeast of Lake Titicaca wrote Barrientos inviting him to visit their colonies. They made similar pleas for water pumps, fencing, machetes, sprayers, shovels, and affirming that while they would honorably "continue with our activities in these distant places of the country [even though] at the present we have not received help of any nature."[92]

Like the state, petitioners often linked these specific demands for material support to broader ideas of patriotism and social transformation. From Roboré, on the plains to the east of Santa Cruz, the regional Director of Education wrote

President Barrientos explaining that locals were demanding the "bricks, roofing materials, professors of education, agriculture, hygiene, and practical skills," promised by the MNR, "so that progress can be made in all regions of the country [and] new generations will be formed with a spirit of work, discipline, civic duty and honor."[93] In the highlands, frustrated Andeans threatened to immigrate to Argentina if state aid was not forthcoming, while in San José de Chiquitos, a small city on the eastern plains, the mayor wrote that residents were fleeing to Brazil—another "pernicious exodus" from a region with its own "natural resources in abundance." "They see themselves as abandoned" he continued, requesting that the government provide agricultural credit and extension services.[94]

On one of the final pages of a colonization pamphlet produced by USAID and the MNR in the early 1960s and titled "The Route to Development," a captioned photo proclaimed "the colonist will be absolute owner of his land with titles of domain and possession after the third year of work."[95] In 1966, Bolivian President René Barrientos received a letter from the Colmena Cooperative in Caranavi, one of the colonization zones pictured in the pamphlet. Its members noted that they had settled in the region in 1953 and rehearsed a standard discourse about their role in national development and territorial integration. Despite complying with their end of the state–settler bargain they still lacked titles to their property, which prevented them from using land as a guarantee to obtaining advance harvest credit from banks. They also extended this critique into a more reflective domain. The lack of titles, they continued, "positions us as a *transitory people* who cannot develop the fullness of our possibilities . . . almost as if we were *minors*, without right to manage our own affairs [emphasis added]."[96] No response to this petition is preserved in the archives although the Colmena cooperative continued to petition the government over the following years. In 1968 members protested the dismantling of a community development service operating in the area, through which their cooperative "and the entire campesino sector has escaped from the ignorance in which we were subsumed."

For a state worried about abandonment, the assertion that these petitioners remained transitory and juvenile should have been a troubling one.[97] Through colonization, the state hoped to first generate and then restrict mobility as migrants became firmly rooted in their new environment. The state also pictured that process as the constitution of independent rural farming families led by masculine fathers. At times state pamphlets pictured abandonment as a failure to comply with masculine honor—conflating abandonment of land and abandonment of family in the hopes of shaming future colonists who might consider leaving the colonization zone. "Women, the force for which man labors," said one.[98] "Family happiness in the zones [of colonization] is a reality" stated another caption accompanied by an image of a man embracing a woman and small child.[99] In the Colmena petition that gendered perspective was inverted. The colonists blamed the state for their protracted instability and in describing themselves as *minors*, sought to call out the hollowness of this gendered ideal of honorable migration.

To the reader, it may seem extreme that after thirteen years in the colonization zone, members of Colmena had still not received legal rights to land. Yet this was a common complaint of colonists from across the lowlands. Why did the state hold back the titling process? One could suggest that by keeping colonists perpetually waiting on property documents the MNR sought to maintain control over its transplanted, dependent subjects.[100] But ascribing cynical intent, may give too much credit to the administrative capacity of the MNR in those years. Securing land title in remote regions without a national cadastral survey was a particularly complicated and maddeningly bureaucratic process which passed through a range of steps from the local office of the agrarian reform to the desk of the president itself where at every turn it might founder.

To establish legal title, colonists were also required to manage their new lands in particular ways. Ownership was processional, a goal that colonists aspired to through the performance of good agricultural citizenship, rather than a legal basis from which they could operate. Building houses, setting fence lines, clearing land and cultivating an increasing portion of the total plot size were all essential elements in fulfilling the social economic function of land that made ownership a reality. As Kregg Hetherington points out for Paraguay, where a large settlement scheme was taking place on the eastern frontier with Brazil in those same years, transforming precarious plots into legal property might include symbolic as well as material elements such as raising the national flag over the land or the planting of perennial rather than annual crops.[101]

In establishing ownership on the ground, colonists also depended on local officials and mobile topography teams who would inspect the improvements (*mejoras*) that colonists had made to their land and verify their physical presence. For officials, the task of maintaining records for colonization plots that, due to high abandonment rates, frequently changed hands without acknowledgement was a daunting one. With thousands of colonists on the move to remote regions, simply locating mobile colonists was often a challenge for the state. In a note from late April of 1959, the Institute of Colonization admitted that 500 hectares of land near Caranavi had been solicited by a group of fifty campesinos but there had been no sign of them since the beginning of the year.[102] For some Bolivian officials, the struggle to control abandonment and illegal occupation was about administrative incapacity. In 1956, Max Molina, an employee of the Ministry of Agriculture, admitted that in the Chapare region of Cochabamba adjudication of new lots had stalled for two years because "the direction of colonization has not centralized the archive," and documentation remained with local authorities. He recommended that no new lots would be given out until a topographical commission could tour the region to inspect "agricultural labors and carry out a survey on the number of families and available lands."[103]

In the absence of strong administrative capacity and reliable centralized documentation, head offices in La Paz were forced to depend on local colony leaders and officials to find land for new recruits. The obvious problem was that most of these individuals, including Max Molina, were colonists themselves, which opened the door to a variety of forms of arbitrariness. Institute officials depended on those individuals but also asked that all materials pass through the central office so that they could register and confirm abandonments. In a letter to the President of Viluyo Colony, the Institute felt the need to "once again reiterate that it is entirely prohibited any occupation of lots that have not been confirmed by this office, any increase in the number of colonists in the colony presided over by you . . . would be illegal and will not be recognized." Yet through the 1960s the Institute of Colonization received frequent complaints that these same colony presidents were illegally selling lots.[104]

In the hopes of addressing issues of arbitrariness, the Institute of Colonization continued to send out teams to survey new settlements. In the early 1960s, Pablo Gutiérrez Sánchez, head of the Department of Colonization and Land, was tasked with sorting through the morass of abandonment denunciations in the Alto Beni region. In 1961 and 1962 he traveled extensively in the zone, reporting to his superiors on denunciations, abandonments, and overlapping solicitations and identifying lands that should revert to the state and others that could legitimately pass on to new colonists. In Corpus Christi Colony, titling appeared to be progressing with sixty-two of sixty-four colonists holding title. Investigating Florida Colony along the Río Coroico, Gutiérrez found that out of sixty-two lots, forty-two colonists had official title, three had been investigated for improper or arbitrary transfer "without the knowledge of the administration," eleven more did not have title for failing to comply with the minimum annual improvements or for arriving spontaneously after the initial colonization contingent. Examples of irregularities included the case of Ernesto Choque who took out loans based on the improvements made by a previous colonist and two individuals that had sold land that they did not technically have title to. Gutiérrez also requested that colony leaders provide reliable documentation including the name of the abandonee, the season and date of the abandonment, fair price for existing improvements, and whether anyone else had subsequently occupied the lot. He also dealt with cases where the entire border between two colonies had not been properly established. Despite these attempts to regulate and structure mobility, *spontaneous* colonists frequently moved to new zones without seeking prior approval from the ministry of colonization and soon outnumbered *directed* settlers in the lowlands.[105]

In the difficult process of establishing ownership over land, colonists were as likely to denounce one another as to denounce the state. When they wrote to the state, would-be colonists often asked for lands that were presumably empty or had been abandoned by their former owners. A particularly rich vein of correspondence exists from the Alto Beni-Caranavi settlement zone in the Amazon basin near La Paz. Some petitioners requested the plots of absentee owners who they claimed lived in the city or those that, to cite one case, had supposedly "abandoned the lot in question without having worked it."[106] In perhaps the most ironic example, Max Molina, the same employee of the Ministry of Agriculture who would later insist on the need to properly archive the records of the Institute of Colonization in the Chapare appears as usurper rather than archival advocate. In 1957, he solicited Lot 58 in Bautista Saavedra colony in the Caranavi settlement region, noting that the original owner Manuela Chacón had not worked the lot since receiving it in 1955.[107] "As a single woman" Molina claimed, "she does not have the slightest capacity to complete the work demanded by our laws."[108] He counterbalanced this with his own masculine role as colonist-provider, explaining that he had brought his family with him to the zone to establish himself permanently. Once again, the line between government-sanctioned official and colonist was entirely blurred.

What did these appropriations look like from the other side? A typical case received by the Institute of Colonization rarely included a response from opposing parties like Manuela Chacón. Fortunately, some colonists accused of abandonment also petitioned the state. These ranged from longtime residents of colonization zones who wrote the Institute of Colonization claiming that they were being displaced by new arrivals to recently arrived colonists who briefly traveled to the highlands and returned to find their new lands claimed by others.[109] Like the usurpers who claimed their plots, the accused presented the state with a justification for their continued claim. In June 1960, for instance, Celestino Roque Chuquimia, noted that he had received Lot 29 as one of an original group of colonists to settle in the Alto Beni in 1956. He had worked it for three years, but the new tropical environment had affected his health and he was forced to abandon the lot and return to the altiplano for several months. After recovering he attempted to return to his plot only to discover that it had fallen into the hands of one Belberto Machaca who, he claimed, had done no work to improve it. "Return the lot to me," he asked, invoking his status as "a founder of the colony dedicated to agriculture work," adding, in a plea for compassion, that his wife and young son had both died while he was in La Paz the previous year.[110] In an undated letter a veteran of the Chaco war—a status that was frequently, though not always successfully, invoked as part of a moral claim to land—complained of a frustrating experience in Caranavi in which he had been accused of abandonment before even settling in the colony! He had been adjudicated lands during the rainy season which he deemed ridiculous in a region in which "one can only work in June."[111] Nevertheless, he avowed that he had made two short trips to Caranavi to inspect his lot but each time the head of the colony was absent and thus unable to confirm his presence. When he finally arrived in the dry season to begin clearing his plot, he discovered that a woman named Luisa Valdés had been given his *abandoned* land. This was just six months after the original adjudication. The frustrated veteran questioned the seriousness of the entire endeavor and suggested that the national government and its local representatives where "doing business" with state lands and "maliciously disrespecting the law."[112] He argued that the woman who had taken over his plot was a merchant and not a farmer and that his usurped plot sat next to her husband's ultimately making it "all just one plot" which the latter was working. The veteran's accusation that Valdés was a merchant and not true a farmer marshaled the moral authority of agrarian citizenship to reinforce a contested claim to land. His related argument that her husband was farming both lots invoked colonization laws restricting maximum plot sizes in the settlement zones.

In the preceding correspondence, colonists might have denounced their neighbors and state shortcomings on the ground but in a modern reworking of *naïve monarchism*, they also appealed to an imagined benevolent president to resolve those outstanding issues. As Richard Turits writes of land reform under Trujillo in the Dominican Republic, colonization created a "novel role for its national state officials and institutions in peasants' daily lives as the mediators of land access, protectors against eviction, and providers of needed resources."[113] Although the petition's immediate effectiveness is not always apparent, by denouncing state failings and detailing their disputes with neighbors, colonists tended to reinforce the role of the state as ultimate arbiter. From the perspective of state building and governmentality, this discourse of disillusionment should be seen as generative. The anthropologist Tania Li, in an analysis of development policy in Indonesia, points out that, "failures invite new inventions to correct newly identified—or newly created—deficiencies," creating additional room for the state.[114] Despite foregrounding the failure of development schemes, James Scott acknowledges that their effectiveness "lies as much in what they replace as in the degree to which they live up to their own rhetoric."[115] If those initiatives pull subjects into a more intimate relationship with the state, reduce the capacity for independent action, or incorporate marginal territories into the nation it matters less that they fail to achieve their more altruistic aims.

Andeans sought to use denunciations of abandonment to appropriate vacant or contested land and to drive the state to provide promised support. Another key element in colonists' petitions framed abandonment in the context of transport and marketing. One news clip from the ICB's early years consisted of a steady, crudely hypnotic, camera shot of the front of a bulldozer leveling virgin forest.[116] In another pamphlet given to colonists, neatly stacked bags of grain sat by the road while, in the background, trucks carried others off to highland markets.[117] The message was clear. With state help, unruly landscapes would be transformed to mesh with the needs of individuals and capital. Yet colonists soon discovered that getting themselves into the settlement zones was easier than getting their produce out. While the government attempted to track settlers who abandoned their plots, settlers complained bitterly of their inability to transport their crops along terrible roads, claiming they had been abandoned by the state.

In 1965, Barrientos received a letter from a group of ex-miners who described their fate, "suffering all kinds of injuries . . . and the inclemency of nature, below the sad roof of our humble huts." The group lacked sanitary posts and complained of snakebites and tropical diseases, but most bitterly of all, that a so-called all-weather road constructed by USAID was consistently impassable in the rainy season. It was blocked by "landslides and the waters of the Río La Paz . . . making it impossible even for animals to pass," precisely in the months when the colony began to harvest crops. In language that resonated with their former occupation, these ex-miners claimed that they "see ourselves enclosed (*encerrado*)" in the forest.[118] With no access to market, their crops would rot, and residents would "remain in misery." El Palmar cooperative in Caranavi similarly petitioned for a bridge that would connect their fields to the highway across the river. Their leader affirmed that members were unemployed before becoming colonists but considering the difficulties bringing their crops to market "they want to abandon their lands and establish themselves in the city again."[119]

Excessive rain and humidity and impassable rivers were not the only aspects of tropical nature that colonists struggled with. Settlers from Huaytú Colony's agrarian union—a form of rural organization that became increasingly common in the mid-1960s—wrote to President Paz Estenssoro throughout early 1964. In earlier petitions from the region, farmers argued that while they had complied with national calls to increase rice production to satisfy national demand, the government had failed to guarantee prices when production exceeded the capacity of limited local and regional markets. In the first few months of the year they wrote letters accusing the Bolivian Development Corporation of "cheating them" and making "false promises."[120] Producers, some of whom had taken loans to increase production, were selling their production at below cost to predatory merchants. In May, Virgilio Zabalaga, elected representative of 175 families totaling nearly 900 colonists, submitted a list of demands for technical support and the establishment of credit offices. "Huaytú is a rice-producing colony," he reminded the government, and should have services including a sanitary post, bridges over small river crossings, and a "properly graveled road." Finally, given that they were approaching their ten-year anniversary, Zabalaga asked that the government build a public plaza and acknowledge the community's role in the March to the East with "a monument to the Bolivian colonizer."[121] In a July letter, they again expressed their frustration and described a public meeting to try and devise a way to combat the giant rats that were destroying their harvested rice and that they had "wasted [their] resources fighting."[122] In appealing to the MNR about their struggles with these pesky rodents, the colony admitted their struggles with a fecund tropical nature evident in the teeming animal life of the lowlands. But the underlying cause was one of frustrated mobility—of roads rather than rodents—evident in stockpiles of rice that could not be brought to market in a timely fashion and for a fair price.

That same year, the neighboring colonies of Cuatro Ojitos and Aroma, also established in the mid-50s, wrote to Paz Estenssoro. "You called us your children [*guaguas* from Quechua]" they reminded the president—invoking the deficient paternalism of the state.[123] Yet these colonists who were the "true campesino workers" were being cheated. The issue again was directly related to marketing—evidence of the dramatic expansion of regional production. The group affirmed that its "2,000 humble families had been brought from the valleys [of Cochabamba] by the revolutionary government to culti-vate the land and live exclusively from our work growing sugar cane and other articles." The government-run Guabirá sugar mill had drastically reduced their cane quota to 25 percent of their total harvest. "We are unfortunate" they stated, explaining that they could not pay the excessive cost of transport-ing even this amount to the mill. They demanded that Paz Estenssoro, as a benevolent father figure, rectify this transport and marketing issue thereby guaranteeing them an outlet for their crops. "Anything else," they concluded, "would be to leave us as orphans."[124]

In a lowland context in which new colonies struggled to market their pro-duction, uncertainty functioned in the case of transport as it had in the case of land. Individual colonies frequently wrote to the government asking that deteriorating roads be refinished, or that bridges be constructed across rivers that became impassable during the rainy season. The lack of government support was as likely to lead colonists to denounce their neighbors as to de-nounce the government. In May of 1964, the residents of General Saavedra (another colonization site to the north of Santa Cruz) wrote to the president. They wanted the government to modify the planned road between Chané and the Guabirá sugar mill. Although the current rough route passed through their town, the projected road would effectively bypass them. Such an act would be nothing less than a "death blow [*golpe de gracia*]" for the town, "pushing forwards its slide into decline."[125] The previous year the "Víctor Paz" colony in the Alto Beni wrote to their eponymous president, "with a small request that is not money." USAID was currently constructing a road be-tween La Paz and the colony of Broncini, linking a zone that encompassed approximately seven thousand families. "We are located at the start of the road," along with the neighboring colony of Incahuara, they explained.[126] Members of both colonies had freely given their personal labor to the danger-ous task of opening a 25 meter right-of-way in the forest. Because of a petition by members of Incahuara, engineers had decided to move the road closer to that settlement effectively stranding Victor Paz colony. They wanted the pres-ident to intervene and restore the road to its original course avoiding, "irrepa-rable future damage," to his namesake colony. The residents of both Victor Paz and General Saavedra understood the centrality of mobility to survival in the Bolivian lowlands and recognized what historians of transport have also pointed out—that the effects of infrastructure are uneven. As they pull two distinct places closer together, they effectively magnify the distance between other locales. In the process some sites would become future nodes, but others would be abandoned. As with the commentators in *El Deber* who complained bitterly about the new train service—even as they made ample use of it—these settler petitions also reveal a shift, demonstrating the degree to which their expectations about mobility and access as a fundamental right had been transformed.

Conclusion: New Repertoires and Old

As lowland settlement expanded in Bolivia throughout the 1950s and 1960s, colonists wrote letters to presidents and government officials pleading, cajoling, threatening, and demanding a range of state intervention. They also described conditions that jarred with the representation of colonization put forward in state-produced propaganda in the 1950s. The challenges they faced were often environmental as when floods, storms, tropical illnesses, and giant rats threatened the stability of colonists. They were also legal and political as when colonists denounced illegal occupations of land and petitioned for outstanding titles to their new properties. Economic understandings of mobility also emerged in complaints about bad roads and lack of accessible markets.

Hope and failure, the respective reigning motifs of the first and second rounds of correspondence I examine in this chapter, may appear diametrically opposed. After reading correspondence from a drought-stricken prospective settler begging for support to travel to the fertile lands of the east, it is difficult to read the subsequent letters of a disillusioned colonist or the damning report of a project official without accepting their explicit narrative that the reality of colonization had utterly failed the hopes of colonist and state. It is then easy to read that failure into a broader critique of "new lands settlement," an initiative that was taking place in developing nations across the tropical world in the mid-twentieth century.[127] This conclusion was increasingly voiced by national and international officials who came to see Cotoca as the first in a series of directed-colonization failures and advocated for a rollback of state support by the end of the 1960s.

But failure is not modernization's opposite, and cynicism and optimism are often intertwined. Settler denunciations harnessed a discourse of failure as an authoritative lament rather than one of passive despair. Through claims to reality, letter writers and development workers diagnosed failure and cried abandonment to demand new rounds of intervention. Colonists also made extensive use of the idea of failure and abandonment as they sought to push the state to provide them with promised resources. That much was already evident in precolonization petitions in which Andeans construct the high-lands as a failed place or wrote of labor exploitation in neighboring Argen-tina. Ensconced in those new colonization zones, colonists sought to critique failings, not for their own sake, but to generate a new round of intervention on the part of the state. Spontaneous colonists, who arrived in the settlement zones without official sponsorship, were as likely to denounce fellow colo-nists who had supposedly abandoned the zone in the hopes of securing their plots.

The diverse repertoire that petitioners employed reflects a unique histori-cal moment. The period of MNR rule in Bolivia (1952–64) was one of broad social and political reform and was thus highly conducive to the sort of state-citizen discourse found in the aforementioned petitions and letters. Despite the overthrow of Paz Estenssoro by René Barrientos in 1964, the following four years were still characterized by a populist, if increasingly authoritarian, approach.[128] This was particularly evidenced in countryside and coloniza-tion zones where Barrientos forged a rural peasant-military pact (PCM) to combat radicalism in cities and mining centers. With distinctive personalist flair, Barrientos encouraged a robust culture of directly petitioning govern-ment officials and the president for access to resources, land, credit, and titles even as the PCM suppressed independent forms of political organizing.

This state-citizen exchange continued in the tumultuous years after the unexpected death of Barrientos in 1968 but took on an unscripted dimension. Two left-leaning military leaders, first Alfredo Ovando and then J. J. Torres, presided over increasing radicalization of laborers, miners, and rural workers leading up to a popular constituent assembly in 1971. Colonists, organizing into agrarian unions by the mid and late 1960s, were no longer content with petitioning state authorities for the delivery of promised resources and, as I show in the following chapter, took more radical and direct forms of action. Instead of complaining about missing roads and incompetent administrators, they began to block roads, occupy public space in Santa Cruz, and even kidnap international development experts to support their demands.

With the coup of General Hugo Banzer in 1971, and subsequent repres-sion, that populist moment definitively ended as it would across much of the southern cone in the face of repressive military regimes. Although petitioners surely continued to write to the president and various ministries, they had less reason to be hopeful that their pleas would be well-received, and they could no longer claim the revolutionary legacy of 1952 to push the govern-ment into action.[129] Even as the state–settler pact unraveled in authoritarian

Bolivia, a new sort of actor took on an unprecedented role in lowland colonization. A range of religious figures that had long been active in frontier development found increasing room to maneuver in Banzer's Bolivia. The subject
of the following chapter, these members of Catholic orders, Protestant missionaries and volunteers with Protestant relief agencies often stood in as
proxies for an absent state or as conduit between colonist and state. In the
early 1970s, in a period in which directed colonization was widely seen as a
failure by the international development community, these faith-based development workers began to administer the largest colonization project in
Bolivian history.

CHAPTER FOUR

To Minister or Administer

Faith and Frontier Development in Revolutionary and Authoritarian Bolivia, 1952–1982

The phrase "faith-based" development conjures a linguistic contradiction. How can economic development, a symbol of this-worldly, material improvement of science, and of progress, be based on faith?

—Erica Bornstein, *The Spirit of Development: Protestant NGOs, Morality, and Economics in Zimbabwe*

In some quarters it is an article of faith that the great forested heartland of South America can and must be utilized if Latin America is to reach its development goals.

—Michael Nelson, *The Development of Tropical Lands: Policy Issues in Latin America*

Another time when I was working [for USAID] in Costa Rica, [Bastiaan] was there . . . and he had invited me for Sunday dinner at his house and his friends asked what did I do, and I said . . . "well, rural development." Bastiaan said "Hell, tell 'em the truth Harry, tell them what you really are!" And I said, "well, what do you think I am?" [He responded,] "well you're a god damn missionary, that's what you are!"

—Harry Peacock, *Methodist missionary to Bolivia*

As Harry Peacock recalls, his USAID supervisor, the bombastic Bastiaan Schoutten, seemed to revel in any opportunity to *out* him as a missionary rather than a development practitioner. Yet he also insists that Schoutten, who he had first met in Bolivia while attempting to secure USAID financing for a massive colonization scheme known as the "San Julián Project," had a "deep Christian orientation" and was sympathetic to the settler orientation program that Peacock, along with an unlikely, ecumenical, and alliterative coalition of fellow Methodists, Maryknoll nuns, and Mennonite Central Committee volunteers, known as the United Church Committee (CIU) had pioneered along Santa Cruz's expanding settler frontier in the 1970s. Like Peacock, members of these organizations routinely troubled the boundary between faith and development by engaging in extension work of a *spiritual* as well as *technical* nature. The first MCC volunteers in Santa Cruz's Cuatro Ojitos Colony, for instance, forged a relationship with settlers by digging wells and building houses. They also ran a booth in the local market where they sold an impressive array of spiritual and agricultural implements—from vegetable seeds and insecticides to Bibles and Christian literature—with proceeds invested in magazines for the church library.[1] Maryknolls in Bolivia, like their contemporaries in Guatemala, also moved nimbly between the spiritual and the developmental, by turns establishing catechetical systems and cooperatives in the rural communities they operated in.[2]

Beyond Bolivia, Peacock's personal reinvention does not look out of place in the expanding world of international development. In her study of Protestants in Zimbabwe, anthropologist Erica Bornstein has shown that while the theoretical separation between faith and development, is tightly linked to the logic of modernization theory, in practice, the second half of the twentieth century witnessed an increasing interconnection between the two that was bolstered by strong "political support for faith-based work in the U.S." and a growing "transnational reach of faith-based humanitarian aid." In his critical assessment of the Green Revolution in Asia, historian Nick Cullather also reminds us that, "religious terminology" was routinely used by planners to describe "conversion" to new technologies while "development has been described as a global faith."[3] Given these literal and metaphorical intersections, it is less surprising than we might expect then, that Peacock, and the faith-based actors he worked alongside, regularly transgressed the boundary between ministering and administering as they moved fluidly between religious organizations, university departments, NGOs, and government financing agencies in the name of religion and rural development.

Their prodigious personal trajectories, in turn, profoundly reshaped the migrant landscape of lowland Bolivia. Flying over the plains of Santa Cruz, that imprint is most evident in the dozens of pinwheels forming the San Julián Project. Its unique spatial design, in which forty radial settler plots emanate from a central point make a sharp visual contrast with the linear *penetration roads* and accompanying *piano-key* settlements of an earlier generation of Bolivian colonization. In a further break from those prior schemes, the planning, administration, and orientation of San Julián settler *nuclei* was turned over to the CIU, the faith-based NGO that Peacock, Maryknoll sisters Mary de Porres Pereyra and Maureen Keegan, and MCC volunteer Marty Miller created in 1971.

Over the following decade, as CIU-run settler nuclei spread across the forested landscape of Santa Cruz, word of San Julián also traveled through the

An aerial view of San Julián's unique settler nuclei. Photo by author.

international development community. The unique colony attracted the attention of a range of planners and academics who arrived to assess the project and often labeled it one of Bolivia's—and Latin America's—few successful colonization schemes. When the SUNY Binghamton–based Institute for Development Anthropology (IDA) commissioned three anthropologists to study the colony in the early 1980s, their final reports joined a growing body of graduate theses, reports, and assessments on everything from the colony's cooperative systems, orientation program, gender relations, and NGO activity. The IDA would prominently include the San Julián Project in its 1981 state-of-the-art evaluation of new land settlement initiatives across the global south.

These positive pronouncements emerged through a transnational constellation of faith-based, NGO, academic, and state actors that were critical to the project's continued financing. Yet strangely, as academic study of San Julián proliferated, the religious underpinnings of the colony were gradually obscured. This chapter aims to return to the role of Protestant faith-based workers in the development of colonization in lowland Bolivia as a window on the *Protestant Boom* throughout Latin America. As such I devote greater attention to the activities of Methodist and Mennonite faith-based development workers than to the Catholic Maryknolls they worked alongside.[4] Like other transnational actors analyzed in this book, these Protestants carried out their activities immersed in a web of interconnections. In linking local projects to national programs, international financing, and global faith communities, they served as intermediaries and were often required to justify their activities to an array of stakeholders including colonists, U.S. and Bolivian officials, and funding constituencies back home. On the ground, the interactions between settler and missionary are also revealing. In standing in as proxies for an absent state along the nation's frontiers, these faith-based development workers were often the primary authorities to engage with settlers. Beyond Bolivia, the evolution of the CIU, also speaks to the proliferation and transnational reach of nongovernmental organizations in development modernization.[5] That shift would become fully apparent as democracy was restored to Bolivia in 1983 and the state further stripped itself of direct responsibility for lowland development. The trajectories of many of the individuals discussed in this chapter embodied that transition as they transferred their repertoires from church organizations to NGOs and international development agencies in the *lost decade* of the 1980s and in the subsequent shift to neoliberalism—an economic order that, based as it was on a dramatic reduction of the state, proved uniquely conducive to NGO-led development. The reputation of San Julián also changed over those years. A prominent success story in the 1970s, the image of San Julián shifted again during Bolivia's 1980s hyperinflation when its settlers turned to those same radical strategies employed by older colonies in Santa Cruz more than a decade earlier, cementing their reputation as a rebel people.

This chapter begins, like the preceding three, in the revolutionary period by tracking the emergence and growth of faith-based rural development after 1952. Curiously, postrevolutionary secularization in Latin America often meant a dramatic *increase* in Protestant activity in the service of the state. Bolivia was no different and the MNR quickly invited a range of Protestant organizations to work in the lowlands. The second section narrows in on a key moment of transition—beginning with an environmental crisis (a devastating flood that swept through lowland colonies) in 1968 and closing with a political one (the Banzer coup) in 1971. Those few years were defined by the development of new flood-inspired models for colonization on an expanding frontier *and* an increasing radicalism in established colonies. Mennonites and Methodists found themselves at the center of each of these seemingly bifurcated processes.

The final section of this chapter turns to the work of the CIU in the San Julián project after the 1971 Banzer coup. Banzer's seven-year rule (the "Banzerato") was part of a tide of repressive governments that swept across South America in the 1960s and 1970s. Yet bureaucratic authoritarianism—as it was characterized by Guillermo O'Donnell—was enthralled by grandiose development initiatives, especially when they turned attention away from human-rights abuses.[6] Brazil's military government was a prime example, expanding road-building and settlement in the Amazon basin while engaging in urban counterinsurgency tactics. The Bolivian case reveals similar intersections between authoritarianism and environmental change with contradictory results. Banzer, a native of Santa Cruz, suppressed radical settler organizing yet also pushed forward development in the lowlands, expanding the MNR's vision of the March to the East. His allocation of large landholdings to political cronies is well known, but his regime also carried out massive new colonization schemes that brought thousands of new Andean settlers to the tropical frontier. For Protestants, Banzer also signaled continuity *and* change. Some faith-based development workers found themselves exiled or imprisoned by his government, but others successfully negotiated the transition between revolutionary nationalism and bureaucratic authoritarianism to take on an unprecedented role in Bolivia's largest colonization endeavor during his reign.

Death Trails, a Land of Decision, and the Bolivia Mystique

Italian Methodist Sante Uberto Barbieri remembers a conversation with a tearful Paz Estenssoro shortly after the 1952 Revolution, "when the President was telling us of those [lowland] regions and people, of their tremendous need, and inviting us to go there to open up churches, schools, and social centers."[7] As the MNR looked to new rural and frontier regions that it wished to incorporate into national life, a resource-poor revolutionary government welcomed the independent initiative of churches in education, health care, and community building. Protestants rushed to fill the gap. This invasion of the sects in post-1952 Bolivia, was part of a boom in Protestant activity across Latin America and particularly in nations that boasted large indigenous populations.[8] Although Barbieri's discussion with Paz Estenssoro hinged on service provision, other groups arrived in Bolivia to engage in proselytization. A comparison of the activities of the staunchly evangelical World Gospel Mission with two service-based organizations, the Methodist Church and the Mennonite Central Committee highlights these alternative conceptions of *ministering* and *administering*.

When prospective colonists petitioned the state they typically claimed that the lands they were asking for were empty or free from competing claims. These assertions concealed the fact that the nation's frontiers were not vacant but home to more than thirty indigenous groups, distinct from the Spanish, Aymara, and Quechua speakers who were arriving as colonists. Some lowland indigenous peoples, like the Guaraní and Guarayo, had a long history of interaction with Spanish culture—primarily through mission stations—but many others, including the Ayoreo and Yuqui, lived at the margins of settled areas and had long resisted outside incursion. Violence between settlers and indigenous communities intensified during the March to the East providing opportunities for organizations like the Chicago-based World Gospel Mission (WGM).

WGM missionary Carroll Tamplin had been conducting missionary work in Bolivia since the early 1930s, translating the Bible into Aymara in the highlands before moving to lowland colonization zones. When fellow WGM missionary Robert Geyer arrived in La Paz in 1945, he joined the Tamplin family at the small settlement of Guanay in the Amazon basin, "a base from which advances might be made to reach the Indians of the forest."[9] In a subsequent book recounting their missionary experience titled, *From Death Trails in Bolivia to Faith Triumphant*, Geyer provided harrowing descriptions of the descent from the Andes to the Amazon along precipice-lined mule trails that typify the heroic missionary narrative. But after several years and with thousands of dollars invested, the missionaries abruptly abandoned the project. "There were programs. There were plans. There were prayers. There were dreams. And then we lost it," Geyer obtusely reflected on the failed initiative. Undeterred, by the "hasty comments and withdrawn confidence" of their superiors in Chicago, the pair turned from this failure in the northern Amazon basin to a new potential mission site near Santa Cruz.[10]

In April of 1950, Tamplin wrote to the Ministry of Colonization with what was surely intended to be an alarming claim. He explained that near the colony of Yapacaní, where the government had settled Chaco War veterans in

1938, "this year four citizen-settlers ... were killed with poison arrows."[11] According to Tamplin, the handful of settlers faced attacks from the Sirionó and Yuqui and he proposed that the WGM be given exclusive license to contact and convert the perpetrators. The timing of Tamplin's request was clearly critical for the suspect missionaries whose own organization was raising serious doubts about their capabilities. Yet his petition also arrived at a seemingly inopportune moment for the Bolivian state. The previous year the government had secularized and dispersed the last Franciscan missions in the region which local elites and state authorities had long reviled as a lingering colonial institution whose corporate privileges, large landholdings, and control of indigenous labor they saw as a barrier to regional development.[12] That the state would evict one missionary organization and welcome another, speaks to the unique image of Protestantism in mid-twentieth-century Latin America. In contrast to the established Catholic orders, Protestant missionaries were able to present their work in a decidedly modern light that harmonized with nationalist aims. In Mexico, Protestants became missionaries of the state as they mastered indigenous languages through the Summer Institute of Linguistics— an institution that some Bolivian missionaries trained with.[13] In Guatemala, missionaries offered a "cultural package ([with] its emphasis on 'development,' as evidenced by medical clinics, schools, and translation projects) [which] endeared missionaries to the liberal government."[14] Across Latin America, postrevolutionary secularization entailed an expanded role for Protestant organizations.

In Bolivia, Tamplin spoke to national authorities in the language of civilization and citizenship, assuring officials that his mission among the Sirionó would be directed entirely toward, "their subsequent incorporation into the life of the nation ... in strict accordance with the provisions of the Political Constitution of the Republic."[15] He invoked the colonial distinction between barbarous indigenous groups outside of state control and *gente de razón* to differentiate the seminomadic Sirionó—a tribe of "authentic barbarians"—he sought to convert from the "partially civilized" Guarayo population the state was dispersing from Franciscan missions. In their search for "uncontacted tribes," Tamplin and Geyer had no interest in the latter, sharing with many anthropologists also circulating in the eastern Andean lowlands a disinterest and, at times, a barely masked disdain for corrupted indigenous groups, living near settlements and working in extractive industries.[16]

Lost in the tumult of the 1952 revolution Tamplin's petition was approved in 1955, evidence of the MNR's embrace of evangelicals whose activities appeared alongside other images of lowland development in ICB bulletins.

Geyer and Tamplin felt they had a "divine commission to the Indian of the forest, savage though he was."[17] The MNR gave them a state commission defining a sector over which they had exclusive religious jurisdiction.[18] The two missionaries immediately began their search for *untouched* Indians in northern Santa Cruz but were soon disappointed. Unable to locate more than a handful of Sirionó, their first successful convert was an ex-army sergeant they referred to as Don Ángel. With his help and the aid of a Cessna 180 they dubbed "Wings of Peace," they pushed further into the jungle searching for the Yuqui.[19] Although presented to home churches as selfless acts of potential martyrdom, such expeditions often had the feel of hunting parties in which well-armed missionaries were as likely to shoot, as to proselytize, so-called hostile tribes.[20] While building an airstrip in the forest in August of 1959, Don Ángel was killed by a Yuqui spear, falling victim to the very thing that Tamplin had claimed his arrival would prevent. Searching for a "return to the present day dynamic demonstration of the spiritual power of primitive Christianity," Tamplin had found his first martyr—a foundational moment for any struggling mission.[21] When another convert was gravely wounded in a second attack the missionaries positively rejoiced.[22] In pursuing U.S.-based support, these fatal and violent encounters furnished the missionaries with an irresistible narrative. In the millenarian language of Geyer, who traveled the United States sharing the story of Don Ángel and collecting donations from church groups, there was a clear trajectory, "from death trails in Bolivia to faith triumphant." Similar missionary organizations continued to flock to the region as expanding settlement further displaced lowland indigenous communities. By the end of the 1960s, more than twenty evangelical missions were operating in Bolivia, seeking to outdo one another in frontier exploration.

Like the WGM, the Methodist Church established operations in Bolivia in the early twentieth century. Facing hostility from the Catholic Church, Methodists won few converts, but their American Institutes in La Paz and Cochabamba became prestigious schools for the children of the elite. By 1960, a full third of sitting members of Congress were alumni.[23] In the mid-1950s, under the leadership of Director Murray Dickson, the Methodist Church sought to dramatically expand its operations and promote Bolivia to its membership in the United States. In 1956, just before its annual general conference, the Methodist Board of Missions decreed that Bolivia along with three other locations—Sarawak on Malaysian Borneo, South Korea, and the Belgian Congo—were "lands of decision," for global Methodism. While acknowledging the diversity of these locales, the board justified its selection by invoking the ideological climate of the Cold War as well as broader notions of primitive society and modernization. Bolivia (in the midst of a social revolution), Korea (emerging from a devastating war), and Malaysia and the Congo (where growing anticolonial movements threatened European rule) were all "at a place in their history when they must decide which way they will go."[24]

Dickson personally toured U.S. churches in support of the "Land of Decision" campaign giving impassioned speeches, passing out pamphlets, generating donations, and recruiting young missionaries.[25] Reflecting on their new agenda in spatial terms, Argentine Methodist missionary Sante Uberto Barbieri agreed: "Our work in a way has been limited to the high places. It is now time to go down to the plains, to the jungles to the midst of new and throbbing communities."[26] Recruits would serve in the church's new rural development and colonization programs. While the WGM, targeted uncontacted tribes living along the frontier, the Methodist church saw the dislocations produced by the internal migration of indigenous Bolivians as ideal entry points for missionary activity.[27] At the personal invitation of Paz Estenssoro, they linked their work directly to the national project of eastern expansion where the state presence was weakest. "People are moving in by the thousands" wrote Dickson, "and the church is being urged to come in because these immigrants need spiritual and moral help."[28]

Dickson's tour took him through the Methodist heartland of south Texas where Harry Peacock, a young church member from Brownsville, was captivated by the Land of Decision campaign.[29] In 1961, Peacock and his wife Patricia, headed to Bolivia to join missionaries Robert and Rosa Caulfield who had established a high school and agricultural training institute in Montero. As the terminus of the Santa Cruz-Cochabamba highway, the small city, forty miles north of Santa Cruz, was something of a development hotspot in those years. The U.S. government had used Point Four funds to build a machinery pool and technical station on the south side of town, and the Salesian mission also established a school in the area before taking control of Point Four's operations. Montero was also a supply center for nearby national colonies including Yapacaní, Cuatro Ojitos, and Aroma and for Okinawan colonists living along the Río Grande forty kilometers to the east.[30] The Methodist Rural Institute, the Caulfield's imagined, would serve as a regional base, an "axis of [Methodist] operations [from which] teams of agricultural extension, social work, education, medicine, and literacy headed out to settlements of camba, kolla, Japanese and Mennonite origin."[31]

Peacock originally taught languages in Montero but like Dickson, and many other Methodist volunteers, he came from a farming background and was fascinated by the region's colonization projects.[32] In the mid-1960s he took a

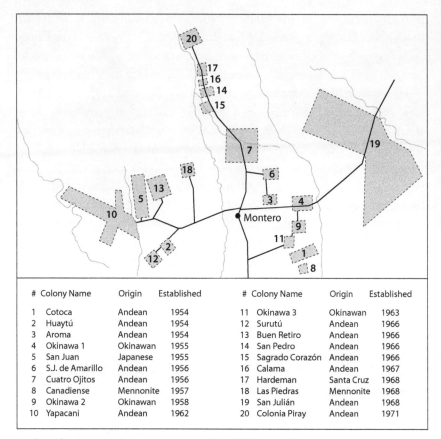

#	Colony Name	Origin	Established		#	Colony Name	Origin	Established
1	Cotoca	Andean	1954		11	Okinawa 3	Okinawan	1963
2	Huaytú	Andean	1954		12	Surutú	Andean	1966
3	Aroma	Andean	1954		13	Buen Retiro	Andean	1966
4	Okinawa 1	Okinawan	1955		14	San Pedro	Andean	1966
5	San Juan	Japanese	1955		15	Sagrado Corazón	Andean	1966
6	S.J. de Amarillo	Andean	1956		16	Calama	Andean	1967
7	Cuatro Ojitos	Andean	1956		17	Hardeman	Santa Cruz	1968
8	Canadiense	Mennonite	1957		18	Las Piedras	Mennonite	1968
9	Okinawa 2	Okinawan	1958		19	San Julián	Andean	1968
10	Yapacani	Andean	1962		20	Colonia Piray	Andean	1971

Settler colonies near Montero. Map created by Bill Nelson. Base layout based on a map in Kenneth Graber, *La vida agrícola en las colonias: estudio económico de diez colonias al norte de Santa Cruz, Bolivia* (La Paz: Iglesia Evangélica Metodista en Bolivia, 1976).

two-year furlough in the United States. Along with church tours, graduate studies were a central element of Methodist furloughs that might just as easily introduce young missionaries to rural sociology as theology. Peacock focused on social psychology, a discipline he found particularly applicable to the environmental and cultural shocks experienced by new settlers in the tropics. From 1966–68 he was back in Bolivia serving as director of the Rural Institute but spending most of his time traveling through settler colonies.[33]

Foreigners like Peacock were not the only Methodists active in lowland colonies. Membership of the nascent national church grew in the 1950s and 1960s and missionaries sought to recruit Quechua and Aymara speakers from the highlands to serve indigenous settlers in rural colonization zones. Ana

Fajardo, a native Quechua speaker from Oruro, was living in Cochabamba when she made a chance encounter with Methodist missionaries. An initial commission to sew a banner for the city's American Institute led to membership in a Methodist youth group that provided the church with a stream of volunteers to join its new initiatives in the Alto Beni, Chapare, and Santa Cruz.[34] Because of her knowledge of Quechua, Fajardo was asked to go to Yapacaní Colony, fifty miles west of Montero, where a large contingent of fellow Orureños had settled. For a young Fajardo, as for many highland–lowland migrants, it was an impressive but daunting opportunity. Her father agreed to let her go because two of her cousins had already migrated to Montero and lived down the street from Methodist missionary Harry Peacock. In Yapacaní, Fajardo worked as a home economics agent and translator along with two North American Methodists. Traveling by bicycle along the colony's piano-key jungle penetration roads, she stopped at rustic palm shelters to demonstrate hygiene techniques, food preparation, and sewing. Like the MNR, the Methodist church also embraced the didactic power of film and Fajardo assisted a church-owned mobile cinema truck showing informational films throughout the colony.

Jaime Bravo, a member of the Redentor Church in La Paz, was another highlander who migrated to the east with the expanding Methodist church and would play a key role in its operations. "With the road that Paz Estenssoro opened in the early 1950s, a great movement of people began," he recalls, "and I was among them." At sixteen he moved to Montero where he enrolled in the Wesleyan seminary, a theological school attached to the Methodist Rural Institute, and "got to know a lot of Texans" like Peacock, Jim Pace, and Bob Caulfield.[35] For Bravo, "the Methodist church had the virtue of combining social action with evangelization, but the stronger side of the church was its *envangelio integral* (the whole or total gospel), a salvation that had to do with hunger, poverty, justice . . . with doing well for the poor, the widows, the orphans, and the migrant."[36] He "went out with this mentality" after completing his studies at the Rural Institute and was soon director of the church's new operations in the Alto Beni colonies where more than 1,500 colonists and their families had settled since the early 1960s with Alliance-for-Progress funding.

Describing his work to North American readers of the Bolivian Methodist newsletter *Highland Echoes*, he wrote of these new migrants (Aymara speakers like himself) in words that mirrored his own transformation. "When altiplano Aymara Indians move into the lowlands," he explained, "this change in geography notably affects their personalities. Their reserved nature becomes more open. Their facial expressions soften, and frequent smiles denote a new happiness."[37] Bravo worked in agricultural extension in the region where four Methodist congregations had recently been established. Their names—New Zion, New Santa Fe, New Israel, and New Jerusalem—expressed the dual sentiment of pioneering and renewed faith that the Methodist church saw in Alto Beni. He also managed a new generation of Methodist volunteers, known as "Latin America 3s." As with Fajardo's experience in Yapacaní, North Americans were paired with Bolivian Methodists. The teams received a three-month training program in cross-cultural communication, and community development led by a Methodist anthropologist, John Hickman, a Cornell University doctoral student, who had just completed research in Aymara regions of neighboring Peru. Teams would serve as intermediaries, wrote Hickman, "to fill the vacuum of contact between the people of a given colony, and between the colonists and the church and government services available from outside the community."[38]

As Hickman's assertion makes clear, Methodists quickly became a third party in what—from a planning perspective—might have seemed limited to national agencies and international financers. Settlement projects funded by the U.S. government through the Alliance for Progress in the 1960s were supposedly administered by the Bolivian Development Corporation and later by the National Institute of Colonization. But in practice, aside from road construction and land surveying, much of the day-to-day service provision was informally farmed out to the Methodist church and led by mobile foreigners and nationals like Hickman, Peacock, Fajardo, and Bravo.[39]

For the Bolivian state, missionaries and laypeople were the perfect go-betweens. They received small salaries (paid by the church), were connected to North American congregational funding, and (as part of their sense of ministering) were willing to live and work in remote regions eschewed by qualified nationals. This skill set was as appealing to U.S. funders as it was to the Bolivian state albeit for different reasons. The former engaged with missionaries as trustworthy fellow Americans. Methodist missionary Jim Palmer, remembers one official arguing that, "with men like Murray Dickson managing the work of the Methodist Church in Bolivia, they get twenty times as much result from a dollar as we get."[40] The statement is compelling not simply because it suggests missionary work was cheaper and more effective than U.S. development assistance, but for its underlying assumption, that development work and missionary work were essentially equivalent in nature. Methodist missionaries also spoke to a state discourse of abandonment which had emerged in the first decade of colonization. As the previous chapter argued, planners attributed the abandonment of new colonies, albeit dubiously, to settler dependency caused by excessive state paternalism. Writing to a friend, Dickson explicitly invoked the *infant industry* critique of state-led development claiming—in what is now a cliché of the development industry—that "I am convinced that my job, in fact the job of every missionary, is to work himself out of a job. We are not here to perpetuate the need for our being here, but to train, guide, and so prepare the people that if we were to step out at any time, they could carry on, not as a handicapped infant institution, but as the instrument of good which we intend."[41] When Hickman and Bravo trained the Alto Beni teams in the mid-1960s they advised participants not to impose a model but instead to let projects emerge from consultation with their host communities. Whether these arguments about low-cost and collaborative work prevailed or that an underfunded, understaffed state simply had few alternatives, foreign and national Methodists became intimately involved in the quotidian operation of lowland colonies.

While the WGM entered the Bolivian lowlands looking for uncontacted tribes and the Methodist church arrived to help in the moral and social instruction of Andean transplants, the Mennonite Central Committee initially came to Santa Cruz with an entirely different goal—namely, to help their fellow brethren in need. The MCC was a faith-based relief organization that had been founded by American and Canadian Mennonites during the 1920s to support Russian Mennonite refugee resettlement in Canada and Latin America. Mennonite sociologist J. W. Fretz had worked with Mennonites in northern Mexico and the Paraguayan Chaco in the 1940s and 1950s on its behalf. Through his contacts in Paraguay he learned of the first small-scale migration of Mennonites to Bolivia and in 1960, decided to visit Santa Cruz. Fretz's trip, which he wrote about for the magazine *Mennonite Life*, introduced North American Mennonites to the struggles of these new settlers and provided the impetus for the Mennonite Central Committee's Bolivia program.[42]

For the first decade, MCC-Bolivia consisted of little more than a health clinic at the edge of the Tres Palmas Mennonite colony. However, it gradually departed from its simple mandate and was drawn into the nexus of Bolivian colonization with the scope of MCC activities paralleling, and directly overlapping, those of the Methodist Church. Designed to serve a few hundred fellow Mennonites, the Tres Palmas clinic began providing care to neighboring Bolivians (covering a service area of approximately ten thousand individuals) including those from nearby Cotoca Colony whose Medical Center, according to colonist petitions, even lacked the beds for patients.[43] From the outset, MCC was also involved in settler extension. When Fretz arrived in 1960, he toured Santa Cruz with MCC South America director Frank Wiens

who had been "loaned by the MCC to the Point Four in Bolivia." Having supported Mennonite colonization across the Chaco for more than two decades,
MCC was experienced in resettlement and connected to U.S. financing. Wiens
had obtained and administered Point Four funds to Mennonite colonists in
Paraguay and, much like the Methodist church, MCC was viewed by Point
Four officials as a reliable and efficient organization. He was soon working
with the national and Okinawan settlers on behalf of the U.S. government.[44]

Several factors including the increasing movement to urban centers, the
growth of North American Mennonite colleges like Goshen, Bethel, and
Bluffton, and a growing outward-looking evangelical trend in the church encouraged young Mennonites with rural backgrounds to volunteer with the
MCC or foreign missions.[45] Geopolitics also played a role. As members of
one of the historic peace churches (also including Quakers and Brethren),
many Mennonites registered as conscientious objectors (COs) to avoid the
U.S. military draft. In an unexpected twist, the same current of postwar mili-
tarization that drove the construction of military bases on Okinawa also provided a steady stream of COs that took part in the PAX program—eventually
recognized by the United State Selective Service Program. Many of these
PAX men, volunteered for expanding MCC operations in Bolivia and elsewhere. With an ample supply of personnel (but lacking programs beyond the
Tres Palmas clinic) the MCC initially loaned out its volunteers to the Methodist Rural Institute in Montero. They also began to place volunteers in rural
communities without schools through their Teachers Abroad Program (TAP).
By the late 1960s, MCC-Bolivia was on the verge of completing a transition
from a small health clinic to an integrated development agency with its own
programming.[46]

Pulling in young North American Mennonites for three-year stints in a
country in which they had no prior experience, the MCC attracted a mixture
of enthusiasm and naivety that melded practical farm knowledge and church
life. Russ Stauffer, who would work in the San Julián project in the 1970s,
remembers himself as an impressionable youth inspired to join the MCC
through his encounter with returning volunteers "who were the example of a
model citizen of the community, he could doctor sick chickens, teach a course
on hog care, or build a schoolhouse and after his term give slide shows in
church that the whole congregation showed up for to boot!"[47] Wide-eyed enthusiasm aside, Stauffer's recollection suggests that the type of faith-based
work that MCC volunteers undertook in Bolivia played to their experience in
rural areas and small towns across America's agricultural breadbasket. At the
height of the Green Revolution, as USAID was attempting to translate North

American agricultural knowledge to the tropics, the MCC imagined itself as engaging in a similar practice writ small.

The Methodist church attracted missionaries in the late 1950s with its argument that Bolivia was a Land of Decisions. By the end of the 1960s there was something of a "Bolivia Mystique" emerging in Mennonite circles in the United States and Canada, with volunteers and administrators alike ecstatic about the limitless opportunities in the country.[48] For both the Methodist Church and the MCC, Bolivia—closely followed by Zaire—would eventually become the largest of their expanding global operations. In early 1968, MCC delegates from across Latin America met in Bogotá, Colombia. That same year, a meeting of the Latin American Episcopal Council in Colombia would lay the tenets for Catholic liberation theology and base ecclesiastical communities. The MCC—while far more conservative than that radical collection of priests—was also thinking through the implications of rural development. Writing in an MCC-Bolivia newsletter a few years later, one volunteer would claim that "Anabaptists believed, practiced, and taught a Bible-based 16th century liberation theology," centered on "the autonomy of the community."[49] Back in Bogotá, MCC-Bolivia representative Arthur Driediger was on hand to discuss MCC's future in Santa Cruz. He advocated, "the selection of a frontier community to develop a rounded program under MCC administration." Yet if community development could draw inspiration from Mennonite history, MCC activities were not to be linked to evangelization. According to Driediger, their objective "would be to minister to the total man without a predecision to build or not to build a Mennonite church."[50] The timing was fortuitous. As MCC delegates were gathered in Bogotá from February 12 to 18, a disaster near the small settlement of Cosoriocito in Santa Cruz would provide them with an unprecedented opportunity.

From Disaster to Dictatorship

In February of 1968, floods without equal in modern Bolivian history devastated the lowlands and much of the nation. A very wet month was punctuated by several days of intense rain during which the Río Grande east of Santa Cruz spilled its banks and flooded nearby colonization zones forcing evacuation in Aroma, Cuatro Ojitos, and Okinawa Colony. It was followed by the Río Piraí on the west side of Santa Cruz and, days later, by record rainfall in Cochabamba turning Bolivia's second largest city into a disaster zone. On February 13, *El Deber* reported that "eight people have perished from drowning in the turbulent waters of the Río Grande" near Cosoriocito where the river had risen within one meter of homes and fields.[51] Over the following week, rains continued and floodwaters threatened the city of Santa Cruz.[52] In addition to affecting urban spaces and new colonization zones, the disaster struck at central symbols of regional progress that had been constructed in the previous decade. The recently completed rail bridge over the Río Grande was washed away while the Cochabamba-Santa Cruz highway was closed due to dozens of landslides.[53] On the night of February 22, President Barrientos declared a national emergency. The following day, the situation only worsened. Another flooding river was causing the state oil company (pictured in the film *Los Primeros*) to lose 2,000 barrels of oil a day.[54] Further afield, in Rurrenabaque, the rain caused landslides that placed the entire town in jeopardy. The mayor reported that half the town's water system—whose heroic construction had been fictionalized in the ICB film *La Vertiente*—was destroyed.[55]

The Río Grande flood of 1968 displaced thousands of new colonists in the agricultural areas around Montero. In fact, the geography of the disaster, as depicted in maps in Bolivian newspapers, was nearly contiguous with the landscape of settlement established over the past two decades. Settler refugees poured into the city quickly filling the Methodist Institute and the Salesian mission. Missionary Harry Peacock requested permission to use an abandoned road work camp (owned by the Hardeman Company) to house additional victims and, as principal, enlisted the help of his students at the Methodist High School. The MCC also lent its volunteers to the effort. For several weeks in late February and early March, MCCer (as MCC volunteers were known) Elwood Schrock managed *Hardeman Camp* alongside several Methodists, Maryknoll nuns, and Peace Corps workers. Refugees organized themselves into firewood brigades, built brick ovens, and dug latrines.[56] By March 20, most had returned home but 650 people—primarily those from Cosoriocito—whose lands and possessions were destroyed, remained in the camp. "We are hoping" wrote Schrock, "that the churches that worked so well together during the refugee phase can continue working together on a colonization program to help the people build a community better than the ones they've had to leave."[57]

Cruceños welcomed this initiative. Initial gestures of support notwithstanding, displaced peoples have routinely been seen as a liability and the presence of a mass of impoverished refugees produced predictable anxiety in Montero.[58] By March, as the floodwaters subsided, there were rumors of a gang of thugs running rampant in the city, with "the operation of the chicherías" to blame.[59] On April 6, *El Deber* worried that rations for the refugees would last less than three weeks and warned of the "activities of agitators that are operating among the victims" inciting them to illegally occupy lands.[60] Two weeks later, *El Deber's* editor Pedro Rivero Mercado warned of "people without scruples . . . extremists, agents of foreign governments . . . sowing discontent, taking advantage of the tragedy."[61] He suggested that while the material needs of refugees could easily be addressed, the "spiritual and soulful rehabilitation of the [flood] victims" was essential. President Barrientos also linked faith and the flood calling for "Christian strength" to aid in postcrisis rebuilding.[62] Yet, the National Emergency Committee (CEN), he established lacked the funds to even conduct a full study of damages caused by the flood. Perhaps motivated by this incapacity, Juan José Torres, head of the CEN, was soon suggesting that damage reports produced by affected communities, "lacked in seriousness [and] registered exaggerated damages about innumerable losses distant from reality, which had created expectations of aid that the [CEN] could never satisfy."[63] Like cruceños, he alleged that flood victims were trying to take advantage of the disaster.

With hostile local elites demanding spiritual rehabilitation and a bankrupt state calling for Christian strength, the United Church Committee—as the alliterative ecumenical coalition of Methodists, Mennonites, and Maryknolls was referring to themselves—was uniquely qualified to respond. With several hundred colonists left at the Hardeman camp in mid-1968, the CIU produced a report on a proposed resettlement with the hopes to "develop a self-governing community free from reliance on the [CEN], integrated with other developments [in the region and] in contact with the government."[64] As Peacock recalls, the colony would mirror the camp, with the organizations that Hardeman refugees had developed through tireless planning committees guiding the initial settlement process. Methodist and Mennonite volunteers would be on hand part time to supervise progress, and two Maryknoll sisters would remain in the colony on a permanent basis to lead educational programs and foster community development. Yet decision-making would be entirely in the hands of the colonists, who planned to work cooperatively for the first two years. Unlike the Andean colonists that Methodists were accustomed to working with, the displacees from Cosoriocito were lowlanders with an intimate understanding of the local environment. This was a revelation for Peacock and others who remembered that these were people that "knew the jungle . . . they got in and in one year they put themselves in a situation where highland colonists, those who survived, it would take them three years to get to that point."[65]

In a report written only three weeks after the colonists had arrived at Hardeman Colony, committee members noted evidence of rapid progress with buildings, "all very orderly in neat rows."[66] The greater achievement, as Barrientos and *El Deber* had hoped, was a psychological one, as "all visitors have commented on the completely different outlook and spirit here." The subjectivity of the former victims had been transformed, the CIU claimed, affirming that "the 'refugee' atmosphere is gone and in its place is vitality . . . even the small children manfully trotting to and fro."[67] A subsequent report contained a similar affirmation, noting "the most outstanding feature of the story so far has been the really remarkable spirit of community which has been generated among the people themselves." The passive phrasing in both statements (*had been* transformed, *was* generated) begs the question of who (or what) had produced that impressive shift in spirit—missionary, colonist, or the flood itself.

Even as Methodists, Mennonites, and Maryknolls acted as proxies for an absent state, channelers of foreign assistance, and as hands-on participants in the affairs of Hardeman Colony, they were hesitant to claim much responsibility—or long-term accountability—for the project. Rather than establish themselves as a permanent institution, these faith-based relief workers, vigilant against the first signs of settler dependency, sought to work to eliminate the need for their presence.[68] With settlers established on the land, the CIU considered the operation a success and, wary of paternalism, looked to "phase ourselves out of the present operation . . . a tapering off process that must be explained to the settlers, lest it be construed as *abandonment*"[emphasis added].[69] Insisting their role was to remain, "in the background to advise and assist," they proudly stated that the CIU "now has no powers of coercion or control whatsoever over the community," outnumbered by colonists five to one on the executive group.

It became evident, as the CIU withdrew from Hardeman, that members were beginning to conceive of this unique experience as far more than a one-time response to a national disaster. What had appeared ad hoc was beginning to be redefined in CIU project reports as a deliberate and coherent vision for rural development. In the wake of the flood in Santa Cruz, these religious development workers were soon sketching the outlines of a broader model to apply toward "integration and nation-building over the whole local development area."[70] The problem, members argued, was that although many directed and spontaneous development initiatives were underway in Santa Cruz, "all of them are going their own sweet ways regardless of the others." What was needed was a "unifying force to build that sense of community which the whole area needs if it is going to be successful."[71] The foundational Hardeman Colony would be "one small but focal part."

Over the following years members of the CIU expanded their philosophy of settler orientation. This implied an increasingly refined understanding of the problematic of lowland settlement, a rethinking of the challenges of Santa Cruz as a limitless frontier, and a response to prior constructions of abandonment. In an MCC summary of rural development, the author reminded its own volunteers that Santa Cruz was "a terrible land as well as a land of promise," one in which infant mortality reached 95 percent in new settlement sites.[72] The MCC presented the typical highland migrant not as a hypermasculine pioneer but as a hapless victim wielding unfamiliar tools and combating a hostile and unforgiving environment. He faced 120 acres of virgin forest, "and has never cut down a tree . . . swarms of insects and he has never before been bitten."[73] In what President Barrientos called "a titanic struggle to organize and dominate our land" the highland migrant needed help to combat the unruly forces of nature.[74]

In their first attempt to achieve this aim, MCC worker Marty Miller, Methodist Harry Peacock, and Maryknoll nuns Mary de Porres Pereyra and Maureen Keegan proposed the creation of a new "colonization orientation center" in 1971.[75] Drawing on the model of the Methodist Rural Institute, they would offer courses in "basic shop, horticulture, poultry, farm machinery, swine production, animal nutrition, dairy [and cooperatives]," giving new colonists, "critical experience to confront their new environment."[76] The group planned to receive 400–500 newly arrived settlers in 1970, and twice that number in 1971, promoting "skills, understandings, and attitudes necessary to maintain at minimal rates, colony desertion, and mortality." Although the MNR had long hoped to end the exodus of Bolivians to neighboring Argentina, by the early 1970s tens of thousands of highlanders were working as braceros in Santa Cruz's sugar, cotton, and rice harvests, often in equally exploitative conditions to those they had faced abroad. The CIU planned to target "these men [who] come year after year hoping to earn money to buy their own land only to find that their agricultural experience in the highlands ill fits them for life in the jungle. Some fail economically, some lose their health, some simply cannot endure the isolation," explained Peacock.[77] Critical for a state with limited funds, the CIU promised to do all of this at a mere cost of $18 per colonist.[78] Before implementation, Peacock presented the plan to several residents and community organizers from long-established colonies. Ruben Baldivieso from Yapacaní thought the project was good even if it "wouldn't work," explaining to him that "people do not come down [from the highlands] to go to school they come down because they want land." He and Marcelino Limachi (a leader from Cuatro Ojitos colony) suggested that they should bring "orientation directly to the point of settlement," and the CIU quickly began to sketch out a new model for onsite orientation in newly formed Piray Colony.[79]

As the annual flow of harvesters arrived in Santa Cruz from highland towns in March and April the CIU was poised to act. Signs and radio ads targeted new arrivals and advertised 180 available spots. On July 5, the first forty colonists arrived, and the new missionary run colony was "inaugurated with great solemnity." After colonists and CIU staff made speeches, everyone sang the colony anthem, "Let's go to Piray."[80] While MCCer Marty Miller managed onsite construction, Peacock oversaw technical details and the Maryknolls Pereyra and Keegan—fresh from prior service and education work in Hawaii and the Bronx—would distribute rations. However the "most crucial aspect of the program fell on the six, experienced colonist orientators," one of whom had written the colony anthem.[81] Aware that the rapid success of Hardeman Colony was in part due to its local or camba composition, the CIU had recruited longtime colonists or native residents of the Montero area that would facilitate the implementation of local environmental knowledge among Andean settlers. Some of them, like Dardo Chávez, were former students at the Methodist school and had taken part in the 1968 disaster response.

The belated recognition of local knowledge reflects a curious blind spot in prior decades of directed colonization. National authorities had previously maintained physical distance between highland transplants and locals in the hopes of avoiding conflict. Moreover, officials regularly discounted cruceño agricultural experience based as it was on nonintensive production for local markets, viewing the horse-powered sugar presses and heavily laden oxcarts of cruceños as relics of a backward agrarian economy. In contrast, CIU members like Jaime Bravo, argued that their goal, "was to achieve this dialogue between two worlds, the vision of the brave, laboring man, the miner, that wanted to work . . . and the vision of the local, native the original farmer of the jungle that calmly achieved his life, earned [enough] to eat and was not in a hurry to destroy the jungle and shoot animals."[82] The modest aim of the latter, he continued, was simply "to survive . . . for a long time." Referring to high infant mortality in new settlement zones, Harry Peacock also emphasized the CIU's primary goal was simply to keep settlers alive and healthy.[83] By reinterpreting *survival* as *success*, the CIU sought to initiate a revolution of lowered expectations among highland migrants many of whom, in the words of Jaime Bravo, sought to "farm as they had mined"—with expectations of extracting great wealth from the land and moving on. The CIU's hands-on presence at the outset of a colonization venture would allow them to prevent many mistakes, rooted in a fundamental misreading of the landscape, that fed a cycle in which settlers, "in their desperation clear, plant, harvest, and then *abandon* the land, . . . leaving it useless [emphasis added]."[84] In those first crucial days, missionaries found settlers to be paradoxically suspicious because they "had been exploited before" and—in the absence of established institutions—uniquely pliable and open to change.[85]

In early reports on Piray, the CIU claimed that through the work of the local orientators, new colonists, "men with no experience . . . emerged from their discontent and suspicion and gradually acquired a quiet confidence in their own ability."[86] As in Hardeman Colony, the orientation program ended as quickly as it had begun. By October, only the Maryknoll nuns—accustomed to long-term residence in service areas—remained to give continuity to the project. In locating their role in orientation, and insisting on their withdrawal from the community in short order, Protestant members of the CIU defined a more conscribed range of operations than their Maryknoll counterparts. They might provide instruction in marketing, securing credit, and the formation of cooperatives but would ultimately be gone long before the success or failure of these endeavors was apparent. In short, the CIU saw its role as something like that failed but pioneering colonist it sought to help—as the vanguard of lowland colonization "moving onward year after year to virgin territory."[87] Indeed departing CIU members were already looking to their next venture, the first major colony in the "Agricultural Expansion Zone" to the east of the Río Grande.[88]

Even as faith-based development workers pioneered a strategy for new land settlement in remote locations across Santa Cruz's expanding agricultural frontier, they continued to work in colonies that had been established a decade, or more, earlier. The former presented members of the CIU with two elements that planners prized above all: miniaturization and a blank slate.[89] In the latter, missionaries faced not the challenge of pioneering but the frustrated expectations of established colonists. Many had spent years in settlement zones and still lacked titles, agricultural credit, and extension service, and felt, as they made clear in frequent petitions, they had been abandoned by the state. Between 1968 and 1971, as the CIU's camba-inspired orientation program took shape in Hardeman and Piray, these colonists, along with landless braceros from the highlands, organized to challenge explanations about limited state resources.

After more than a decade of active extension work in settler colonies, many Methodists were sympathetic to the demands of these colonists. Their support was in part a product of personal contact through service provision, but it also meshed with a transnational theological strain in Latin American Methodism. The two were linked by the movement of individuals such as Bolivian Methodist Jaime Bravo who recalls that "we fell inside a wave of deep radicalization within the church [which] repeated what was going on in its social milieu."[90] Harry Peacock's furlough to the United States, where he studied social psychology in 1966, had helped him conceptualize a new program of settler orientation in Hardeman and Piray. After his work in the Alto Beni colonies, Bravo also went on furlough in 1967–68, heading south to Argentina where he completed graduate studies in theology at the Interdenominational Evangelical Faculty of Theology in Buenos Aires. For the Protestant faith in Latin America, the school was an epicenter of liberation theology, and Bravo remembers being mentored by José Míguez Bonino, author of the canonical *Doing Theology in a Revolutionary Situation*.[91] Other graduates of the seminary included Federico José Pagura, who served as a Methodist bishop in Costa Rica and Panama in the late 1960s and early 1970s, helped Chilean refugees escape the Pinochet regime and protested with the "Mothers of the Plaza de Mayo" as part of the Evangelical Argentine Methodist church.

Bravo returned to northern Santa Cruz in 1968, still committed to colonization but his vision "had broadened in Argentina" and he was interested in deepening the social work of the church.[92] It was a time, as anthropologist Lesley Gill points out, "of economic expansion [in which] new opportunities lured migrants to northern Santa Cruz, while the expansion of agroindustries dashed their hopes of becoming independent producers."[93] Bravo set his sights on the small colony of Chané-Bedoya. When a university commission surveyed the area that year, they found two hundred and fifty people were living in a "subhuman" state having paid local elites for land that ought to have been granted for free through the agrarian reform.[94] Bravo began working with these sugarcane harvesters that, "came from the interior, those that suffered injustices . . . and young people that didn't have health services."[95] He helped organize unions and orient colonists, "so that they could demand their rights" and push for land. With support from the National Bolivian Workers Congress and the Methodist Church, colonists in Chané-Bedoya formed the Union of Poor Campesinos (UCAPO) an organization that began to stage land occupations in the area.

In Yapacaní Colony, Methodists also found themselves at the center an emerging agrarian radicalism. By the late-1960s, Yapacaní contained more than two thousand settlers brought in under an Inter-American Development Bank (IDB) agreement with the Bolivian government. Among them was Juan Espejo Ticoña. An ex-miner who had organized industrial workers in La Paz in the early 1960s, Espejo was forced to flee the highlands after the 1964 Barrientos coup and arrived in the colony clandestinely. With settlers prohibited from forming unions by the Barrientos regime, Espejo and two other colonists from Cochabamba and Oruro organized an innocuous-seeming "Committee Pro-Yapacaní." Their first objective was to secure basic health services as the medical posts established by the Colonization Institute "were empty apart from a few aspirin."[96] With government support lacking, they turned to the Methodist missionaries Jaime Bravo and Brooks Taylor who helped secure a $10,000 grant for the construction of a hospital and a pharmacy in Yapacaní along with the provision of an ambulance and the training of fifteen nurses and home economics personnel (like Ana Fajardo) to carry out extension work. It is telling that in Espejo's written account, the Methodist Church is only discussed in terms of these social services and never mentioned in a separate section on church activity in the colony. As to the latter he insists that the Catholic Church, "did not help at all" and that the belated arrival of evangelical churches only fomented division within the community.[97]

Divorced from the question of faith, the Methodist presence in Yapacaní extended beyond health care and soon placed the church at the center of a conflict over marketing. By the late 1960s, settler colonies in northern Santa Cruz had created an "empire of rice" but as production surpassed national demand, oversupply led to dramatic fluctuations in prices made worse by the limited storage and processing facilities.[98] After traveling rough roads to arrive in Santa Cruz, colonists were forced to accept whatever price the city's handful of buyers was offering. With profit margins slim, the colony's struggling farmers reacted with outrage when the nearby town of Buena Vista—a long-established provincial center strategically located along the highway to Montero and Santa Cruz—attempted to introduce a transport tax on Yapacaní rice. For Espejo, the town's camba "ranching oligarchy" sought to live as "simple parasites" off the work of highlanders. Drawing from his experience as a labor organizer, he rallied a group of colonists, traveled to Buena Vista, and burned the checkpoint. Buena Vista's mayor soon rebuilt the gate and brought soldiers from Montero to police it. After unsuccessfully petitioning Santa Cruz's mayor, Espejo called a colony-wide meeting in April of 1969. Colonists returned to Buena Vista, destroyed the checkpoint, and occupied the town for several hours.

The incident brought condemnation from a regional press already nervous about the increasing militancy of highland migrants. *El Deber* alleged that "twenty colonists in a state of drunkenness [had] carried a noose through the plaza [with] the clear intention of hanging Mayor Percy Antehlo."[99] In an editorial the following day, *El Deber's* editor, Pedro Rivero conjured a timeless image of camba pastoralism, that "until a few years ago was peaceful and as progressive as one could hope," and was now characterized by "road blocks, the taking of hostages as in international conflicts, occupations of peaceful towns, threats of lynching, public demonstrations of the contempt for the law and legally constituted authorities."[100] In calling for the arrest of the instigators, *El Deber* directly implicated the Methodist Church in settler radicalism, identifying not only Espejo, but Methodist minister Ruben Baldivieso as a culprit. The latter was a resident of Yapacaní, close friend of Methodist Harry Peacock, and brother-in-law to Methodist Director of Rural Development, Jaime Bravo.

Protests and land occupations continued over the following months and mirrored an increasingly radical vision of agrarian citizenship that swept the nation after the death of René Barrientos in 1968. With Espejo's organizing experience, Yapacaní's colonists formed an official union (CECOYA) linked to other emerging colonist centers in the Alto Beni and the Chimoré and to Santa Cruz's student population. Earlier that year, Guillermo Capobiano head of the students at Gabriel René Moreno University (UAGRM) in Santa Cruz had brought a group of students to meet Espejo and other colonists near Yapacaní. The two leaders then traveled back to Santa Cruz in a joint student-settler convoy of forty trucks. In the cruceño capital they blockaded all roads in and out of the city, occupied the plaza, and eventually pressured the government to set fair prices for rice.[101] The presence of militant rural folk in an urban setting outraged cruceños who drew explicit comparisons to the invasion of highland militias a decade earlier during the civic struggles.

In July of 1969, under Capobiano's leadership, students of UAGRM also petitioned new President Luis Adolfo Siles directly on behalf of colonists. They had sent a commission to study colonies like Chané-Bedoya and denounced the lack of titles, transport, and technical support they had uncovered. They also leveled what they took to be a damning critique, pointing out that while Japanese colonists received continuous support from the Japanese government, "the only manner of subsistence for the Bolivian campesino is to become a peon of the Japanese colonist."[102] Unlike *El Deber's* editor who nostalgized a respectful campesino cruceño in contrast to raucous highland migrants, the students reminded the President and readers of their open letter that the so-called campesino cruceño was a fiction, "constituted [entirely] of elements that have migrated from La Paz, Oruro, Cochabamba, and Chuquisaca."

The grassroots activities of Methodists like Bravo and Baldivieso, organizers like Espejo and Capobiano as well as organizations like CECOYA and UCAPO precipitated a coordination of colonist cooperatives throughout the nation's lowlands. In February of 1971, settlers from across Bolivia gathered in La Paz for the First National Congress of Colonizers. Delegates created a National Federation of Bolivian Colonists, to address the needs of "the campesino colonizer who feels orphaned in their necessities."[103] The Federation held that the colonizer sat at the vanguard of social revolution due to "their origin and the act of migration [the former] because a grand part of the colonists are displaced miners, unemployed workers, etc. connected with the centers of revolutionary ideological diffusion and [the latter] because it has conditioned important cultural changes that gave them an accelerated awareness of the Bolivia drama."[104]

A few months after the conference, the transport situation in Yapacaní once again led to conflict involving the Methodist Church. An International Development Bank loan was intended to pay for a bridge over the Yapacaní River that would extend the highway from Santa Cruz directly to the colony and avoid the need for costly barge transport. The CBF had contracted the job to a local company but the initial pilings had been destroyed by floodwaters. With progress stalled, Juan Espejo and fellow colonist Marcelino Morales traveled to La Paz to petition the government. A bridge commission visited the colony but "left without doing anything," and as in Buena Vista, Espejo opted for direct action.[105] Drawing inspiration from Uruguay's Tupamaro movement (which was taking wealthy Uruguayans and foreign dignitaries hostage) Espejo decided to kidnap four IDB auditors scheduled to visit the colony and hold them until the government agreed to complete the bridge.[106] On the night of August 8, the operation took place. After a brief struggle, the auditors were sequestered and taken to a safe house while armed colonists, expecting a military response from the government, blockaded the entrance to the colony.

Quite a few Methodists found themselves behind the barricades in Yapacaní. In addition to foreign and national workers like Brooks Taylor, Ana Fajardo, Larry Sanders, and Mary Sayers, on long-term assignment in the area, the church had twenty-nine U.S. student summer volunteers assisting in the construction of a new Methodist hospital in the colony.[107] After the IDB hostages had been dealt with, Espejo went to Taylor, the head of the hospital commission, and asked if the students would be willing to take part as hostages of "a symbolic character."[108] The answer came back that the volunteers stood ready to cooperate in solidarity with the colonists.[109] At the time, Harry

Peacock was working in the new orientation program in Piray Colony to the north of Montero. Returning to the city one evening, he had dinner with fellow Methodist Ruben Baldivieso who lived and worked in Yapacaní.[110] Unaware of the hostage situation, the two decided to make an impromptu courtesy call on the youth group—the majority of whom were from Peacock's hometown of Brownsville, Texas. When they arrived at the Yapacaní River well after dusk, the barge operators warned them about "people with shotguns" guarding the town. As a settler in the colony, who had worked with Espejo during the Buena Vista conflict, Baldivieso was able to negotiate their entrance. The pair of Methodists found the students in the colony school where their frantic chaperone, a Methodist minister, was attempting to convince the youth group that they could not "volunteer to be hostages to pressure the government, this or the other." According to Peacock, a sixteen-year-old girl from Corpus Christi, Texas, the youngest of the group, declared "I'm not sure you have the right to make that decision for us, we came down here knowing that we couldn't do anything really significant to change the lives of these people . . . but they have come and they have asked us to do something and I don't think I'm ready to say no."[111] The group, along with Peacock, elected to remain as hostages and observe a nightly curfew while completing their work on the hospital during the day.

Settler-government negotiations proceeded at the local and national level, and on August 16 a bridge accord was signed and all hostages, voluntary and otherwise, were released. During the conflict, Yapacaní had received the strong support of settler cooperatives from across the lowlands and especially from UCAPO in Chané-Bedoya, the organization that Methodist Jaime Bravo had worked extensively with after his return from Argentina. By adding international profile to their demands, the unlikely union of Methodist teenage hostages from South Texas and radical agrarian politics from eastern Bolivia appeared to have achieved a tangible gain. Peacock's presence among the hostages also blurred the distinction between two very different forms of Methodist organizing in Santa Cruz—new settler orientation on the frontier north of Montero and radical solidarity work in established colonies to the west—that suddenly appeared compatible. The moment was short-lived. Less than a week after the hostages were released, Colonel Andrés Selich Chop, leader of the Montero-based rangers that had captured Che Guevara, launched a coup d'etat. President Juan José Torres—a leftist member of the military—was removed and replaced by General Hugo Banzer. Banzer, a native of Santa Cruz, would orchestrate a definitive shift to the right over his seven-year rule (referred to as the *Banzerato*).

Negotiating the Transition: Faith and Authoritarianism

In colony cooperatives and Methodist circles, the effects of the coup were felt immediately. The week after the bridge resolution, Espejo learned that paramilitaries were rounding up union leaders in Yapacaní. He and a friend swam across the Yapacaní River well downstream of the ferry crossing and wandered through the bush until a friendly Japanese farmer from San Juan Colony directed them to a nearby road. From there they traveled clandestinely through Santa Cruz and up to the highlands before settling in the Alto Beni colonization zone near La Paz. Returning to Yapacaní the following year, Espejo was captured, detained, and tortured in Santa Cruz by agents of the government and spent the next three years in prison.[112]

The coup also impacted the Methodist Church, albeit in uneven ways, thereby revealing points of fracture between national and foreign workers as well as between Protestant missionaries and workers more and less committed to radical measures. In a report on the fallout from the coup also distributed to the MCC, local Presbyterian missionaries James and Margaret Goff explained to their mission board that "during the Ovando and Torres administrations a group of progressive priests had moved far ahead of the [church] hierarchy in advocating and promoting social change [and] the Methodist Church in Bolivia has seen itself as having a humanizing role in the revolutionary process."[113] MCC officials, who worked alongside the Methodists in rural development, but had remained distant from radical politics, also expressed their unease. In a letter to the home office in the days after the Banzer takeover, MCC-rep Dale Linsenmeyer reflected that "in one sense this was just another coup . . . but this one affected us more directly in Santa Cruz and a shift from left to right was a surprise. I hope the Methodists aren't too far out on their limb."[114]

Some were. Although uninvolved in the hostage situation, Jaime Bravo was working with the settlers and fieldworkers in UCAPO at the time, and the organization was targeted by the new Banzer regime. When local radio broadcast a list of people that had been asked to present themselves before the authorities the week after the coup, Bravo's name appeared along with other Methodists, both national and foreign, including Harry Peacock and Brooks Taylor. Authorities arrested both Bravo and Taylor, several priests, and student leaders like Capobiano that had shown solidarity with the colonists. Foreigners, like Taylor, were soon released. Peacock presented himself before authorities—some of whom had been students at the Methodist Rural Institute in Montero—and was not incarcerated.[115] Others went into hiding.

In their report, the Goffs noted that forty priests were in exile while others had taken sanctuary in foreign embassies.[116] Several current and former Maryknoll sisters—such as U.S. citizen Mary Harding—were also among those targeted by the regime.

"I never thought I was a Communist, I was a Methodist Christian with a socialist orientation, but Communist, no" Bravo would explain, but conceded that the ideological climate of the Cold War, "took hold of us Christians like a sandwich" making those distinctions impossible. "After turning himself in to authorities" he was subject to violent interrogation and expected to be killed. Ultimately, the Methodist Church was able to use its institutional clout to secure Bravo's release on the condition he leave Bolivia. In exile, Bravo first went to Lima where he was lodged with the Methodist Church and worked alongside members of the independent new bulletin "Noticias Aliadas." While he was out of Bolivia, right-wing militia broke into his house in Montero—located next door to the Methodist rural health director, Jim Alley—burning his library and screaming at his wife that her husband was a Communist.[117] It was not until a 1978 amnesty at the tail end of the Banzerato that both Espejo and Bravo, two of the many examples of dislocation produced by the Banzer coup, were able to return home.

For those faith-based development workers that elected to remain in the country and work under the Banzer regime, the coup also led to a deep reflection on the radicalism of the pre-coup era and the church's place in politics. Peacock remembers that Bravo and other Methodists were very involved with UCAPO but clarified that the organization ranged from "basically socialist" to those committed to a "violent overthrow . . . on the same page as what was going on with the left in Uruguay and Argentina at the time."[118] Methodist worker Ana Fajardo was far more critical, claiming that Bravo's actions, along with other radical Methodists that "had gotten involved in politics" had fundamentally discredited the church. "They were going in the trucks of the Methodists, taking their Bibles but involved in another project . . . meeting in the church, saying that it was a church meeting, . . . doing all of these things against the government."[119] In her interpretation, politics was something extraneous to the work of the church in rural development that led to a gradual separation between U.S. and Bolivian Methodists and eventual autonomy for the Bolivia Methodists Church. Her narrative reflects waning ties with North Americans she had worked alongside in Yapacaní. "We had a lot of fields, properties, it all fell apart because of the politics; we lost the connection with the gringos," she concludes.[120]

Like Fajardo MCC administrators were critical of the Methodists' political organizing in the period before the coup. In the aforementioned letter, Dale Linsenmeyer may have expressed concern for the Methodists who were "out on their limb," but learning of Bravo's imprisonment, he insisted that "it is without a doubt their involvement in politics earlier and especially their leaning and advocacy of the Leftist movement which was the sole reason for their severe treatment by the new government."[121] He also recognized that, given MCC's vocally apolitical stance, the coup, "should mean that the price of [our] stock in Bolivia has risen."[122] As Linsenmeyer predicted, the MCC experienced a dramatic increase in the scope of its operations under the new regime. That much was evident a year and a half into the dictatorship as MCC Director for Latin America Edgar Stoesz toured the country. "[Nineteen seventy-two] was in many ways a good year," he began his report, praising the stability of the Banzer regime, and agricultural growth in Santa Cruz where "cotton is king." MCC-Bolivia had expanded to thirty-six workers and unlike in the past, when they had farmed out their volunteers to the Methodist church and other agencies, "all but seven are [currently] under MCC programming."[123] They worked in four interrelated fields (education, medical services, colonization, and agricultural extension) each "entirely directed toward either new colonization areas or rural villages."

As with the Methodist church in the 1950s and the 1960s, the MCC filled in for an absent state in the 1970s and 1980s, strategically placing its volunteers in regions where basic services were lacking. Stoesz insisted "the most meaningful area of interaction no doubt takes place at the village level where MCC workers seek to achieve the highest level of interaction with Bolivians . . . a trend that must continue if MCC Bolivia is to earn the right to continue and expand."[124] Lynn Loucks, founder of MCC's Teacher Abroad Program, also celebrated "the exciting aspect of going where there are no educational services," encouraging TAP volunteers "to look outside their classroom more and begin to form a community vision."[125] These MCC officials portrayed their organization in stark opposition to "La Paz-based" agencies like USAID, and MCC took pride in their low overhead and direct contact with rural villagers.

Among numerous zones of activity, MCCers worked in El Torno-Jorochito, that same region that in 1966 had petitioned for the transfer of the Cotoca center and eloquently demanded that the state provide técnicos and not tourists. Wendell Amatutz was carrying out demonstrations of animal traction, drilling wells, publishing a newsletter, leading cooperative classes, and doing extension work in encouraging wheat and soybean production. He had also become a "resource person for the town . . . and surrounding area"—a role that saw him take part in youth clubs, sports programs, the credit union, and the syndicate.[126] In other locations, MCCers shifted nimbly between extension and evangelical work. Calvin Miler, for instance, demonstrated animal traction techniques to colonists, but also enjoyed "helping out the Protestant churches in the area" and found among the new migrants in El Torno, "a really beautiful and active group of Christians."[127] Others like George Reimer and Murray Luft expressed reservations with the overlap between ministering and administering. Reimer was impressed by the dedication of neighboring evangelicals but was also "mindful of the pitfalls in too close an association with labels under suspicion and theologies slightly different."[128] Luft, writing in a newsletter circulated among Mennonite volunteers, promised "not to bring my Christian theology to bear on Bolivians." I say this for two reasons: (a) I do not feel personally comfortable in an evangelizing role whether here or at home in Canada, and (b) I have too many unanswered questions about the culturally specific implications and consequences of evangelizing in rural Bolivia. Hence, I challenge the simple assumption that if I am a Christian, my only job is to communicate the gospel verbally and bring about an individual conversion experience in the heart of a Bolivian peasant.[129] Instead Luft wished to simply "concentrate on ministering in a *limited* way to the material and social needs of Bolivians."

Circular Logic, Secular Logic: Missionary and NGO in the San Julián Project

As the preceding examples make clear, the MCC flourished in the turn to authoritarianism under Banzer. Although its members were persecuted, and its radical activism suppressed, the Methodist Church also survived the coup. For both organizations this was particularly true of work in new settler orientation, which emerged in the wake of the flood at Hardeman and Piray colonies and appeared apolitical to the regime. In 1972, as the Piray project concluded, the CIU (with Harry Peacock as director) signed a contract to administer settler orientation in a new colonization scheme. In the two decades since 1952, settlers and agribusiness had gradually filled the "Integrated North" around Santa Cruz and Montero. This would be the first major colony in the "Agricultural Expansion Zone" to the east of the Río Grande. For those sympathetic to the type of political organizing that had preceded Banzer and who had witnessed close friends jailed and exiled, the choice to work on behalf of the new regime was likely a difficult one. "We had contact with [Banzer's] new director of colonization," Peacock recalls, "and we were invited out to San Julián and then invited to work . . . those of us in the CIU did some real soul-searching, Marty [and I] had the same response, that it would be the easy and clean thing for us to say we didn't want to get our hands dirty working with these people. But who pays? Who suffers? . . . if we're going to get on the field and play, we're going to get muddy. We decided to get on the field."[130]

During a week-long conference held in Santa Cruz from May 22 to May 27 of 1972, it was evident that faith-based missionaries and volunteers would continue to play active roles as proxies for the state in rural development under the new Banzer regime. The meeting brought together a range of institutions, and foreign agencies under the leadership of the Public Works Committee, the regional governmental organization responsible for managing Santa Cruz's oil royalties. The tone was hyperbolic. President Banzer, a native of Santa Cruz, was present to welcome participants. "We believe," he told the crowd that, "[we are living in] an age of progress, an age of development and an age of integration."[131] The President of the Committee Pro-Santa Cruz also spoke, claiming that Santa Cruz was "not just the place where our country comes together, but all of America."[132] The President of the Committee of Public Works recounted the history of neglect that had long characterized the region that now "constitutes the greatest economic potential of Bolivia."[133] Over the following days, experts led panels on infrastructure, public utilities, petroleum development, and technical assistance, but a significant portion of the conference was dedicated to agriculture, including the cotton boom, "the migration of farmers from the interior," and the construction of a new highway linking the highlands with new colonization zones in Santa Cruz and Cochabamba.

A special roundtable on colonization was attended by members of the military and the INC as well as MCC-rep Dale Linsenmeyer and Harry Peacock, Ken Graves, and Bishop Mortimer Arias from the Methodist Church. INC director Ulrich Reyes began by outlining some of the problems faced by the 12,000 campesino families that had made the move to the lowlands since 1952. He noted the lack of tools, draft animals, inadequate parcel sizes, and declining fertility on cleared land. Most grave was the "excessive geographic dispersion of the colonies," inhibiting service provision, transport, and marketing.[134] Attempting to link Bolivian colonization to global theories and practices of resettlement, the conference organizers had brought in two experts from Israel. The agronomist Shai Arazi explained the evolution of the kibbutz system over more than sixty years, which like Bolivian colonization had employed community organization to secure national frontiers, resettle populations, and develop agriculture. His colleague, Y. Gazit provided members with a proposed budget for establishing a cooperative agricultural colony on the kibbutz model.[135]

Representing the United Church Committee, Harry Peacock spoke to their new role in the San Julián project. "Catholic, Methodist and Mennonite," he began, "through an agreement with the INC and with financing from the World Council of Churches and the Collaboration of the World Food Organization have created a capacity-building program (capacitación) for colonists that will settle the zone of San Julián."[136] Peacock's speech drew the practices that the CIU had cultivated in Hardeman and Piray into a coherent strategy. With its Mennonite and Methodist volunteers experienced in outreach and extension they intended to teach courses in sanitation, land clearing, hygiene, and nutrition. In addition to project technicians drawn from the ranks of the CIU's member organizations, the orientation program would rely on the orientator model pioneered in Piray in which camba orientators would impart local environmental knowledge. Beyond these technical elements, their orientation had a broader sociopolitical element which would include the idea of solidarity. The proposal closed with a budget of $1,900 per colonist, substantially lower than earlier INC initiatives.[137] At the 1972 conference, San Julián was imagined as a unified project of infrastructure, service provision, settler orientation, and community building, managed in a partnership between a religious NGO and the state.

The kibbutz model put forward by Arazi and Gazit was not adopted in San Julián, but the project's design took a radical spatial approach to address similar questions about community formation and Reyes's concern with geographic dispersion and service provision. As opposed to the piano-key formations of earlier colonies in which settlers were strung out along access roads, colonists in San Julián were concentrated in nuclei of forty families. In a unique format, that is particularly striking when viewed from above, pie-shaped plots radiate out from a central point. These individual nuclei were grouped in blocks of nine (referred to as a NADEPA), connected by principal and arterial roads, with the central nuclei containing a high school, a clinic, and other services. The circular logic of San Julián was that settler's homes, community centers, wells, and schools would be drawn to the center of the pinwheel, creating an inescapable solidarity and reducing the universally high abandonment rates of previous colonies.

Critically, missionaries and volunteers from the United Church Committee offered arriving settlers an intensive three-month orientation program. Peacock had recently completed a U.S.-based furlough during which he studied social psychology at the University of Texas-Houston. The training was

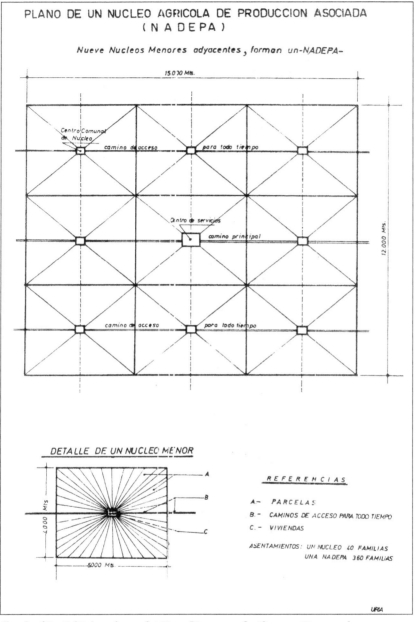

Sketch of San Julián's settler nuclei. From "Programa de Alimentación para el asentamiento de colonos en el Proyecto San Julián" (La Paz: INC, 1975).

evident in his characterization of the settler who, "at the moment of occupy-ing his new territory, because of the need to change his methods, or for being outside of his environment and because his social relations are different is much more receptive to change."[138] It was a moment, according to Peacock that "is the best to implant new ideas," and lasted for a few months followed by "stages of resistance and passivity." Essentially, the orientation program sought to re-create the experience of the 1968 flood response at Hardeman camp. Settlers would form self-governing committees to construct shelters, prepare food, and clear land. It also contained an environmental component. Experienced camba orientators would be on hand to impart local knowledge and survival strategies in a meticulously structured fashion. This included Alejandro Araus, a native of Montero who had worked beside Methodist missionary (and future Heifer Project director for Bolivia) Walter Henry as an orientator in Piray colony. Araus joined the CIU as an orientator at San Julián, spending every day for three months with the colonists assigned to his nucleus. He remembers a minutely planned program of daily and weekly objectives from committee work to planting. The nucleated settlement pattern proved con-ducive to settler surveillance as well as solidarity. When one enterprising colonist sowed cash crops in space reserved for subsistence gardens, Araus was on hand to force their removal. Amid this seemingly contradictory mix of orientator-driven regimentation and settler-based initiative, the CIU ran con-tinual assessments of the orientation program to be applied in future nuclei.

By the end of 1974, the CIU had established nine nuclei in San Julián form-ing the first NADEPA. In a report of that year, the INC praised this new form of "semi-directed colonization" as a radical departure from past practices.[139] Earlier schemes had failed, according to the authors, because, "the politics, like the plans, projects, services, and other activities of colonization moved between margins that were too broad." Abandoning its lofty former high modernist goals of incorporating the peasantry, engaging in massive popula-tion transfers, improving the balance of trade, and making the nation agricul-turally self-sufficient, the INC resolved that it was "of great urgency to delimit [future schemes] with more precision and locate their execution within our reality."[140] That reality, the report acknowledged, included the misappropria-tion of tens of thousands of dollars of funding for earlier projects. But unwill-ing or unable to address a rampant culture of corruption, the INC claimed that a culture of paternalism was the most significant obstacle to success as it, "increases the incapacity of the colonist" to engage in basic actions from the formation of cooperatives to mutual aid.[141] The idea seems astonishing given the INC's inability to administer basic services in earlier colonies. Yet a fictional dependency induced by phantom state largesse served as a convenient pretext for further reductions in INC support.

In the 1974 report, we see a perpetually absent state engineering its future disappearance through NGO-led colonization. At the same time, the INC maintained a broader cynicism about Andean peoples who, they pointed out, displayed "a certain negligence and lack of initiative to learn new and more lucrative cultivation techniques."[142] This "low cultural level" was apparent in such presumably financial decisions as their inability to consume industrialized products to their imbalanced diets. It created a dilemma for the INC in which untrustworthy colonists, as they saw it, could not be left to their own meager devices any more than they could be showered with excessive, paralyzing support. Partnering with an NGO in San Julián was one practical step in this progressive narrowing of ends and diversification of means. INC officials were understandably excited about the project given that the CIU had taken control of many aspects previously in the domain of the state. They were particularly enthusiastic about the orientation program, under which settlers were creating mutual aid organizations that were "the fruit of their own labor and initiative and not something imposed from outside that they have no responsibility for."[143] Read the other way, it was the INC expressing relief at its own lack of responsibility, financial or otherwise, for the new colonists.

As the first round of settlement drew to a close, state officials were not the only ones imagining San Julián as a model for the future. MCC volunteer Mark Epp, worked with the CIU in San Julián from 1972–74 while the orientation program in the first nine nuclei was underway. Returning to the United States, he transferred his experience into a Master of Professional Studies in Cornell's College of Agriculture. A condensed version of Epp's 1975 thesis, "Establishing New Agricultural Communities in the Tropical Lowlands," was then translated into Spanish and provided to project orientators as a guide when Epp returned to San Julián to take over directorship of the CIU from Peacock. Like the INC report, Epp conceded that many of the structural and everyday aims of resettlement were unrealistic, but he preferred to think of the CIU as a buffer between colonists and a heavy-handed state bureaucracy. The CIU's pedagogical role extended beyond instructing colonists. Epp felt an equivalent need to encourage more "humane considerations" on the part of national institutions privileging dignity and community as much as the rational provision of services. Though carefully admonishing the INC, Epp also shared its fear that, at the moment of settlement, the "seeds of paternalism and dependence are planted." With a touch of the theatrical he warned that the incautious NGO is soon, "cast in the role of patrón, a position it can neither fill nor painlessly retreat from. The settler, on the other hand, has begun to look to the outside for his help. The scene is set for disaster."[144] In the decade preceding neoliberal reform in Latin America, Epp was already advocating for decentralization as a state paradigm. Scaling back the state would carve out space for the emerging NGO, freeing the colonist from the "patrón–client" matrix, he claimed, and transforming them from "follower to decision-maker."[145]

The CIU's initial orientation program relied on a three-year Oxfam grant (another reason it was appealing to Bolivian officials), and as the term ended, project participants scrambled to identify new funding. As the head of the CIU, Harry Peacock was often in La Paz working with the national office of the INC. The rural development office of USAID was across the street, housed in the same building as the Ministry of Agriculture, and he spent much of 1974 crossing and recrossing *avenida* Camacho, courting AID funding. In 1975 the project was sent to USAID's home office for final approval. Peacock, back in the United States on a church furlough, traveled to Washington to support the review. After more than a decade of U.S.-supported resettlement initiatives in Latin America, such programs were "not a popular thing . . . within AID," he recalled, "they had figured out that colonization didn't work."[146] Ultimately AID officials only agreed to fund a new round of settler orientation on one firm condition. Peacock would be "directly responsible to AID and not to the Methodist church or the [CIU] or Oxfam." He accepted, officially leaving the church after thirteen years as a missionary.

Peacock's abrupt decision to exchange missionary work for a position in international development reflected a growing tension for many Protestants in Bolivia. Back in 1969, Nelson Litwiller of the Kansas Mennonite Mission Board had toured the country and pointedly asked, "How it is possible for MCC to be active here for ten years and give only the life, and not the word, or in other words, establish a Mennonite church." Yet a few years later, MCCer Murray Luft, who was active in the San Julián project, wrote a piece in MCC-Bolivia's newsletter, to "challenge the simple assumption that if I am a Christian, my only job is to communicate the gospel verbally and bring about an individual conversion experience in the heart of a Bolivian peasant."[147] As members of a relief agency (albeit a faith-based one), it was easier for MCCers to resist the conflation of ministering and administering. In contrast, Peacock found that as a Methodist missionary working with Mennonites and Catholics in colonization he "wasn't making Methodist churches and that began to be a problem."[148] Even as he moved further into development, the nationalized Methodist church was focusing ever more on evangelization. Furthermore, a new core of Aymara leaders within the church were gradually assuming responsibility, recalls Jaime Bravo, "and various missionaries," some advancing critiques of dependency in the relation between Bolivian and U.S. Methodists strikingly similar to those espoused by the CIU, "were thinking the time had arrived to withdraw."[149] Uruguayan Methodist Mortimer Arias would also resign as bishop of the Bolivian Methodist church at this time and even Bravo, himself Aymara but positioning himself as "not too Aymara, but also not very Gringo" left the church.

Although Peacock's move to USAID might appear as a subtle secularization in which ad hoc religious organizations and missionaries were gradually giving way to an emerging professional development industry, the reality was far more paradoxical. Even if the USAID funding agreement was conditional on Peacock's departure from the church, a shared faith also made the agreement possible. He confided the deal "never would have come about had it not been that . . . Bob Moffat [director of rural development for USAID in Bolivia] was a Methodist, Bob Moffat was from South Texas. We had not known each other . . . but we were from the same part of the country."[150] Moffat visited the nuclei in that first orientation program and while they had "never discussed religion" directly, Peacock felt that the fellow Texan's rural Methodist background provided the framework for his appreciation of the CIU and their Christian commitment to development, just as he would latter claim about his future USAID supervisor Bastian Schoutten.

With USAID funding secured, San Julián would quintuple in size over the following three years. A consequence of that expansion and of Peacock's recruitment by USAID was that this remote colonization scheme along the Bolivian frontier was pulled further into dialogue with parallel forms of rural development and new lands settlement projects across the globe. With its unique spatial layout, orientation program, and low abandonment rates, San Julián soon gained a reputation as a success story, appearing to observers not just as a model for future colonization in Santa Cruz, but for global tropical resettlement as a whole. The project would appear in studies and edited collections as visiting sociologists, anthropologists, and agronomists documented its progress and passed through the small bunkhouse the CIU had built in the first phase as those distinctive pinwheel-shaped nuclei spread across the forested landscape.[151]

In those years, the intertwined logic of the Green Revolution and the politics of decolonization were driving a range of studies in tropical agriculture, resettlement, land reform, and colonization directed at identifying exportable models for future projects. Within Latin America, Craig Dozier's 1969 case studies of Peru, Bolivia, and Mexico, and Michael Nelson's 1973 study of colonization projects across South America, stand as prominent examples.[152] Other organizations and individuals were looking further afield. Cornell University, where CIU member Mark Epp had completed a master's in 1975 had pioneered global research on applied anthropology and agricultural economics, including a prolonged case study in the Vicos Region of Peru. At Caltech, anthropologist Thayer Scudder had made a career of studying Africa's dam-driven resettlement projects, beginning with the study of the relocation of 100,000 Nubians during the construction of the Nile's Aswan Dam and the displacement of nearly 60,000 Tonga due to the construction of the Kariba Dam on the Zambezi River. He would go on to support a generation of research on human ecology and resettlement. In 1976, just as the second round of settlement was underway in San Julián, Scudder joined fellow scholars Michael Horowitz and David Brokensha to form the Institute for Development Anthropology (IDA) based out of SUNY Binghamton. Through the IDA, a network emerged peopled by a range of specialists and experts that traversed the globe—often commissioned by USAID to assess the success or failure of rural development initiatives.

Having taken U.S. furloughs to pursue graduate studies, CIU members were aware of this emerging body of research. Peacock was inspired by Nelson's *The Development of Tropical Lands: Policy Issues in Latin America*. Exploring twenty-four case studies of directed settlement in the humid tropics, Nelson proposed that new colonies passed through three stages: pioneering, consolidation, and growth. With the first round of USAID funding about to expire in 1978, San Julián had expanded dramatically from nine to forty-seven nuclei but the road beyond núcleo eleven remained unfinished. The members of the CIU typically left settlement schemes in the early stages but at Peacock's urging, they decided to pursue a new round of USAID funding that would, in Nelson's formulation, consolidate San Julián.[153] As part of the consolidation proposal USAID also brought a seven-member interdisciplinary review team to Bolivia in 1978. The group included anthropologists Allyn MacLean Stearman (who had spent four years as a Peace Corps worker in Santa Cruz in the 1960s, where she had worked in San Julián), Nelson (personally chosen by Peacock), and a young PhD candidate in Anthropology from Cornell named David Hess. At the time, the IDA was embarking on a global review of new lands settlement in cooperation with AID's Office of Rural Development. In addition to reviewing existing literature, this included funding for new doctoral research and Hess would receive USAID-IDA support to write a dissertation on the San Julián Project. Nelson, Stearman, and Hess left "favorably impressed with the progress of the colony and recommended that, following Nelson's schema, a 'consolidation program' should be undertaken."[154]

The pursuit of additional funding pushed forward the process of secularization that had begun with Peacock's move from the Methodist Church to AID. The CIU was formally disbanded and its members created FIDES (Inter-American Foundation for Development) an appropriately secular-sounding NGO (but with an identical composition to the CIU) to apply for the grant.[155] Jaime Bravo frames the move as a response to the incompatibility of their work in rural development with an evangelical-minded Methodist church that, after nationalization in the early 1970s, "didn't have the same vision," arguing that with the shift to FIDES, "we were more autonomous and independent of the church, and we were much more pluralistic, we could work with the Catholic church more openly and channel resources from foreign governments which before . . . working between church and USAID [was difficult], but with the NGO and [USAID] it was easy because we were a 'civil organization' and not a religious organization, and not doing work of proselytization, USAID would not give us help any other way."[156]

Despite the secularization of the CIU, faith again provided the unspoken framework for the financing. USAID's manager on the ground in Santa Cruz was Gary Alex who had previously worked in rural extension in Cambodia for seven years with the Quaker-founded International Volunteer Service. Although Alex's superiors remained unsympathetic to colonization, thinking that "it was sloppy, and we couldn't control it," he was impressed by the project and brought in an experienced grant writer to assist FIDES with the pitch.[157] With Alex's support, and the Nelson-Stearman-Hess positive review, USAID approved the proposal. The result, was "the largest cooperative agreement grant ever given to a private volunteer organization in Latin America: $1.5 million USD to be disbursed over a five-year period."[158] On the verge of the *lost decade* when state-sponsored development entered a profound crisis, a central shift in development financing in Latin America was underway with the nascent NGO at its center.

Roadblocks on the Path to Development

The promising period of USAID-FIDES financing began inauspiciously as Banzer attempted to rig the 1978 elections in his favor and was overthrown in a military coup after which a series of short-lived presidencies and military dictatorships stripped state companies of resources. It ended as a dramatic hyperinflation gripped Bolivia. Flora Gómez, a former colonist who had become a Methodist-trained FIDES health worker in nucleus settlement, remembers the 1980 coup by Luis García Meza, "broke the colonization [program] and I withdrew" eventually opening a pharmacy in San Julián's small but growing urban center along the highway.[159] In addition to political and economic crises, environmental disasters also threatened the colony. In May of 1983, a massive flood devastated several of San Julián's settler nuclei, causing the evacuation of more than a thousand families and the loss of the entire harvest. As in 1968, settlers formed emergency committees to respond to the disaster. But while the '68 flood in Santa Cruz had provided the foundational moment for the creation of the CIU and its unique orientation program, the 1983 flood seemed to bookend that unique period. With annual inflation approaching 300 percent, Harry Peacock, whose children divided their lives in Bolivia into *before* and *after* the 1968 flood, also left the project to join a USAID initiative in Costa Rica. In an assessment blending principle and pragmatism, Peacock confided that "[for] twelve years I had been there, I was getting to be too much of an institution and the whole idea of dependency and so on and so forth . . . and besides that I was broke, and they were offering me a good job."[160] "Everyone suffered" remembers Jaime Bravo of Bolivia's record-setting hyperinflation. With inflation over 20,000 percent, in 1985, he also left Bolivia, returning to the United States, where he had spent his exile, to work with Latinx migrants in California.[161]

In addition to producing an exile of project workers, the events of the early 1980s shook the academic image of San Julián as an unqualified success and produced an increasing radicalization among San Julián's settlers. This much was evident in the differing findings of two evaluation teams that visited the colony in 1983 and 1984. The first, led by Richard Solem of USAID arrived to assess the progress of the consolidation program and to unofficially determine whether a new round of funding should be extended that would take the colony into a growth period—the third of Michael Nelson's stages of settlement. Given the difficult situation faced by settlers, Solem was remarkably positive. He seemed to accept the CIU's original premise, that San Julián's success lay in a resilience grounded in subsistence practices that could survive economic (hyperinflation) and environmental (flooding) disasters. While many USAID officials saw FIDES extension work as nonsubstantial or *do-gooder*, Solem privileged the "intensive, sensitive, and ever adaptable" relationship between faith-based planners and settlers. His report is punctuated with quotes from anonymous FIDES employees such as, "[success] depended on a mystique—a social compromise with the campesino that came from religious roots . . . salvation had to be reflected in all aspects of life, not just in spirit," and Solem endorsed continued funding.[162]

The year after Solem's visit, IDA anthropologists Michael Painter and William Partridge traveled to San Julián. Partridge had previously studied dam-induced settler relocation in Mexico's Papaloapan Valley and their visit reflected the continued importance of comparative global approaches to re-settlement. In a series of provocative studies and meetings through the early 1980s, the IDA attempted to bring together the lessons of these projects around the world. Participants in one 1982 meeting in Binghamton discussed "Latin America and Colonization," "Large-scale Settlement Schemes in Tropical Africa," "Transmigration Experience in South Asia," and "AID Experience with River Basin Development and New Lands Settlement." The IDA was not alone in its enthusiasm for comparative research into pioneer settlement in the tropics. In September of 1985 a group of UN-supported geographers and other social scientists gathered in Kuala Lumpur for five days to discuss resettlement in the humid tropics of Latin America, West Africa, and Southeast Asia before publishing an edited volume on the subject.[163]

During their 1984 visit, Painter and Partridge were tasked with a follow-up assessment on what IDA-founder Thayer Scudder had referred to in 1981 as "one of the few examples of successful settlement in the Amazon." While acknowledging San Julián's unique orientation program, spatial layout, and high retention rate "represent noteworthy advances," Painter and Partridge claimed that such "project-level innovations do not address the longterm problem facing San Julián and other settlement areas." Planners, they continued, preferred to see "settler's problems as technical in nature," while eschewing attempts to "insure [settlers] win a [greater] share of development resources" as "too time-consuming or politically unrealistically." Yet the latter strategy was critical, they continued, as settlers "are not alone on the frontier," where loggers, ranchers, and regional elites, "dominate access to truck transport, physical infrastructure, credit sources, and scare resources." This much would have been apparent for many FIDES workers, whose only access between Montero and San Julián—aside from bicycle, horseback, or foot—was to ride precariously on the top of fully loaded logging trucks that passed through the colony. Painter and Partridge concluded that the same dependent conditions that had driven Andeans to migrate from the highlands had simply been re-created in the tropics, a result, that by USAID's own logic, could not justify the environmental impact of new lands settlement.

That the Painter and Solem teams could draw such different conclusions about the success of San Julián only a year apart may speak to the surreal nature of Bolivia's hyperinflation—which had increased from 300–20,000 percent between 1983 and 1984. Bolivians were soon using wheelbarrows and empty rice sacs to transport money in a fashion reminiscent of Weimar Germany. Amid these bleak conditions it was impressive that colonists could even maintain a basic subsistence base, and Painter would later reflect that an accurate assessment of the colony was simply impossible at the time. Despite their differences, Painter and Solem were unanimous in their praise for the solidarity shown by San Julián's settlers. As Solem explained, the newly created Federation of San Julián Colonists was "taking an active political role in expressing their needs to outsiders." Examples abounded, from the collective response to the 1983 flood, to a staged occupation of INC headquarters that same year. Solem's team had experienced such settler solidarity firsthand, on a particularly bad access road near nucleus 47, when his evaluation team was "detained by a dozen settlers who had strung a heavy cable across the road and, thinking we were from one of the organizations concerned with road maintenance, politely but firmly led us to the community center to discuss the problem." A year later, Painter also championed settler solidarity and acknowledged that the FIDES orientation program had "built upon the experience of highland peasants in organizing themselves," so that at present, "settlers rely heavily upon community organizations to define and defend their rights."

Painter and Partridge left San Julián in August of 1984 questioning whether planners could address the broader socioeconomic forces—manipulation by wealthy ranchers, the abuse of local infrastructure by logging companies, and increasing ethnic tensions within the department of Santa Cruz in the context of highland migration—that threatened the viability of San Julián. Only two months later, San Julián's Federation of Colonists would provide its own answer, mobilizing to carry out a dramatic roadblock that—like the Yapacaní hostage-taking more than a decade earlier—drew together these issues while catapulting the colony into the regional media. In October of 1984, a tense situation boiled over when the regional director of the INC Juan Terrazas requested that road equipment from San Julián be taken south to serve another project. Colonists, suspecting the machinery would not be returned, refused. An irate Terrazas arrived in the colony to demand the machinery in person and was taken hostage by San Julián's residents. Just as Yapacaní's colonists had done, the San Julián colonists then sent a commission to La Paz with list of demands. They also established a roadblock along a nearby highway providing the only access to the neighboring department of Beni as well as the Brazilian border. Over the next few days, tensions mounted as foreign contraband, agricultural goods, and logging trucks began to back up. Through the "Special Federation of San Julián Colonists" settlers maintained a constant vigilance, organizing shifts and food supplies for those involved in the blockade, and prominently displaying Bolivian flags on their barricades. On the day that a commission was expected to arrive from Santa Cruz, violence broke out when a group of drivers tried to break through the barriers and fired into the crowd of colonists. Despite several casualties on both sides, it was colonists that were portrayed in local media as ignorant "brutes" and "savage kollas."[164] Several thousand cruceños gathered for a march to "reconquer the orient" from these "invaders." Throughout all of this, the Federation of San Julián Colonists held fast and was able to negotiate an agreement with the government.

The roadblock (*bloqueo*) is notably absent from academic reports on San Julián even as it forms a foundational moment in the present memories of colonists.[165] For foreign observers, perhaps the community's participation in this widely publicized roadblock/kidnapping appeared an act of desperation that fits within a larger narrative of the *failure* of this unique colonization endeavor. However, the phenomenon of cooperative roadblocks—so integral to the future of grassroots resistance to neoliberal politics in Bolivia—might be read in an entirely different light, as an indicator of community solidarity and organizational capacity. In fact, amid increasingly pessimistic anthropological reports on San Julián during the debt crisis, popular accounts of the protest as recorded through personal testimonies escape from the success/failure paradigm of planners. As in Yapacaní, a decade and half earlier, blocking roads and taking hostages were strategies of resistance for colonists that had been abandoned by the state.

Through the organizing of committees, cooperative kitchens, and labor teams during the blockade, settlers reworked the very practices employed in the CIU's orientation program toward a new end. At the 1972 Public Works Conference in Santa Cruz, Peacock had explained that the CIU would take part in capacitation at San Julián. He seemed to imply a technical bundle for confronting the lowland environment. The apolitical terminology passed without scrutiny at a conference presided over by President Hugo Banzer, who had made every attempt to crush radical political organizing after his 1971 coup. Yet, Peacock had also referred to solidarity with colonists. Methodist Jaime Bravo, who returned from his exile under Banzer to work with FIDES and Lutheran World Relief as a health planner in San Julián, embraced this more politicized version of capacity building. Over the years he also obtained funding for the capacitación of lowland colonists through the NGO Lutheran World Relief. He remembers that religious aid had been key in political organizing in San Julián and elsewhere, that through "the NGOs be they Catholic or Protestant or civil like FIDES, we have helped to prepare the road," for radical politics. In the case of Lutheran world relief, its Canadian funders "did not know that money went to those that were blockading. If they had known they would have cut me off but I allocated the funds saying, 'We are in active capacitación.' What is active capacitation? Blocking roads [*bloquear*]!"[166]

A compelling aspect of faith-based development was its ability to contain both these qualities: a technical package of service provision and planning that appealed to authoritarian leaders like Banzer and a form of rural mobilization that meshed with the needs of campesinos. Such radical expressions of agrarian citizenship—as much as the meticulously planned orientation of new settlers—were ultimately the legacy of faith-based development in San Julián by the mid-1980s just as they had been in the early 1970s in the north of Santa Cruz when the colonists and field hands of Yapacaní and Chané-Bedoya also blocked roads, took hostages, marched, and formed peasant unions to secure their rights. Two seemingly bifurcated strategies—solidarity and orientation—had come together again.

Conclusion: Trajectories in Development

The history of CIU, FIDES, and San Julián illustrates the early growth of NGO-led development, a system of reallocating state responsibility to third parties, which has had a powerful legacy for Latin America and much of the developing world (better described as the world as imagined through the lens of development modernization). In the mid-1980s, the Bolivian state was only at the beginning of a long process of divesting itself of responsibility for colonization along with many of the other revolutionary projects of the MNR. As with other Latin American nations, most notably Argentina, Bolivia began to privatize state institutions and sell off state property in response to the debt crisis. For the Bolivian highlands, the privatizing of the state mining agency COMIBOL in 1985 was the definitive moment of betrayal in neoliberal transition but the disbanding of the CBF, the initial organization responsible for lowland colonization and infrastructure, in the same year was far more relevant for Santa Cruz and the March to the East. In the absence of state institutions, voluntary organizations rushed to fill the gap, offering their services across Bolivia. It is no surprise that when anthropologist Nathan Wachtel returned to the highland community of Chipaya in 1989 for the first time since conducting field work seven years earlier, he found the community awash with NGOs and evangelicals.[167] As the FIDES contract indicates, USAID and other international funders were also increasingly likely to rely on these organizations, in lieu of state agencies, to administer development. Despite Painter's negative assessment of San Julián in 1984, the following year, he was proposing a new three-year program to bring the colony from consolidation to growth, the final of Michael Nelson's three stages of pioneer development. When political scientist Allison Ayers visited San Julián the following decade, she found that the academic curiosity engendered by the CIU—and fed by IDA publications—had led to a flood of NGO activity in the colony. Ubiquitous NGOs were openly competing with one another for institutional space, going so far as to undermine local settler organizations in the process while remaining oblivious to the total absence of support in neighboring settlements outside an intense but narrow sphere of donor interest.[168]

For the workers and visitors that traveled the well-worn physical and intellectual routes through San Julián in the 1970s and 1980s, the colony provided a manageable, miniature space to reflect on modernization theory, dependency, and cultural transformation. It also proved a springboard for highly mobile careers. Although distinct from the travels of Jorge Ruiz, displaced Okinawans, frontier-seeking Mennonites, or border-crossing Bolivians, their prodigious trajectories also linked the lowlands to regional, hemispheric, and global development in the second half of the twentieth century. Jim Hoey, who came to Bolivia as a "Latin America 3" Methodist missionary in the 1960s transitioned to a career with Heifer Project International, an organization that had worked extensively with colonists throughout Santa Cruz. For several individuals, Cornell University, with its early adoption of applied anthropology and robust agricultural economics program, and SUNY Binghamton's Institute for Development Anthropology also proved uniquely conducive to these movements. While Cornell researchers had been engaged in community development in locales as diverse as India, Thailand, Canada, and, most famously, Peru's Vicos region in the postwar era, the IDA took on a growing role in global rural development in the 1970s and 1980s with its close ties to USAID and other financing institutions. CIU member Mark Epp, completed an MA at Cornell, before returning to direct the second phase of San Julián's colonization. He later became head of MCC-Latin America (which maintains a large Bolivia program). Peter Leigh Taylor who studied the cooperative system in San Julián for his Master's in Development Sociology at Cornell went on to study tropical forest management in the Maya Biosphere between southern Mexico and Central America. David Hess, who also completed a PhD dissertation on San Julián at Cornell in the late 1970s with research support from the IDA, went on to work for USAID, a career that took him from Bolivia to Peru, the Ivory Coast, and India. While managing a US$2 billion USAID development loan to Tanzania, he still remembered San Julián as formative moment, arguing that "there is no replacement for me, for having ... an experience I had doing my anthropology research in the jungles of Bolivia."[169]

For some Bolivians, connection to the Methodist church or the CIU also provided opportunities for mobility or advancement on a local or even hemispheric scale. Dardo Chávez, the high school student who had volunteered in the Hardeman camp and later at Piray and San Julián with the CIU, runs the Montero Office of the Rural Andean Health Center. Flora Gómez, a San Julián colonist-turned-health worker, received training at the Methodist hospital in Montero and runs a successful pharmacy in San Julián. Ana Fajardo, who worked in Yapacaní as a home economics agent for the Methodist Church in the 1960s, still lives in the colony and organizes women's circles while her son works in the municipal government. Leaving Bolivia after the Hugo Banzer coup in 1971, Jaime Bravo's trajectory was far more expansive. Mortimer Arias, Methodist Bishop in Bolivia, met with an exiled Bravo in Lima and negotiated for his travel to New York, location of the Mission Board headquarters. The abrupt transition from the periphery to the heart of United Methodism's global operations still makes Bravo laugh. "After sleeping on a tiny little bed [in Montero] and being in a jail sleeping on the floor with some old newspapers in Santa Cruz, I just arrived in New York and they put me up in a Hilton Hotel [with] all the luxuries."[170] Like returning missionaries on furlough, the exiled Bravo traveled the network of U.S. churches speaking to congregations about the situation in Bolivia. Still blacklisted at home, the Mission Board eventually sent him to Dallas to study theology at Southern Methodist University where he was finally reunited with his wife. As an indigenous Latino, he proved remarkably adept at engaging with the civil rights struggles of the 1970s, from the United Farm Workers to the American Indian Movement. Bravo, who had worked with migrant field workers in Bolivia soon found himself involved with Mexican and Chicano migrant laborers in Texas. "I started to work with Julio César Chávez, el Brown Power," he recalls. In Bolivia, Bravo had also taken part in a growing rural Aymara movement within the Methodist church that sought to challenge the control of urban mestizo Methodists. Out of place in Texas he, "was also confused with the North American *indígenas,* and various said, 'hey brother' and I was 'red power' as well."[171] Texas was not Santa Cruz but the organizing skills that

Bravo had cultivated working for the church in Bolivia proved transferable. The Río Grande conference of the Methodist church offered him "a large salary, house, car" and communities to work in. Although he decided to return to Bolivia after an amnesty in the late 1970s, he went back to the United States several times. After leaving the Methodist Church he established connections with other faith-based development agencies like Lutheran World Relief, acting as a broker between North American rural church communities and the Bolivian settlers that they supported through the debt crisis.

Some foreign faith-based development workers never left Bolivia—or later returned. Milton Whitaker, who ran an MCC agricultural program is still living in the small town of Santiago de Chiquitos, where new settler colonies continue to expand a short distance away, with his family nearly half a century later. Harry Peacock, one of the central actors discussed in this chapter, grew up in South Texas and came to Bolivia as part of Murray Dickson's "Land of Decision" campaign. His transition from missionary to an employee of USAID eventually brought him out of Santa Cruz to Costa Rica in the early 1980s. There he worked in another rural development project along the Nicaragua-Costa Rican border for the duration of the Contra war before returning to Bolivia with USAID in the 1990s. Along with several other San Julián alumni—including Whitaker, and UW-Madison trained anthropologist Henry Sanabría—he took part in alternative cropping programs in the Chapare, the counterpart to the DEA's coca eradication campaign in the early 1990s. In this colonization zone near Santa Cruz, U.S. funding was offered to push coca growers like future Bolivian president Evo Morales toward other crops. Disillusioned with the program, Peacock eventually returned to San Julián where all project workers, himself included, had received a fifty-hectare plot in the colony. Into the early 2000s, he farmed in San Julián before turning the property over to his son Oliver. The joke in Methodist circles, that Peacock had been a CIA agent throughout this entire period points to the tight relationship between rural development and counterinsurgency during the Cold War. Similarly, USAID director Bastiaan Schoutten's reminder that Peacock was not engaged in rural development but was in fact a "god damn missionary" illustrates a related connection between development modernization and faith-based mission work. This chapter has suggested that the divide between the latter two was not always clear cut in colonization zones in the Bolivian lowlands, where the activities of missionaries, relief workers, and development workers blurred the boundary between ministering and administering along with other divides such as those separating university and mission field and agrarian radicalism and bureaucratic authoritarianism. Although their particular history has been subsumed by the activities of the hundreds of NGOs that began to operate in Bolivia in the 1980s and the continued rapid growth of San Julián, these faith-based development workers, as proxies for an absent state, were central to the expansion of lowland colonization, providing a surprising strand of continuity as Bolivia moved from revolutionary to reactionary regimes.

CHAPTER FIVE

A Sort of Backwoods Guerrilla Warfare

Mexican Mennonites and the South American
Soy Boom, 1967–Present

It seems to us an illusion, but here we see that, incredibly, all is possible
even in this age of the atom, the twist, psychedelics and LSD.
—"The World of the Mennonites," *Excelsior* (Mexico City), May 27, 1968

[Old Colony] Mennonites have a proven means of economic and cultural
survival. Their existence is sustained at the culturo-ecological, agricultural
stage just inside the borders of the industrial threshold. Be it the pioneering
environment of the Ukraine, Manitoba, Mexico, or Bolivia, the Bolivian
Mennonites have successfully eluded the encroachment of industrialization . . .
through a series of immigrations. Always they were successful in acquiring
a new habitat to continue their secluded agricultural way of life. Whether there
are remaining places in this world for future migrations remains to be seen.
If there are such "vacant niches," we can be assured the Old Colony Mennonites
will find them.
—Menno Wiebe, "First Draft Bolivia Report," March 20, 1975,
MCC-Bolivia Files, 1975, MCCA

In the densely forested region to the south of Santa Cruz de la Sierra, two independent parties of foreigners were on the move on January 18, 1967.
Accompanied by Bolivian guides, both groups were scouting the area. The
first, a small party of revolutionaries, led by Ernesto "Che" Guevara, had recently arrived in a transitional zone (or ecotone) where the last mountain es-
carpments on the eastern edge of the Andes overlook the plains of the eastern
lowlands. Slowly establishing their forest base, they were surveying the surrounding land and making occasional covert supply runs to nearby Camiri
and Santa Cruz de la Sierra. On that particular day, Guevara's plans were cut
short by heavy rain. His diary report notes, "Started out cloudy, so I did not
inspect the trenches."[1] Two members of a scouting party, aliases Braulio (a
Cuban) and Ñato (a Bolivian), arrived in camp with the news that the rest of
the group would not be returning that night because Inti, another operative,
had lost his rifle and been badly bruised after a fall in a flooding river.

Unbeknownst to Guevara, the second party of foreigners a short distance below him on the forested plains was also affected by the rain. Led by Oscar Rivera Prado, a member of the Bolivian Institute of Colonization, this scouting party included three Mexican Mennonites—Peter Bergen, Peter Fehr, and Abraham Peters—and their topographer Mario López. In search of vacant arable land, the five men traveled by jeep south from Santa Cruz along the old highway to Camiri. Their route paralleled the rail line from Santa Cruz to the Argentinean border. Along the way they passed the vestiges of the region's expanding oil industry—periodic exploration roads (*brechas*) that intersected the main road every few kilometers. At brecha 5.5, occasional small farms and ranches gave way to dense bush extending over an area of 60,000 hectares. This was the land the Mennonites had their eye on. However, as the party reached brecha 13, it started raining and the dirt road quickly turned to impassable mud. Slowly heading home after completing a sixty-five kilometer excursion, the group unanimously agreed that the land possessed excellent prospects for farming.[2] Several days later, on January 23, while Guevara tended to a feverish member of his party suspected of suffering from malaria, Rivera, accompanied by two other Mennonites—Isaac Fehr and Francisco Wall—returned to finish their aborted inspection.[3] This time they conducted a more detailed survey of the land, noting the quality of the forest, the presence of several species of trees suitable for construction, and occasional grassy clearings in the bush. With improved weather conditions they reached the settlement of Mora, nearly ninety kilometers from Santa Cruz, before returning home.[4]

To the reader it might seem little more than an intriguing coincidence or a convenient hook to begin this chapter with the simultaneous presence of the two parties of foreigners, Mexican Mennonites and Cuban revolutionaries, in Santa Cruz in early 1967. Yet, it is worth exploring as more than happenstance. The logic that drove Guevara from Havana to the Bolivian frontier in the late 1960s was, in a sense, shared by the small Mennonite exploratory commission from the state of Chihuahua in northern Mexico. Both groups were guided by an underlying environmental rationale that viewed the region's dense undeveloped bush as a fugitive landscape rife with opportunity just as they had once viewed Cuba's Sierra Maestra and the eastern valleys of Chihuahua's Sierra Madre, respectively.[5]

Reflecting on Bolivia's lowland landscape back in 1942, residents of Santa Cruz had invoked a fragile territoriality, that "the uncultivated bush [and] the absence of roads, of railways and of population are an assault on our nationality."[6] Mennonites and Cubans understood state sovereignty in similar terms but invoked it to opposite ends. For Guevara, the transnational revolutionary, frontier underdevelopment offered a unique opportunity to establish a guerrilla force that could capitalize on the lack of surveillance and control in a region where the state was absent. Enshrined in his 1961 work, *Guerrilla Warfare*, it was a logic replicated by a range of guerrilla groups across Latin America in the following decades. For transnational Mennonite farmers like Bergen, Fehr, and Peters, whose coreligionists had already settled in frontier regions of Mexico, Paraguay, Belize, and Bolivia over the past half century, and whose parents and grandparents had engaged in similar pioneering on the prairies of Canada and the Ukraine, frontier underdevelopment afforded a similar opportunity for autonomous action. In their case it involved extracting generous concessions for cultural, religious, and social practice from the state based on their conversion of borderlands into breadbaskets.

Given that the regional development engendered by Mennonite settlement often eroded the basis for those privileges and exemptions, this was a tenuous balance. In an introduction to a 1969 study of Old Colony Mennonites in Mexico, University of Chicago sociologist Evert C. Hughes referred to them as fighting, "a battle . . . a kind of backwoods guerrilla warfare against an ever-expanding world."[7] Steeped in the assimilationist perspective on ethnicity that he and his colleagues at the Chicago School of Sociology would pioneer, Hughes harbored little hope for the long-term survival of traditional, ethnic communities like the Mennonites.[8] Yet it was the Mennonite's *guerrilla* vision, and not Guevara's, that ultimately prospered in Bolivia. Within the year, Guevara's party was wiped out by U.S.-trained Bolivian rangers, whereas over the following decades Mexican Mennonite settlement in Bolivia expanded dramatically, converting the bush land south of Santa Cruz and much of the eastern lowlands into prosperous farming colonies. As another more sympathetic commentator pointed out, in their search for the "vacant niches" of the world, Mennonites were capable of "revolutionizing native economies."[9]

This chapter begins by exploring the ways that arriving Mexican Mennonites were perceived by the Bolivian press, government officials, and their religious brethren in the Mennonite Central Committee as they established themselves on their new lands in the 1960s and 1970s. Throughout this period, I emphasize a persistent contradiction—namely, the impressive and varied forms of mobility that were often invisible to these commentators even as they made Mennonite settlement, and subsequent expansion, possible. This much is evident in backroom discussions with lawyers and government officials, titling processes recorded by the agrarian reform, import registers, personal production data, and farmers' recollections (through oral histories and diaries) of long voyages, bad debt, and bad weather. These sources reveal a

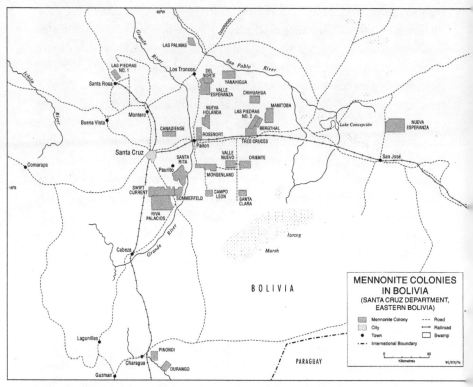

Map of Mennonite colonies in lowland Bolivia by William Schroeder. William Schroeder and Helmut Huebert, *Mennonite Historical Atlas*, 2nd ed. Collection donated to Mennonite Heritage Center, Winnipeg, Canada. Used with permission.

community performing an image of productive but humble agrarian isolationism even as some of its members led remarkably transnational lives.

During that initial period, Mennonites were still model farmers in search of a model crop. In previous migrations to Russia and then Canada they had established themselves as large-scale wheat farmers. Arriving in Mexico in the 1920s, Mennonites turned to traditional crops like corn and beans while pioneering dairy production in the region. The latter would become their mainstay during the decades-long midcentury drought, and they would transplant their dairy industry, along with the Holstein cattle it depended on, to semitropical Santa Cruz. However, in Bolivia, Mennonites' success also depended on a novel crop—a temperate-zone migrant like themselves—that would become their star commodity. The second half of the chapter centers on the role of Mennonites in the South American soy boom. As it spread over former forested regions of Brazil, Paraguay, Argentina, Uruguay, and Bolivia, that humble bean would profoundly alter the economy and ecology of the South American interior. Soy transformed landscapes and livelihoods even as an economic crisis took hold of Latin America, and the experiences of Mennonite soy producers (*soyeros*) provide a unique ethnic perspective on the broader regional contours of hyperinflation, the debt crisis, and the turn to neoliberalism.

These phenomena, while emerging across the hemisphere, were particularly extreme in the Bolivian case. Furthermore, the rapid privatization of the country's large public sector—a legacy of the 1952 revolution that had been maintained by the authoritarian governments of the 1970s—came with a painfully ironic twist. It was none other than the former leaders of the Nationalist Revolutionary Movement, and Victor Paz Estenssoro himself, that oversaw the privatization of the revolution's defining institutions—from the state mining company to the Bolivian Development Corporation. Although neoliberalism signaled the abandonment of the statist legacy of the 1952 revolution at a national level, in Santa Cruz, the neoliberal turn resulted in continuity as well as change. This was most apparent in the intensification of the revolution's enduring policy: The March to the East. As the highland mining economy crumbled in the mid-1980s, the frontier once again beckoned as a space free from Bolivia's economic turmoil. As I show, Mexican Mennonite producers were well positioned to capitalize on the disconnect between national crisis and regional boom.

"We Did Not Invite Them in Order to Socialize"[10]

The land that Rivera and the Mennonites surveyed in January 1967 would become the site of two of the largest Mennonite colonies in Bolivia, Riva Palacio and Swift Current. These names referenced the Mexican and Canadian migration histories of Old Colony Mennonites. Between 1967 and 1969, the first 288 Mennonite families, totaling 1,329 individuals, arrived in Bolivia. Some took chartered planes directly from Mexico City, but one group of 167 followed an exhausting route of two months over land and water. This included traveling by bus and train through Mexico and Central America before boarding an Italian steamer in Panama that passed through the canal before skirting Colombia, Ecuador, and Peru on its way to the Chilean port of Arica. From there, Mennonites boarded a train to La Paz, where they again disembarked and finally, took another long bus ride to Santa Cruz.[11]

Unlike the small-scale migration of Paraguayan Mennonites in the previous decade, these new arrivals gained the immediate attention of the Bolivian press who were on hand when the beleaguered group arrived in La Paz on

February 28, 1968. In an article in *El Diario* the following day, "campesinos Nemonitas [sic]" the paper reported on the curious scene, noting that there were already six hundred colonists settled in Santa Cruz and three hundred more to arrive shortly. They are a "working people" of "healthy customs" the reporter claimed, noting that Mennonites had paid for state lands and brought machinery and capital. Turning to the migrants' appearance, he described women who "cover themselves to the ankles in long dresses, preferring dark color [and] long blonde braids tied back tight against the neck," and men that, "seem like a Texas rancher."[12] While *El Diario*'s coverage was generally positive, this gendered distinction revealed a tension that would appear frequently in the evolving relationship between Mennonites and the Bolivian public. Were Mennonites modern farmers, as the "Texas" reference surely implied, or were they a traditionalist, and perhaps disconcertingly backward people, evident in the retrograde fashions of their women? With their long dresses and tightly woven braids, the latter maintained a style of dress closer to indigenous Andean women than *mestizo* elites. Elsewhere, the author noted that Mennonites "consider themselves German," positioning the new arrivals as the European migrants Bolivia had been unsuccessfully courting for decades.[13] Yet this implied desirability, with its racialized association of whitening, was tempered by the news that the Mennonites were establishing a colony "where they will be contained without the need to interact with the eastern population."[14]

These muted questions reappeared with more force, and from distinct regional and national perspectives, in two subsequent dueling articles that actively debated the racial, economic, and social merits of Mennonite migration. The first, in Santa Cruz's *La Crónica* was entitled, "What Benefits Does the Mennonite Immigration Bring?" The author denounced the settlement plans and claimed (incorrectly) that Mennonites were expelled from Mexico for disobeying the law and that their lands had been turned over to Mexican nationals. "As always," the author continued, "the Department of Colonization has talked to us of 'desirability,'" sarcastically linking the current migration to previous plans to bring Koreans into the country.[15] This was followed by a series of (correct) assertions that Mennonites, "form separate nuclei within our country, refuse to speak our language, and shun us Bolivians," as well as that they "refuse military service, dress in their own way and . . . do not attend Bolivian schools."[16] The article also alleged that the land in question had previously been occupied and ended on a dire note, warning that soon, "Bolivians will not own a single inch of our native land."

Two weeks later, Roberto Lemaitre, director of the INC, fired back in an editorial in *El Diario*.[17] He also took the opportunity to defend Japanese migrants who faced similar nativist attacks, albeit in a curious fashion. His defense of the Japanese was steeped in a deeper tradition of racial pessimism in Bolivia dating to works like Alcides Arguedas's *Pueblo enfermo*.[18] Noting that his country was "basically composed of Indians or mixtures in which the Indian blood is predominant," he concluded bleakly that "the argument that the Bolivian race could be made worse by mixing with other races can hardly be accepted," going even further to stress that "there is no actual Bolivian race" and hedging that assertion with the claim that even if that were not the case, the small Asian immigration, "could never produce fundamental changes in the racial, mental, and spiritual characteristics of our people."[19]

Turning to the Mennonites, ostensibly deemed *white* by the editorialist, Lemaitre acknowledged that religious and cultural rather than racial issues were the problem. As with his support for the Japanese, his defense of Mennonites moved simultaneously in two opposing directions. On the hand, he assuaged fears by downplaying the extent of the migration. "At best, or at worst as some people may say," he explained, "as many as 2,000 families with a total of 12,000 people may be authorized . . . what influence could 12,000 have on 4 million . . . when one considers Bolivia's potential ability to support 100,000,000 inhabitants?"[20] On the other hand, he emphasized the degree to which Mennonites, and their capital, could transform Bolivia and proclaimed, "this is only the beginning of an immigration which we have reasons to believe will gain momentum day by day."[21] Bolivian citizenship was also linked to mandatory male military service—in the past settlers had often been drawn from the ranks of young conscripts—and Lemaitre faced objections about the exemptions that pacifist Mennonites had received.[22] In response, he invoked an idea of agrarian citizenship that Mennonites would put forward repeatedly over the following years. By feeding the nation and the army, and thereby freeing more men to go to war, Lemaitre claimed that Mennonite farmers supported the military strength of the nation even if they refused to serve. He drew on the example, a bitter one for many Bolivians, of Mennonite settlers in the Paraguayan Chaco, where their position proved crucial for Paraguay's defeat of Bolivia during the Chaco War. He explained that it was this reputation as key frontier producers that had led "Paraguayan civil and military authorities to encourage Mennonite colonization in the Chaco," in the 1920s.[23] As to the cultural characteristics of the migrants, Lemaitre insisted Mennonites were "by tradition, but also by religion" bound to the land and none had ever abandoned the countryside to take jobs in the city. This was an accusation that was continually evoked against potential Jewish, Japanese, Korean, and Chinese immigrants in Bolivia.[24] "What does it matter if they use their own language (while learning ours), that they wear their own kind of clothing or even that . . . they do not assimilate" he scoffed, "we did not invite them [here] in order to socialize with them, . . . the important thing is that they work and make our land produce."[25]

The dueling articles in *La Crónica* and *El Diario* provide a differing perspective on Mennonite settlement and ideas of race, national sovereignty, and modernity. On one hand, Bolivians might choose to criticize the extension of considerable privileges to foreigners. They might challenge the modern quality of Mennonite migrants and their racial desirability as Europeans by pointing to backward cultural traditions evidenced by the long dresses and head coverings of Mennonite women. On the other hand, they could follow Lemaitre and dismiss those claims by insisting that the only *culture* that mattered when it came to Mennonites was a culture of agrarian production evident in a singular resume of frontier development in Russia, Canada, and Mexico. As with other nativist responses to immigration, the polemic did not necessarily reflect the process. Two years before the editorials appeared, Mennonites and the Bolivian state had quietly completed a series of land negotiations and the contentious set of privileges that *La Crónica* condemned had been passed back in 1962 as Supreme Decree 06030. Intense behind-the-scenes negotiation had preceded the inflammatory, but ultimately fleeting, public discourse on Mennonite settlement. In those private negotiations the terms of debate were not race, culture, and sovereignty but hectares, investment, and agricultural knowledge.

In accord with Bolivia's Agrarian Reform Law of 1953, Mennonites were required to justify the social-economic function of their settlement, and although they may have been silent on how they fit into the racial and ethnic hierarchies of Bolivia, Mennonite delegates excelled at framing their migration in agroeconomic terms. On February 10, 1967, only one month after Rivera and the Mennonite delegation had surveyed the land, the five Mennonite delegates presented the Institute of Colonization with a "Minimum Plan of Work" for Riva Palacio.[26] The plan included a schedule for the immigration of 395 Mennonite families and the potential for more to come. Entries would be staggered, with the immediate arrival of twenty-five families and thirty more to arrive every two months. "Our colony works together" they wrote, while explaining that early arrivals would clear roads and open areas for subsequent settlers.[27] What followed was an impressive list of capital to be invested by the colony in houses, schools, well drilling (for which the colony had contracted the oil company, Equipetrol) land clearing, wagons, furniture, tractors, tools, and other goods. Each item was carefully monetized,

and the delegates claimed a total minimum investment of 5.7 million bolivianos (Bs) or just under half a million U.S. dollars. As modern farmers, the
colonists also promised to build a "center for agricultural experimentation,
research and meteorological study," with the aim of providing a detailed ecological study of the soil and the prospects for farming.[28] These Mexican dairy
farmers also noted their special interest in animal husbandry, and promised
that within three years, once pastures had been established, each family
would have at least ten head of cattle.

At the tail end of the "decade of development" when U.S. funding for colonization initiatives was waning under Nixon, the financial weight of the Mennonite plan bears emphasizing.[29] The proposal was also distinct from early
small-scale Mennonite migration from Paraguay in the 1950s, where colonists
arrived with scarce capital for large-scale farming and land clearing. This
was apparent to some members of the INC who wrote to director Robert
Lemaitre expressing their excitement. In words that would provide fodder
for his subsequent defense of Mennonites in *El Diario*, they reminded him
that "it is of vital importance to observe the investment of capital that [the
Mennonites] will make in the country and the tendency to its gradual increase
and development."[30]

While supporting the Mennonites' "Minimum Plan of Work" two members of the INC, General Valenzuela, head of the Division of Colonization,
and Epifanio Rios, head of the Department of Promotion for Spontaneous
Colonies, had some reservations about the immigration. Valenzuela felt that,
given the large number of children in the average Mennonite family, future
migration beyond the original 395 families should be capped until it was
possible to determine if the colony truly served as a "motor of development"
in the region.[31] More critically, he argued that 40,000 hectares was excessive
for the needs of the colony. He proposed a dramatic reduction to only 17,650
hectares with individual families receiving parcels of ten, thirty, and fifty
hectares according to the quality of the land and the distance from the Argentine railway on the colony's western flank. He also advocated a higher price
for land—up to 500 bolivianos per hectare along the railway and 150 in the
interior—than the 100 bolivianos per hectare proposed by the INC, bringing the cost to over US$40 per hectare. Although still a bargain by North
American standards, it was high price for undeveloped land in Bolivia, making the standard fifty-hectare Mennonite plot that the colony proposed cost
upward of $2,000.

In addition to the restrictions advocated by Valenzuela, INC member
Epifanio Rios challenged the claims put forward by the agronomist and

topographer Mario López who Mennonites had commissioned to provide a survey of the property in question. While López had confirmed that "no campesinos were within the area," Rios asserted that a portion of the land was currently occupied, a point that one dissenting INC reader of the original document (perhaps the sympathetic Lemaitre or the Mennonite booster Carlos Zambrana) had highlighted and then scrawled "no" alongside of.[32] Rios also accused the colony and their topographer of deliberately distorting the size of the parcel, insisting that it was in fact 60,000 hectares rather than the 40,000 they claimed. In a country without a cadastral survey in which petitioners generated land documents for the state, this was a common habit, Rios pointed out, in the request for public lands. It allowed the petitioner to pay less tax and usurp state land, later claiming the boundaries indicated on their inaccurate sketches as inviolable.[33]

Beyond these issues, Rios went on to cite a series of advantages of the lands in question including their proximity to Santa Cruz, the railway and the old road to Camiri and the Río Grande, and the fact that the entire area was intersected by exploration roads developed by the oil company. These all combined to make the site "one of the most advantageous in the country" far better than colonization zones in the Alto Beni.[34] If, Rios concluded, the INC sold these lands to the Mennonites, once again, prime land, would be "in the power of a few hands [resulting in] the sad situation of being unable to give quality lands to the mass of campesino farmers that wait to be relocated." The latter, "in their condition as nationals and owners of the lands, should have the right to larger parcels than foreigners."[35] Like Valenzuela he advocated small parcels because, "a family cannot attend to more than twenty-five hectares" and would have to bring in workers, "that would place the [Mennonite] colonist in the category of the *patrón*." "In Bolivia there are already too many *patrones*," he concluded, "and it would stand against reason to import more."

Lemaitre passed the objections on to Carlos Zambrana, the INC official directly responsible for the Mennonite settlement. Zambrana might have challenged Rios and Valenzuela on the question of parcel size pointing out that the recommendation for smaller lots failed to appreciate the degree to which Mennonite settlers were not arriving as small-scale pioneers who might carve out a few hectares for subsistence but as highly mechanized producers of cash crops. Instead he chose to simply remind Lemaitre that the lands in question, while possessing quality soil, lacked water, and "that only businesses with significant capital can settle on them with success."[36] The case carried on into mid-1967 with INC official Armando Torrico recommending a new survey of the zone. Despite the suggestions for an increase in land price to 300 Bs, the subsequent study vindicated Zambrana and the INC set the land value at 20 Bs per hectare with Mennonites agreeing to pay a final price of 800,000 Bs.[37]

The Mennonite Embassy

With their land purchase confirmed, the Mennonite delegates began to prepare for the mass migration of their coreligionists from northern Mexico. In addition to contracting for well drilling, they began to purchase construction supplies from local companies and contract drivers to travel back and forth from the colony. As word spread in Santa Cruz de la Sierra that the Mexican Mennonites had money to spend, enterprising cruceños flocked to the only identifiably Mennonite institution in the small city—the headquarters of the relief agency and NGO, the Mennonite Central Committee. Understandably, they were ignorant of the profound denominational differences between Mexican Old Colony Mennonite settlers and young North American Mennonite volunteers. For cruceños, the MCC offices logically appeared an informal "Mennonite embassy" for these new arrivals. As Arthur Driediger, head of MCC-Bolivia, explained in an exasperated letter sent back to the head office in Akron, Pennsylvania, he had already spent considerable time, "telling the people that these immigrants, though they are called Mennonites, are coming to Bolivia under their own sponsorship."[38]

Like the aforementioned editorialist and INC representatives, MCC officials had mixed feelings about this new migration. By 1968, MCC was involved in numerous aspects of colonization and development, including the flood relief activities described in the preceding chapter, in addition to its small health clinic in the first Paraguayan Mennonite colony. The sudden arrival of thousands of Mexican Mennonites in Santa Cruz raised the possibility of a new sphere of operation for the NGO. But these traditionalist colonists were rightfully suspicious of MCC initiatives given that in Mexico they had come bundled with aggressive evangelical outreach programs targeted at converting members of the Old Colony faith. With the potential for conflict as well as cooperation, MCC officials were slow to approach Mexican Mennonite colonists. Yet they maintained a close watch on these new arrivals thus making MCC archives a detailed, if problematic, source for Mexican Mennonite settlement. MCCer Elwood Schrock was assisting Methodists and Maryknolls in the flood zone around Montero in 1968, but he took time to dutifully translate the editorials about Mexican Mennonites that appeared in *El Diario* and *La Crónica* into English before passing them on to his superiors. The quiet but obsessive cataloging of these activities resonates in striking ways with the activities of USCAR officials that also translated newspaper articles regarding Okinawan dissent and made quiet inquiries about the opinions of Bolivians regarding the new arrivals.

The awkwardness of the MCC–colonist relationship was apparent in a clandestine visit that MCCer Alfred Kopp made to the new colony of Riva Palacio in mid-1968. Aware that his presence might be unwelcome, Kopp did not travel as an official MCC representative. Instead, he accompanied a German agricultural extension agent who was interested in working in the region and lied to the Mexican colonists, claiming that he was a Paraguayan Mennonite from the nearby Tres Palmas Colony. Kopp and the extension agent traveled in a jeep "through deep forest" with small clearings carved out along the roadside. With Kopp translating from the low-German spoken by colonists to high-German spoken by the extension agent, the two had a pleasant day. They found the colonists friendly and listened to their plans to begin a dairy industry and experiment with cotton that was experiencing a boom among farmers in the Santa Cruz area. Comfortable that his ruse had succeeded, Kopp decided to ask one colonist—a "Mr. Loewen"—whether the colony was interested in working with the MCC. Loewen quickly replied that "no . . . he didn't think that would work at all" adding that "some had [in the past] but that was another reason they had left Mexico." "He was obviously referring to the MCC and General Conference mission work that is going on in Mexico," Kopp concluded.[39]

Kopp's findings confirmed his fears, that future cooperation between the NGO and the colonists was unlikely, or at least unwanted, but MCC continued to express a sense of duty to "our Mennonite brothers from Mexico"— over the following years. Rebuffed in person, the organization decided that it should become involved behind the scenes at the national level to intercede on *behalf* of colonists in their relationship with the state. When Latin America director Edgar Stoesz was planning his inspection of MCC-Bolivia operations in late 1968 he explained to local reps that he would like to arrange a meeting with government officials, "to discuss the immigration."[40] Stoesz justified taking this liberty because Mexican Mennonites "were always misunderstood [in Mexico] and . . . never did enjoy a good relationship with the government," and he hoped it would "be possible for us to *interpret* the Mennonites [to Bolivian officials] in such a way that there would be more understanding [emphasis added]" confident that a brief presentation of facts could serve a "preventative" function and forestall future conflict between colony and state.[41] A few years later MCC representative Menno Wiebe noted, "the tie between the individual colonies and the government of Bolivia is tenuous and haphazard."[42] He felt that although MCC could not justifiably act as official spokesman, becoming the Mennonite embassy that many Bolivians assumed it to be, the MCC could still play a critical intermediary role. Wiebe had already been doing just that, engaging in dialogue with state officials and symbolic gestures that asserted, like Lemaitre's editorial, that Mennonites were producers for the nation. For instance, while preparing several MCC reports on colony Mennonites, Wiebe wrote to Minister of Agriculture Alberto Natusch Busch in advance of a formal meeting between the two in late 1974. He informed Busch that he had recently completed a visit to the colonies, "and [was] impressed with their agricultural progress."[43] He confirmed that they were exclusively farmers and that they had cleared thousands of hectares of bush and were producing a variety of goods for the local market. He looked forward to a more detailed discussion of the agricultural progress of the colonies and presented Busch with a "token of gratitude . . . fresh cheese" from one of the Mennonite colonies.

In addition to demonstrating the importance of Mennonite production to state officials, the MCC also attempted a more ambitious program—to transfer Mexican Mennonite production technology to surrounding communities. "The colony Mennonites have much to contribute" wrote Menno Wiebe in 1975 only two years after E. F. Schumacher's *Small is Beautiful* popularized the notion of "appropriate technology" in development.[44] "They are a local source of intermediate technology . . . which is useful to their Bolivian neighbors."[45] After all, Mexican Mennonites had arrived in Bolivia with the promise that they would become model farmers for surrounding communities. The minimal plan of work crafted by Mennonite delegates in 1967 promised an agricultural research station would be constructed in the colonies.[46] A follow-up survey in late 1969 by the INC—a condition of obtaining full legal title to the land—noted with satisfaction that the colony had "broadly satisfied its promise" in most areas. However, no research center had been constructed and officials requested its speedy conclusion, although they extended full title to Riva Palacio despite this omission.[47] Without it, how were Bolivian neighbors to benefit from Mennonite farming prowess? MCCers, who were implementing an animal traction program in the San Julián Project, hoped that Mexican Mennonites would share their expertise in animal care, handling, and the manufacturing of harnesses and other equipment with Andean settlers. The former may have employed tractors (with steel wheels)—beyond the purchasing power of most Andean settlers—but also made regular use of horse-drawn buggies and wagons for personal transport and hauling the harvest.

The MCC attempt to speak for Mexican Mennonites and manage public perception was not unlike the role played by Jewish relief agencies in Bolivia and Brazil two decades earlier.[48] The Mennonite case was far less dire, but it was, like that earlier experience, rife with contradictions that revealed profound differences within the diaspora. MCCer John Friesen began one sympathetic report by acknowledging that he had previously thought of colony Mennonites, as "close-minded, conservative, obstinate, and cruel [and] unresponsive to outsiders."[49] Yet the MCC asserted that its objective was to transform that relationship, to "relate to the Mennonites in Bolivia and Latin America in such a way that we can invite their participation . . . and service to those in need." Whether this involved demonstrating horse-drawn farming techniques for fellow Bolivians or advocating for a cooperative marketing system that would tie together production among the nine Mennonite colonies in Santa Cruz, the MCC continually insisted on its right to intercede. After all, Wiebe concluded, colony Mennonites were a people "whose rural, agrarian culture . . . makes it difficult if not impossible for them to perform this function."[50]

Border-Crossing Bishops and Transnational Traders

Through the early 1970s Mexican Mennonites made few responses to MCC overtures. One reason was that, just as colonists had explained to Alfred Kopp during his clandestine visit, MCC activities, even economic initiatives like the creation of a co-op, always came bundled with evangelical ones. Although the MCC did not officially endorse evangelical activity, a range of North American missionaries used the institution's expanding presence across the developing world as a convenient staging ground for their own operations. By 1976, the Evangelical Mennonite Mission Church from Manitoba, Canada, was already active in Santa Cruz having purchased land near Riva Palacio colony for "young Christians" expelled from the Mexican Mennonite colonies.[51] When they did speak with MCC officials, colony leaders clearly insisted, "disunity within the church and colony is everywhere the result of [these] unwanted preachers from the North."[52]

Although this may have been the most contentious cause for nonparticipation in MCC activities, another more compelling reason was that Mennonite colonists, despite their rural, agrarian culture, were far more adept at negotiating with local and national governments than the MCC gave them credit for. Notwithstanding the image of horse-drawn buggies, restrictive clothing, and a simple farming life, colonists quietly demonstrated a remarkable cosmopolitanism. This unexpected mobility was evident whether they were challenging new taxes, engaging in the cross-border cattle trade, or jumping between the northern and southern hemisphere to import farming equipment. A select few mobile entrepreneurs engaged in migrations, negotiation, and route-building that allowed the diasporic community-at-large to benefit.[53] Those practices, already established in Mexico where Mennonites had engaged in seasonal labor migration and equipment importing during the midcentury drought, emerged in the Bolivian case in a range of sources, from diaries, to government papers and oral histories.

The diary of Mennonite bishop Johan Wiebe, a founding member of Riva Palacio colony, highlights these practices. Covering a period from 1966 to 1983 his account is an exhausting (and exhaustive) chronology of travel and meetings. Wiebe's work began in Mexico where he and other colony leaders made the 1,500-kilometer bus journey from Chihuahua to Mexico City dozens of times to arrange travel documents and logistics. In the city they would usually check in to the Hotel Principal (a few blocks from Mexico City's Alameda) and the following day make inquiries at the SRE, the Bolivian consulate, and shipping companies and airlines as they attempted to arrange documentation and travel for groups of several hundred individuals with both Canadian and Mexican citizenship. After only a few days they would board buses back to the colonies in Chihuahua.[54] At one point in 1967, frustrated by an endless string of delays at the Bolivian consulate in Mexico City, Wiebe and several other colony leaders circumvented the bureaucratic snag by boarding a multileg flight through Colombia, Ecuador, and Peru to La Paz. There, with help from Carlos Zambrana, they received promises that their paperwork would be expedited directly from the Ministry of Agriculture and the Treasury. After only three days in La Paz, they were back on another grueling trip to Mexico City where they returned to the Bolivian consulate and presented their guarantees to astonished officials before taking a bus back home to Chihuahua. Their interventions were successful, and Wiebe would eventually guide the colonists that arrived in La Paz in March of 1968 and whose exploits by bus, boat, and train appeared in *El Diario*.

The travels of Wiebe and other Mennonite leaders on behalf of the colony did not end with their arrival in Santa Cruz. In the absence of a national cadastral survey, land transactions were rarely straightforward in Bolivia, and the purchase of Riva Palacio was no different. As colonists were arriving on their new land, they soon discovered that the property was barely more than 30,000 hectares, not the 40,000 they had purchased. Having uncovered the error, Mennonites balked at the INC's request to pay for a new survey, communicating that, "the colony cannot pay two times for the same work." With help from Carlos Zambrana they were soon back in the INC offices, adding to their growing land case file, as they petitioned the government to reduce the sale price by nearly 200,000 pesos.[55] Two years after the purchase, the Mennonites received an official agreement to recognize the lower amount. Scarcely had they succeeded when a wealthy cruceño arrived in the colony claiming that he held deed to more than 5,000 hectares of their land. Again, Wiebe and others were called into action in a dispute that lasted an entire year.

Throughout the 1970s, Wiebe and others continued their negotiations of behalf of the colony and succeeded in having sympathetic officials intervene on their behalf. In Bolivia, Mennonites benefited from a clause in their 1962 Privilegium that guaranteed exemption from import duties on agricultural equipment and personal effects. With demand for imported livestock and machinery among new colonists and the continual arrival of more migrants from Mexico, this was a privilege they guarded carefully; when a 1969 decree appeared to revoke that privilege, colonists went to the INC officials again for help. As the head of the INC, Robert Lemaitre wrote to José Luis Roca, the Minister of Agriculture in March of 1970 on behalf of Mennonite settlers. He reminded Roca of the 1962 decree guaranteeing "importation of materials and diverse equipment designed to help their settlement in the country." Secondly, Lemaitre informed Roca that the representatives of Riva Palacio, Sommerfeld, and Swift Current, the three largest Mennonite colonies in the country, had personally come before him to let him know of the potential conflict with two recent 1969 decrees relating to import duties. According to Lemaitre they feared that with the annulment of the right to free imports, other conditions of Supreme Decree 06030, such as freedom from military service, might also come under threat. He included a copy of the decree as well as a report indicating that Mennonites had fully completed the "Plan of Work" they had proposed for the region.[56]

Even as they attempted to work through the sympathetic Lemaitre, the bishops of the Mennonite colonies also petitioned the president directly. They began by thanking the president "for the freedom of religion that the previous government, in total trust, has given to us."[57] They explained that they felt, "a deep gratitude to God and our government" and that in each religious assembly they prayed for the Bolivian government and for their continued success, "in this country where we live." They quoted Jeremiah 29:7, to emphasize their commitment to the mutual success of the colonies, Santa Cruz, and the nation. The passage, "Seek the welfare of the city where I have sent you into exile, and pray to the Lord on its behalf; for in its welfare you will find your welfare."[58] This summarized, not only the Mennonite sense of self as diasporic or *exiled* community, but also the way in which as *outsiders* they might justify their conspicuous presence through their role as regional producers. In their letter, the bishops nimbly turned from theological justifications to legal ones by reminding the government of the promises they had been given for free import. Although they were quick to point out that, "it is not against our faith or conscience" to pay the tax, they framed the issue as critical to their continued immigration and explained that many more Mexican Mennonites wished to come to Bolivia, but the cost would be prohibitive were they charged for their personal goods. Whether it was Lemaitre's intercession or the bishops' letter, the plea was successful. In June of 1970, President Ovando signed a decree granting another year of exemption with the justification that the Mennonites had "increased the agricultural productivity" of the zone, "introduced new methods and techniques of cultivation," and that the prior guarantees should be respected to ensure future immigration and maintain the good faith of the state.[59]

Wiebe and the bishops also intervened in other ways on behalf of their fellow colonists. In 1971 they worked with bank officials and local authorities when a colony man was jailed for failing to pay his debts.[60] They made a series of trips to Cochabamba and La Paz in 1972 and again in 1975 when it appeared they would have to pay additional fees to register their children and receive identification cards. They also actively negotiated over the Mennonite Privilegium when the Bolivian consul in Mexico City attempted to charge fees for processing migrant visas and more critically when a 1975 supreme decree by President Hugo Banzer threatened to strip Mennonites of all their privileges under Bolivian law. While Wiebe often writes as a reluctant representative and hoped to leave his official position with the colony in 1977, he continued to work into the early 1980s until the decree was officially repealed. In the meantime, from 1975–82, he and other Mennonite leaders continuously negotiated with Bolivian authorities to make sure that most of its provisions were never implemented.

In addition to fighting for existing guarantees, Mennonite leaders were also investigating new land purchases for incoming colonists in those same years. In 1974, Wiebe inspected lands in nearby Camiri, and in 1976 he helped a colony of Belizean Mennonites purchase the land to found Nueva Esparanza colony in eastern Santa Cruz department. In this way Mennonites became familiar sights at El Trompillo airport in Santa Cruz as they boarded planes to La Paz, checked in to the Hotel Sucre, and traveled up and down Avenida

Camacho visiting the Ministries of Agriculture, Migration, and the Agrarian Reform offices with their Bolivian lawyers. Although the activities described involved almost incessant negotiation with a range of officials, rarely do we find details of the defenses employed in struggles over taxation, privileges, and land. Rather Weibe's diary reads as a dizzying, or perhaps stupefying, succession of times and places in which all activities from flagging down a cheese truck bound for Santa Cruz to the delay in boarding a flight to La Paz or breakfast at the Hotel Sucre in La Paz stand next to the precise time the delegation arrived at the Ministry of Agriculture or the number of hours they waited for an absent official. It is bereft of the compelling justifications we might find in other sources, like the editorials on Mennonite migration in *La Crónica* and *El Diario* or INC memos, but Wiebe's relentless travels are worth recounting in detail because they demonstrate the impressive delegation of mobility by a community for whom most members rarely traveled beyond nearby Santa Cruz. For the average colonist, MCC statements about the incompatibility of "rural agrarian culture" and high-level bureaucratic negotiation might have made sense. Yet most Mennonite colonists did not have to board buses, trains, and planes and make innumerable trips to embassies, ministries, and lawyer's offices precisely because a select few individuals, like Wiebe, did so on their behalf.

While their bishops and *vorsteher* were busy negotiating with state officials, other Mennonite colonists were crossing national borders and jumping between hemispheres to supply the nascent colony with the improved stock and agricultural machinery that would enable Riva Palacio to become a key regional producer. The colony man whose release from prison Wiebe had secured in 1971 was Johan Guenther. In his diary, Wiebe never explains how Guenther had accumulated such a large debt with the national bank in such a short time. The answer was likely apparent to him and other Mennonites who knew Guenther as an enterprising cattle importer. In their 1967 Plan of Work and in conversations with MCC officials, Mennonites had made it clear they hoped to establish a dairy industry in Santa Cruz like the one they had left in Mexico. Yet the extensive ranching techniques and *creole* cattle found in lowland Bolivia were poorly suited to the task with a daily production of a paltry 2.5 liters. A study by the British-funded Center for Tropical Investigation in 1969 found that of 130 small dairy operations within forty-four kilometers of Santa Cruz, only ten boasted any Holstein or Swiss cattle (breeds capable of producing more than ten times the amount of creole stock).[61] In the absence of local stock for sale, the entrepreneurial Guenther stepped in. His activities are evident in Ministry of Agriculture archives where Guenther's name appears in the ledger alongside large-scale importers who were bringing thousands of *zebu* beef cattle from neighboring Brazil and Okinawan migrants who were importing cotton seed from the Alexandria Seed Company in Louisiana. In 1968, the Ministry of Agriculture had authorized Guenther to import 500 heifers and twenty bulls of *holandés* or Holstein breed from Argentina subject to a sanitary inspection at the border. He received an extension to continue importing cattle in 1969.[62] The following year Guenther also received permission to import up to 1,000 horses for his buggy-traveling brethren. He was not alone. Two other colonists, Peter Fehr and Wilhelm Martens, were authorized to import horses and cattle (both Holsteins and Brown Swiss) from Argentina in 1969.[63]

Along with ministerial records, oral histories also provide evidence of this vibrant cross-border trade. Enrique Siemens was a young boy of eight when his family arrived in Bolivia in 1968 with little to their name. His strongest memory of that first year in the bush is drinking canned milk diluted with water because there were no dairy cattle in the colony. The following year his father went with a neighbor to Paraguay and helped him bring back a load of cattle, crossing the Chaco by truck on an old wartime road. "They went and returned in 40 days," Siemens remembers, and in exchange for his help his father received the family's first cow. "After that we were happy, then we had milk [and] my father knew about the business of selling cattle . . . they went again and this time he got two cows out of it."[64] In total, Siemens's father made three excursions to Paraguay. His wealthy neighbor continued to travel to Paraguay (and likely to Argentina) to purchase cattle for sale at auctions in the colony. This transborder trade soon resulted in a thriving Mennonite dairy industry, evident in pastures full of imported Holstein cattle and blocks of *queso menonita*—like the one that MCC representative Menno Wiebe would present to Minister of Agriculture Natusch Busch in 1974 as evidence of Mennonite's success.

For those transnational pioneers that engaged in it, the cattle trade could be both lucrative and risky. Guenther had to request an extension from the Ministry of Agriculture in 1969 noting that he had only been able to purchase eighty-one heifers and one bull.[65] Two years later he was imprisoned for failure to pay debt and ordered to auction off his possessions by the court. Guenther's creditors planned to sell his imported livestock for a paltry 10 pesos a head, but Wiebe arranged an internal colony auction in which attendees paid an elevated 230 pesos per animal, providing funds to pay the entirety of Guenther's debt. It was a colony-based debt-relief strategy Mexican Mennonites would repeat nearly two decades later but on a much broader scale.

Mennonites were not only importing livestock into Santa Cruz from Argentina and Paraguay. Next to the record of Guenther's import authorization we find permissions for colonist David Reimer who was authorized by the Ministry of Agriculture to import two 1950s Allis-Chalmers combines and two Case corn harvesters.[66] Although not specified in the file, much of this machinery was likely brought in from as far away as the United States, building on a long-standing international importing business practiced by Mexican Mennonite farmers. Siemens remembers his father purchasing their first tractor, an old two-cylinder model, from Peter Friesen. Along with his brother Johan, Peter Friesen had been traveling between the United States and Mexico importing older machinery for use in the Chihuahua colonies for years. In his 1969 study of Old Colony Mennonites in Mexico and Belize, sociologist Calvin Redekop noted that such practices were a veritable "[cottage] industry prevalent in all the colonies," with at least twenty individuals who made anywhere from two to ten purchasing trips a year.[67] The Friesens and other border-crossing Mennonite importers benefitted from the rapid technological change and consolidation of farming in the U.S. in those years.[68] Many U.S. farms modernized and consolidated, a process known as the Great Disjuncture. A host of derelict machinery, considered obsolete on the post-war U.S. Great Plains but desirable in Latin American frontier contexts where manual labor and draft animals were the norm, awaited these eager entrepreneurs.

The Friesen brothers brought this acumen to Bolivia. Johan Friesen's son-in-law Johan Fehr remembers his father-in-law as "very smart. He spoke English, was born in Canada . . . it was his business in Mexico . . . he knew how to do that . . . he always brought machinery to Mexico. When he arrived here and figured out what we were lacking he would travel to the U.S., take apart the machinery, put it in wooden boxes . . . and ship it down."[69] In a further transnational twist, Mexican Mennonites that had left Chihuahua to settle in Seminole, Texas, in the mid-1970s provided the Friesens with a diasporic home base to scour the neighboring Great Plains in search of equipment as well as labor to build wooden shipping containers. In those first years, the machinery that the Friesens sent to Bolivia was often old and underpowered. Fehr and many other colonists recall driving two-cylinder John Deere tractors. Nearly obsolete in the U.S. market, they ran on gas rather than diesel and the machine did not do well in the humid conditions of the lowlands. Peter Wall, another Riva Palacio colonist, who had lived in Canada, Mexico, and Belize before coming to Bolivia, marks time in the colony not in years but in a succession of those machines. "[At first] we hardly did anything with machinery . . . it was rough. Then I had a chance to buy a little Allis

Chalmers . . . traded off for a Minneapolis . . . I could sell that . . . I bought a John Deere G, then I [really] started working . . . and then all of the sudden I had a chance to trade for a John Deere A."[70]

In his personal recollection that marks time by moving from tractor to tractor, Wall provides a personal technological narrative that reflects a broader theme. Through the 1970s, owing to the transnationalism of individuals like the Friesen brothers and Johan Guenther, the colony of Riva Palacio became increasingly mechanized and home to a variety of improved stock. A 1975 MCC report on the colonies framed the issue as one of Mennonites "revolutionizing native economies." Old Colony Mennonites and their Bolivian neighbors, "stand in contrast because of a different approach to the use of land and resources," the author wrote. The latter were "not market driven," but Mennonites "farm for profit," producing "eggs, dairy products, meat, and grain . . . necessitating marketing means, more land and a more successful bush clearing technology."[71] In those same years, the struggles of national colonists in Yapacaní and across northern Santa Cruz demonstrated that the issue was one of access to resources rather than immutable cultural differences. Andean settlers eagerly sought markets and profits through farming. Yet, the consequence of that differential opportunity was increasingly apparent by the mid-1970s. In a 1976 diary entry Johan Wiebe wrote, "it is ten years since we came to Bolivia and much has changed. Before half the land was wild and now has been cultivated into fields." According to the MCC, Bolivians were also becoming aware of the scope of this endeavor, and "making their observations about Mennonite colonists."

In 1978, more than a decade after Mexican Mennonite settlement began, agronomist Jesús Bolívar Menacho prepared an extensive report on "The Mennonite Colonies: Support and Participation in Regional Agricultural Production," which he submitted as his thesis at Santa Cruz's regional Gabriel René Moreno University. His findings revealed the consequences of the impressive import networks described previously as well as to the degree to which Mennonites were already established as key regional producers. "We could say that their level of mechanization is extremely high," Bolívar pointed out.[72] Even this was an understatement. According to his survey, the Mennonites of Santa possessed 1,074 tractors, an average of slightly less than one per family. In contrast, the entire remaining agricultural operations of Santa Cruz—a department with a population of approximately 300,000, only made use of 429 tractors. Mennonites were using that mechanical superiority to cultivate 9 percent of the total land under crops in the department. In addition to mechanized farm labor, the importation of cattle by Johan Guenther had led to a surging dairy industry producing 40,000 liters of milk a day. When Johan Wiebe had arrived in Bolivia in 1968, he and his family had slept in his friend Martin Epp's empty chicken coop while they cleared their land. By 1978, formerly empty coops were full to overflowing as Bolívar reported a daily production of 80,000 eggs.[73] Mennonite eggs, dairy, hogs, horses, and cattle, flowed into a regional and national market, thereby creating a constant movement between colony and city. William Dietrich served with the U.S. consulate in Bolivia in the early 1970s and clearly remembers the curious image of "Canadian Mennonites . . . selling butter from horse-drawn wagons in the streets of Santa Cruz." Johan Fehr also recounts his father's weekly travels to Santa Cruz "in the first years" to sell chickens, eggs, and cheese.[74] Riva Palacio colony soon established contracts with buyers in the city who came directly to the colony. Tellingly, in bishop Johan Wiebe's frequent trips to La Paz in the 1970s to negotiate with government officials, he and other Mennonite leaders regularly hitched to Santa Cruz on trucks leaving Riva Palacio laden with cheese.

Mennonite dairy and eggs found a ready market in a region in which production lagged far behind demand. When colonists had informed MCC worker Alfred Kopp of their plans to start a dairy industry back in 1968 he had encouraged them, noting the high cost of quality cheese in Santa Cruz.[75] Bolivian officials and editorialists also frequently bemoaned the low consumption of dairy among poor Bolivians. Milk, in their minds, was truly a revolutionary product capable of physically and culturally transforming undernourished peasants. The Bolivian Development Corporation, the institution in charge of several lowland colonization and road-building projects, also ran a milk processing plant in Cochabamba. In 1967, officials wrote of their struggle both in terms of demand, "before a market that does not possess the habit of consuming dairy," and a lack of supply.[76] They sent commissions to Santa Cruz hoping to find opportunities to increase production there. Seeking to build a model for milk production in Bolivia, Luis Barrón, head of the Cochabamba plant, traveled as far as Europe, touring dairying regions of Holland, Germany, and Switzerland and attending a dairy conference in Munich. In 1968, Jack Wuhl (an expatriate agronomist who had lived and farmed in Santa Cruz since the 1950s) wrote a letter to *El Deber* proclaiming the "superiority of milk" and insisting on the need to increase its consumption among cruceños.[77] A decade later, Mennonites, with only 27 percent of the dairy cattle in Santa Cruz, were producing 47 percent of the department's milk.[78] By lowering the cost of these once-expensive and scarce goods they seemed to be effecting a revolution in consumption among the average Bolivian.

Bolívar's 1978 report was a testament to the fact that Mennonites, as promised, were living on the land, feeding the nation, and thus fulfilling the conditions of their 1962 Privilegium and the goals of the March to the East. He
applauded the idea that Mennonites were not simply creating national self-
sufficiency in products like rice, sugar, and beef that Bolivians already consumed but also introducing new food ways among the population and
"developing nontraditional crops in the region."[79] Yet he could not help but
contrast a Mennonite "social and cultural life almost totally isolated from the
national community" with an economic relationship "through the market [in
which they had] distinguished themselves for the development of a highly
mechanized and diversified form of agricultural production."[80] The result
was an enervating paradox. The very "social religious system" that produced
Mennonite wealth "makes it difficult to understand their experience and
systems of production, and to pass on these understandings to the national
producer."[81] To add further to this irony, the most open and progressive Mennonites in Santa Cruz (those Paraguayan Mennonites that had arrived in the
1950s and early 1960s and used cars, electricity, and placed less restriction on
personal behavior) were fewer in number and far poorer than their conservative brethren from Mexico. If Old Colony Mennonites were model producers
for the region, it thus remained unclear to Bolívar how that model might be
transmitted to neighboring Andean colonists. Because of this, and because of
the dramatic growth in the Mennonite population, Bolívar concluded by recommending measures that had been discussed back in 1968—namely, further
restricting Mennonite settlement to the frontier well beyond the integrated
subregion with its easy access to Santa Cruz.

Like Bolívar, Mennonites had some reservations about their position in
the regional economy after a decade in Santa Cruz. By the late 1970s they
marketed an important quantity of eggs and dairy. But these products of
physical family labor, as important as they were for subsistence and covering
daily expenses, were meant to supplement rather than replace intensive farming of a lucrative commodity. In Russia and Canada, Mennonites had been
key producers of wheat. In northern Mexico, they farmed corn, oats, and
beans. The mechanization that accompanied their early years in Bolivia also
presupposed the large-scale cultivation of a marketable cash crop. Japanese
and Okinawan migrants had carved out a niche as high-quality rice producers
in the 1960s. Many of the Andean colonies in the north of Santa Cruz, such as
Aroma and Cuatro Ojitos, were dedicated to sugarcane, rice, and corn. By the
late 1960s, however, these markets were flooded to the extent that regional
sugar mills operated on a limited quota system and rice, not competitive

outside of Bolivia, had already exceeded national demand. Mennonites might have turned to cotton. The first Paraguayan Mennonites arrived in Bolivia in 1954 with a contract to grow cotton for the Empresa Algodonera. But with a limited market they quickly switched to corn and castor beans, neither of which were high value.[82] In contrast, when Mexican Mennonites arrived more than a decade later, the market conditions had changed. International cotton prices were rising and jumped further after 1973 when spiking global oil prices increased the cost of synthetic-based alternatives. The farmers of Santa Cruz, including many Japanese and Okinawan colonists, responded to the cotton boom by rapidly converting forest into new fields and abandoning sugarcane, corn, and rice on existing farms. Cultivation skyrocketed from 7,000 hectares in 1969 to more than 60,000 in 1972.[83] The expanding harvest generated a dramatic labor migration with more than 70,000 Andean migrants making a seasonal pilgrimage from the highlands to pick cotton along with thousands of Guaraní recruited locally in Santa Cruz. When it outpaced the abilities of labor recruiters in 1975, the army and even school children were recruited to assist in the harvest. Although they arrived at the beginning of this boom and indicated to MCC worker Alfred Kopp their interest in cultivating cotton, Mexican Mennonites were still in the early phases of colony land clearing, road construction, and well drilling in the boom years. In the early 1970s they were planting corn by hand between the stumps of their newly cleared fields. By the time significant bush had been cleared in the colony, the boom was over. 1975 signaled the zenith for cotton in Santa Cruz and for the remainder of the 1970s prices declined as quickly as they had risen. Hastily cleared and poorly managed cotton fields were quickly abandoned and in the absence of ground cover, topsoil blew away and many fields were reduced to shifting sand dunes.[84]

By the mid-1970s, Mennonites in Bolivia were thus model farmers still searching for a model crop. The first decade of their settlement had been characterized by a paradoxical degree of mobility. Mennonites took carts of eggs and cheese into Santa Cruz to supply local demand. Others traveled further hopping on domestic flights to La Paz where they were regular fixtures in the halls of government. Still others crossed borders into neighboring Paraguay and Bolivia to import livestock or took international flights back to the United States to purchase old machinery. Although often invisible to outsiders, these overlapping movements (local, national, regional, and transnational) provided the economic, legal, and technological basis for their initial success. Over the following two decades, another transnational, nonhuman migrant would further alter Mennonite livelihoods, the landscape of Santa

Cruz, and the long-term goals of the March to the East. It had already reached their fields by 1977 when Bolívar was conducting his research and found a "nontraditional crop" covering 1,300 hectares, a small but not insignificant amount of the 14,000 hectares of cultivated land, in Riva Palacio.[85] From this tentative foray, Mennonites would soon embrace soybean cultivation as the economic engine for their expansion.

Mennonites and Soybeans in Neoliberal Bolivia

"Soy! Recipes for the Whole Family," proclaimed a 1984 pamphlet produced by the National Bank of Bolivia and Radio Loyola of Sucre.[86] Its authors introduced readers to a variety of soy-based recipes so that "families, above all those with few resources, can prepare, at a low cost, whole meals of high nutritional value." The versatile bean they were pitching could replace milk, contained twice the protein of meat, and did not require refrigeration. They assured potentially nervous consumers that soy could also be refashioned into any number of reassuringly familiar foods like *humitas*.[87] Ending on a nationalist note the pamphlet reminded readers that, "soy is grown [here] in Bolivia, and can be bought in many stores and markets."[88]

Written amid Bolivia's record setting hyperinflation that saw price increases of 20,000 percent from mid-1984 to mid-1985, the authors of "Soy!" likely hoped to encourage soybean consumption as a low-cost option for an increasingly desperate populace. In the colonies, faith-based development workers were also encouraging new settlers to produce soybeans for internal consumption. The legume was indeed versatile but was rarely prepared and consumed in the intimate domestic way the authors of "Soy!" recommended. Rather, soybeans passed through an industrial system in which they were crushed, separated into oil and a cake soymeal by-product. Although they might still enter Bolivian stomachs through edible oil, and thus contribute to food security, over the following years soybeans emerged, not only as part of the *basic food basket*, but as a star agricultural export crop that circulated in crude or unprocessed form in a global economy of feedlots, biodiesel, and multinational grain conglomerates with futures set in Chicago. Within the Bolivian context, soy would thus represent a fundamental departure from the rice, corn, cattle, and sugarcane that had been the focus of earlier waves of agricultural development and colonization in Santa Cruz. Those earlier initiatives and their attendant commodities were explicitly linked to fostering food security in which settlers were portrayed as "feeding the nation"—a narrative of agrarian citizenship embraced by farmer and state alike. The pamphlet

cast soy in that familiar mold, but by the 1980s the goal of agricultural production in the eastern lowlands had definitively shifted from food security to
generating export income.

The pamphlet's passive construction "soy is grown in Bolivia" was also
silent on the "who." According to annual reports published by ANAPO, a
new national oilseed producer organization, Mennonite farmers in Santa
Cruz, were driving this re-envisioned, extractivist March to the East. In 1984,
they produced the majority of the nation's soy and had been doing so since
the tentative forays first recorded by Bolívar in the late 1970s.[89] The rise of
Mennonite soyeros took place within a national economic context characterized by the debt crisis and subsequent neoliberal reform. It also occurred
within a broader surge in soybean production across the region. From Brazil
to Paraguay, Argentina, and Uruguay, this migrant bean reshaped the economy and ecology of the South American interior as forest was converted into
farmland. Commentators have referred to this combined area, which now
produces the majority of the world's soybeans, as the "United Republic of
Soy" or "Soylandia,"—terms that stress the transnational spread of agrarian capitalism.[90] Tracing the origins and evolution of Mennonite soy production
speaks to that broader narrative, while also responding to recent calls for an
ethnographic, actor-centered approach that situates the soy complex in "specific ecologies and social relations," rather than portraying it as an anonymous
wave of capitalist expansion.[91] The Mennonite soy boom depended upon a
convergence of factors from a local alliance with an immigrant entrepreneur,
agricultural knowledge from Brazil, an environmental collapse in neighboring Peru, and Bolivia's own spectacular period of hyperinflation. These local,
national, and transnational currents played out in miniature form, with telling environmental and economic results on the newly cleared fields of Bolivia's largest Mennonite colony. Riva Palacio offers a unique early window on
the profit and precarity of soy as it would come to dominate lowland farming.

Mennonites in Riva Palacio colony offer different explanations as to why
they first embraced soy production. "[President] Banzer told us to [grow it],"
explains farmer Johan Boldt, emphasizing their support for national production goals.[92] Johan Fehr remembers it differently. "We were planting corn," he
says of the first few years in Santa Cruz, "and [then] we started to sell soy.
There were people here from Czechoslovakia, and they were very knowl-
edgeable about soy."[93] Fehr does not remember the identity of these "Czechs,"
but he may be confusing them with a prominent Croatian immigrant
family—the Marinkovics. A member of the pro-German Ustaše militia in
wartime Yugoslavia, Silvio Marinkovic, and his wife fled Europe after WWII,

settling in Argentina along with many other Europeans, some with ties to fascist organizations. The couple subsequently emigrated to Bolivia where in 1967 Silvio built an oilseed processing plant Industrias Oleaginosas (IOL, S.A.). Already experimenting with castor bean oil in the late 1960s (a principal crop of the Paraguayan Mennonite colony of Canadiense), he turned to cottonseed during the cotton boom of the 1970s. As production declined after the middle of the decade, Marinkovic began to experiment with crushing soybeans. Satisfied with the results, he actively encouraged regional producers to switch to soy, including many Mennonites.[94]

Whether Marinkovic was the person Fehr had in mind or not, the Croatian and his family soon developed a close relationship with the Mennonite colonies, supporting soy production through seed distribution and harvest credit while providing a ready market for the novel crop. Colonists recall these early attempts at soy farming as a near Sisyphean endeavor. They enlisted their children to sprinkle granular herbicides by hand or fashioned wide makeshift wagons upon which four seated individuals with backpack sprayers could cover multiple rows. "We didn't know how to harvest it," laughs Enrique Siemens. Without combines, he and his brother pulled up stalks by hand and fed them into a stationary degraining machine before shoveling the soybeans onto trucks. However, these tentative manual forays soon gave way to mechanized farming as the Friesen brothers traveled to the United States to acquire used North American machinery, from grain conveyors to combines and spraying rigs, that the nascent soy production depended upon.

The seeds that Marinkovic provided to Mennonites on credit were the product of prior technological innovations and market connections that had emerged across the eastern plains in Brazil. Established in the 1960s in southern Brazil, the Brazilian soybean industry had been expanding in the 1970s and 1980s as the construction of highway networks across the cerrado opened lands for soy agriculture and produced a wave of migration and development in large western states like Mato Grosso. Initially, Brazilian soybean production concentrated on more temperate southern regions, including the states of São Paulo and Río Grande do Sul, but as production spread north into Greater Amazonia new varieties of soy were called for. Farming these new lands, long considered of marginal agricultural value, called for new technology. As Susanna Hecht and Charles Mann explain, that expansion was accompanied by an impressive transnational migration, as aspiring Brazilian agronomists, plant scientists, and geneticists traveled north to the United States for training. "In the 1960s and 1970s state schools in the Midwest and California were awash with Brazilians studying crop breeding, soil science, and regional planning."[95] These human migrations north supported the successful migration of soybeans south in ways not unlike the subsequent migrations of Chileans to California to study grape production in the Pinochet era.[96] But transplanting soybeans from the U.S. Midwest to semitropical Brazil was different than bringing grapes from temperate California to temperate Chile. Many of those Brazilian students returned home to work in research stations administered by the Empresa Brasileira de Pesquisa Agropecuária (EMBRAPA) where they developed new soybean varieties. The varieties they designed, including DOKO and Cristalina, grew in the poor-quality soils of the Brazilian cerrado, and matured well in the shorter daytimes of tropical and subtropical latitudes.[97] They were even designed with a characteristic pod height to support mechanized harvest and inoculated for biological nitrogen fixing which reduced the need for costly nitrogen-based fertilizer.[98] A wave of engineered soybeans soon covered newly cleared lands in western Brazil before crossing the border into agricultural expansion zones in eastern Paraguay and Bolivia. Whereas in the 1960s and 1970s, ranchers in Santa Cruz had slowly joined the Zebu revolution by improving their herds with imported Brazilian bred *zebu* cattle selected by EMBRAPA to thrive in tropical South America, in the 1980s and 1990s, cruceño soy farmers could be found planting Brazilian soybean varieties like DOKO and Crystaline that were bred to meet regional growing conditions.[99]

Although dependent on technology imported from the east, initial demand for Mennonite-Marinkovic soybeans was the product of an unprecedented environmental collapse in the rich Pacific waters off the Peruvian coastline, home in the 1960s to the world's largest fishery. The seabirds that had produced the nineteenth century Peruvian guano boom depended on the Peruvian anchovy. With the guano boom long past, that same marine biomass produced a second natural boom in fishmeal in mid-twentieth century Peru. Doubling in size in the decades after WWII, the global fishmeal industry fattened chicken in the United States and hogs in Germany, substantially lowering the price of once-expensive meat. Incorporating sonar technology, nylon nets, imported processing plants and gear, the Peruvian anchovy industry grew exponentially in those years.[100] From 1962–71 the anchovy fishery was at its peak, reaching twelve million tons in 1970, well beyond the recommendations for a sustainable catch.[101] This persistent overfishing proved disastrous when a severe El Niño system in 1972–73 dramatically warmed Peruvian waters typically kept cool by the Humboldt Current. As the anchovy population plummeted, millions of starved seabirds washed up on the coast. The Peruvian anchovy fishery reached a nadir of under two million tons by 1973. For the next two decades, Peru experienced regular strong El Niño currents and the industry remained a shadow of its former self (never again surpassing four million tons into the early 1990s).[102]

Peruvian anchovies had been the world's largest source of fishmeal and a critical fodder for industrially produced cattle, swine, and poultry. Their decline demanded a new source of protein. Conveniently, even as El Niño conditions disrupted the Humboldt Current along the Pacific coastal shelf, new soy fields were popping up in the South American interior. Working in the soy industry in Santa Cruz in the 1980s, ANAPO representative Guillermo Ribera remembers that, supported by the favorable trade terms of the Andean Pact, Marinkovic, and other soy processors in the department began to forge connections with Peruvian businessmen looking for alternatives to fishmeal.[103] With fishmeal sources in decline, soymeal (the by-product of crushing soybeans) emerged as an effective high-protein alternative. This was part of a global transition toward plant proteins in the factory-farming industry and along with national demand for soy oil, increasing foreign demand for soymeal drove forward the nascent soy industry in Santa Cruz where highly mechanized Mennonites were well positioned to thrive.

Transnational connections from the anchovy crisis, the development of Brazilian soy technology, and the travels of Mennonite machinery importers came together with a local alliance with Marinkovic to set the stage for the Bolivian soy boom. Ironically, the conditions for expanded soy production emerged just as Bolivia was entering a profound financial crisis. Inflation, which began in earnest in 1982, took on an increasingly surreal aspect, best captured in oral history, as it reached levels in the tens of thousands by 1984–85. One Mennonite representative who was tasked with depositing colony money in a Bolivian bank arrived in Santa Cruz with an entire wagonfull of bills stuffed in sugar sacks.[104] He remembers being unable to carry out the transaction because the bank's vault, overstuffed with near worthless bill, could not physically contain any more currency. The effects of hyperinflation were particularly traumatic in the first and largest Mexican Mennonite colony of Riva Palacio. "We lost a lot," remembers Johan Boldt, who was head of the widows and orphans fund (*Waisenamt*).[105] A long-standing Mennonite institution used to administer funds to the needy, and critically, to grant loans within the colony, the Waisenamt was of such importance to Mennonite colonists that they had its existence written into Bolivian law in their 1962 list of privileges and guarantees. During the early part of the inflation crisis, the Bolivian government had declared that it was illegal to conduct business in U.S. dollars. Although a massive black market quickly emerged in which individuals made rapid and temporary use of Bolivian paper money as a means of exchanging stable U.S. currency, Boldt and other leaders opted to convert their dollars into pesos because "we didn't want to be against the government [thinking] they wouldn't let us live here [any more]." The colony initially exchanged their dollars to pesos at 1:1,800. By the time the ban on U.S. dollars had been lifted, the currency sat at 1:1,900,000. "It was a pittance," he recalls, of what remained of the Waisenamt and individual savings after the crash.[106]

The financial strain produced by the debt crisis is not only evident in oral histories with Mennonite farmers. The extreme decapitalization of the early 1980s also changed the tenor of the negotiations between Mennonites and state authorities. Once based on their image as successful producers, the petitions of Mennonite colonists became increasingly desperate over the early part of the decade. As part of a new wave of Mennonite migration in the mid-1970s, Mexican Mennonites had founded Del Norte Colony to the east of the Río Grande near the San Julián Project. Del Norte's founders soon discovered that the land had been sold illegally by the former owners and overlapped several adjacent properties. Facing difficulties legalizing their title, they petitioned the state emphasizing their "social-economic function (FES)" as farmers who lived on and worked the land. Their neighbors in the San Julián Project were part of a radical new settlement design that many rural sociologists and planners, as the preceding chapter has shown, considered to be at the vanguard of Bolivian colonization. Del Norte's leaders were likely unaware that experts in the Israeli kibbutz system had been brought to Bolivia by Banzer at the beginning of the decade to advise in colonization prior to the CIU-led expansion of San Julián. Yet, as if directly addressing that plan, they boldly claimed that their own modern agricultural practices were "the envy of the Kibbutz of the Jews and of the [San Julián] cooperatives that surround us [and to whom] we lend all forms of aid and cooperation."[107] They also threatened to take their case before both the Supreme Court and the Human Rights Commission in Santa Cruz, claiming that delays were the result of a "sick and xenophobic hatred" against Mennonites on the part of members of the Agrarian Reform office.[108]

As their case dragged on from the late 1970s into the period of hyperinflation, the sudden impoverishment offered Del Norte's representatives a new logic which they employed in their legal battle. In 1982, Jakob Klassen filed a denunciation before the Agrarian Reform, detailing the "astronomical sums" invested in colony land that now stood to be taken from them. Klassen also noted that his frequent trips to La Paz to contest the case had cost them 100,000 pesos over three "agonizing years" of negotiation and counternegotiation. The collapse in the value of their U.S. dollars deposited in Bolivian

savings accounts, "makes us as poor as any other Bolivian farmer," he continued, "and unlike them we do not abandon [the land], but live and die on the
earth, making it prosper and produce."[109] Mennonites had long emphasized
their prodigious agricultural production as a performance of successful agrarian citizenship. Here Klassen invoked their decapitalization, which had transformed them from immigrant investors to poor campesinos, as a new sort of
claim before the state. His plea resonates more closely with the petitions that
Mennonites had sent to Mexican authorities during the worst years of the
midcentury drought in Chihuahua that also hinged on peasant abandonment and Mennonite stability, than with the confident projections of Riva
Palacio's Minimum Plan of Work in 1967. In suffering the agony of the debt
crisis, Klassen claimed that Mennonites had "come to consider ourselves as
an inseparable part of the great nation of Bolivia, thinking of ourselves as
your own campesinos."[110]

While Boldt and Klassen described the devastating effects of hyperinflation on personal and colony savings, Bolivia's response to inflation paved the
way for a subsequent boom in two settler export crops. In July of 1985, with
annual inflation climbing to 60,000 percent, a young American economist
named Jeffrey Sachs arrived in La Paz to advise the Bolivian government. Based
on Sachs's recommendations for a fiscal shock program, President Victor Paz
Estenssoro passed Decree 21060 a month later. Bolivia's New Economic Policy (NEP), as it would be known, was an early example of the neoliberal policies that would soon be adopted across the region. Under the NEP, the state
revalued the peso, removed protective tariffs, price controls on food, and export restrictions while privatizing major state corporations.[111] The liberalization
of export policies meant that soy prices, previously regulated by the Bolivian
government in cumbersome annual negotiations between the Ministry of
Agriculture, oil factories, and the soy producers' organization (ANAPO),
were opened to world prices for the first time. With the Bolivian peso now
pegged to the U.S. dollar, agricultural exports increased in value by 40 percent
while the government also introduced a rebate for farmers relying on "imported
inputs" such as seed, agrochemicals, and diesel.[112] ANAPO representative
Guillermo Ribera was part of the group that oversaw the first major export of
Bolivian soybeans to Argentina in 1986.[113] It was a moment of victory from
the perspective of the industry's leading organization and for Mennonite producers who had long watched fellow Mennonites in Paraguay sell soy at
higher international prices. In the eyes of state officials, who had just witnessed the collapse of tin prices in the mid-1980s, Santa Cruz soybeans also
stood poised to replace this traditional mineral export and allow Bolivia's

new neoliberal government to improve its declining foreign exchange. With these incentives, production of soybeans would increase nearly tenfold over the following decade to become Bolivia's largest export crop. In the process, the meaning of soy definitively shifted. Once converted into edible oil—within the logic of food sovereignty—and consumed nationally, soy was now circulating as a raw commodity within a global market.

Soy was not the only settler crop to benefit from hyperinflation. In 1985, sociologist Kevin Healy documented another surprising "boom within the crisis" in a special publication of the proceedings from a conference on "Coca and Cocaine" at Cornell University, the same institution that had produced a wealth of studies on tropical resettlement and San Julián in particular. Healy wrote that the combination of Bolivia's national economic crisis and widespread, devastating droughts in the Andes in 1983 had pushed new waves of spontaneous colonization to the tropical lowlands in the first half of the decade.[114] Responding to demand for cocaine in North America and Europe, these internal migrants were turning to coca cultivation. From the Yungas, to the Chapare and western Santa Cruz, Healy explained, "the rapid expansion of coca leaf production and illicit coca paste and base production during the past five years has become Bolivia's boom within the economic crisis."[115] Preceding those economic and environmental migrants was future Bolivian President Evo Morales who had left Oruro for the Cochabamba lowlands in 1978. He became a key organizer among newly arrived coca-producing settlers or *cocaleros*. Lowland development workers were also shifting their attention to coca production. After their work in Santiago de Chiquitos and San Julián, both MCCer Milton Whitaker and Methodist Harry Peacock would spend time in the Chapare as part of USAID's alternative cropping program which encouraged farmers to switch from coca to new crops and was paired with an aggressive and militarized eradication campaign.

Coca, a product traditionally cultivated and consumed in Bolivia and across the Andes, took on a new illicit meaning and economic value in the context of the cocaine boom of the 1980s. Soybeans in contrast, had a relatively modest history of cultivation in Bolivia before the debt crisis but, like coca, they benefited from the economic shocks of the 1980s; their signification, as a star Bolivian export crop, shifted in the neoliberal era. While both licit nontraditional soy and illicit traditional coca leaf production surged in the same years, the products were treated in remarkably different ways. In stark contrast to the alternative cropping and coca eradication programs, which aggressively targeted coca farmers in in new colonization zones like the Chapare region of Cochabamba and the area to the west of Yapacaní in Santa Cruz, the fiscal shock program designed by Jeffrey Sachs actively supported soy producers. On a regional level, cocaleros and soyeros were also racialized in distinct ways by cruceños even as the income from each of these crops helped Santa Cruz mitigate the worst effects of the *lost decade*. Soy was explicitly linked to modern agroindustrial production and identified with wealthy cruceños, Mennonites, and Okinawan settlers. Indeed, from its earliest annual reports, ANAPO divided soy production into these national, racial, and ethnic categories. In contrast, cruceño newspapers like *El Deber* exclusively linked coca farming and cocaine production to indigenous Andeans like Evo Morales and associated it, through coded references or open accusations, to criminality.

At times the celebration of immigrant soyeros and the vilification of Andean cocaleros came together in a crude juxtaposition. In April of 1987, next to an editorial that proclaimed "Santa Cruz does not surrender in the face of the [debt] crisis," *El Deber* carried an article celebrating the agriculture of Japanese and Mennonite colonists in national territory, "or better put, cruceño territory" they clarified.[116] The article contrasted the beneficial presence of these immigrant settlers with that of "other supposed colonists that come down from the highlands of the country without order, without plans, without possessing physical aptitude to grapple with the rigors of the tropics." Mennonites and Japanese, the writer concluded, "do not come to take land from anyone, they do not come to displace, by any means including violence, the old and traditional cruceño farmer . . . organize unions . . . fight . . . create problems for the government . . . disrupt the order by blocking roads demanding this or that in peremptory or intransigent terms." The rant, drawing on a now familiar script of indigenous highlanders as invaders, was tangential. The article, titled "Protection for the Colonies," was intended to spur local government to establish a police presence in the Mennonite colonies where, according to the author, "criminals without scruples" were robbing and assaulting colonists and stealing property. Increasing thefts in Mennonite colonies in 1987 might have reflected the bountiful 1986 soy harvest being, for the first time, sold at world market prices. Mennonites, many of whom did not trust banks to safeguard their earnings after the peso crash, were likely to keep large amounts of cash at home. Whatever the causes, the author warned that if nothing was done to catch these criminals, Mennonites, "men and women of peace and work," would surely leave the country.[117]

While appearing in industry records and public commentary, surging Mennonite soy production is most evident in the detailed personal account books of Mennonite farmers like Jakob Knelsen who arrived in Riva Palacio from Mexico in 1975 as initial forest clearing had finished and Mennonite production was getting underway. His precise accounting for those years provides an intimate source for understanding how these regional and global factors played out on Mennonite fields in Riva Palacio.[118] As Bolívar had indicated in his 1978 report, colonists like Knelsen maintained a diversified system of production. Knelsen's records show that he sold cattle, pigs, hens, and ducks on the Santa Cruz market along with milk and eggs. However, egg production, a key income generator from 1977–81 declined substantially over the following four years, and then disappeared entirely after 1985, as other colonies, and especially the Japanese colony of San Juan Yapacaní began to flood the Santa Cruz market with egg production on an industrial scale with a plant producing balanced feed. From 1985 onward, San Juan maintained roughly 300,000 laying hens and egg sales, according to reports became the "pillar" of the colony's economic stability.[119] In those same years we can also clearly see Knelsen's move into the soy sector. While production of sorghum, a winter crop and important fodder for his dairy herd remained relatively consistent through the 1980s, the amount of corn he produced steadily decreased in favor of soy. From 1978–81 he was predominantly growing corn on his recently cleared land. It was only in 1982 that he grew soybeans for the first time—taking in twenty-four tons and the same amount again in 1983. In 1984 his production dipped to eighteen tons but then he more than doubled his harvest in 1985 (39 tons) and again the following year (97 tons). His land under soy cultivation increased accordingly from 11 hectares in 1982 to a high of 53 hectares by 1989—equivalent to an entire individual farm for most Mennonites and just above the fifty hectare maximum to still be defined as a "small producer" in Santa Cruz under Bolivian agrarian law.[120] Knelsen's personal accounting was reflected among his neighbors in Riva Palacio who embraced mechanized soy production. "[It was] the best harvest I know. Clean. Easy. Good," Isaak Peters remembers. Cornelio Peters agrees, emphasizing that in comparison to the intensive labor of dairy, or early manual corn harvests, "the soy that we harvested in that time, it seemed like it was done without labor." The weight of these anecdotal accolades was reflected in annual reports from ANAPO. The area under cultivation jumped from barely 20,000 hectares at the start of the 1980s to well over 100,000 hectares by the end of the decade as Mennonites, and the Department of Santa Cruz as a whole, staked their future on soy monocropping.[121]

Debt and Drought on the Mennonite Soy Frontier

The narratives of successful Mennonite soyeros in the 1980s jar with an article entitled "Agricultural Project" that appeared in *El Deber* at the start of the next decade.[122] The author explained that a local Women's Club had begun a project to promote farming at Santa Cruz's Palmasola prison as a source of food and form of therapy among inmates. The trial plot included eight hectares of beans, citrus, and vegetables. Those in charge were "people with agricultural knowledge who are in prison for various crimes." In the two photos that accompanied the newspaper article, the philanthropists looked on approvingly as a group of prisoners—including at least three Mennonite men in their trademark overalls and cowboy hats—labored on the jail's farm. What had transpired between 1987 and 1990 that Mennonites, once celebrated as model farmers and defended as victims of theft, had become inmates in the city's most notorious jail?

While no explanation is offered in the article, the three Mennonite men were almost certainly imprisoned for failure to pay their debts. The origins of this *Mennonite debt crisis* emerged rather unexpectedly out of the excellent soybean harvest in 1986. With the prospect of good international prices some Mennonite farmers looked to expand their soy operations and replenish savings they had lost to the currency devaluation. Yet hyperinflation had also disrupted institutional and individual systems of lending in Riva Palacios and, unlike the corn they had farmed in the early 1980s, soy production required a highly mechanized farm operation reliant on expensive inputs. As a result, many Mennonites looked to off-colony borrowing to support mechanization, the acquisition of new lands, and annual planting costs. Like Okinawans who were also mechanizing in those years, some colonists had actively benefited from the peso crash because of outstanding debts that could be paid at a fraction of their original value and emerged from the mid-1980s crisis with a speculative approach to borrowing that presumed further devaluations of the peso.[123] They found ready lenders in Santa Cruz where, according to a subsequent MCC report "unscrupulous moneylenders were plotting and setting into motion an orchestrated scheme."[124] In early 1987 several Santa Cruz banks were actively encouraging Mennonite farmers to take out five year loans at 13 percent annual interest (a low rate for Bolivia). When Mennonites arrived at these banks, they were told the funds were not yet available and were redirected to external lenders where they were offered interim high-interest loans. Others unwittingly signed blank checks or forms acknowledging receipt of funds that had never been given out.

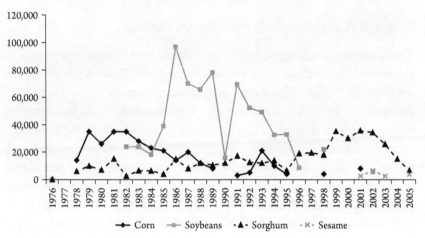

Jakob Knelsen's annual harvest (in kilos). Produced by author from Knelsen's account book.

With a soybean harvest similar to 1986, colonists might have repaid these predatory loans. Unfortunately, the following years in Santa Cruz were characterized by increasingly dry conditions.[125] Once again Jakob Knelsen's detailed accounting is revealing. While he increased his area planted in soy year after year, the yields never reached the 3.8 tons per hectare of his bumper crop of 1986. For 1987–89 they declined from 1.8 to 1.4 tons. In 1990, he took in a paltry 0.3 tons per hectare.[126] Reporting on that 1990 harvest, *El Deber* warned that drought-stricken small producers in the south of the department, the location of the largest Mennonite colonies, had lost everything, and demanded debt relief from the government.[127]

Droughts like these were a recurring phenomenon in Santa Cruz which sits at a transitional point between humid Amazonia and the semiarid Gran Chaco. But the environmental disaster of the late 1980s in Riva Palacio was also directly linked to Mennonite soy farming practices over the previous decade. When Jesús Bolívar provided his survey of the colony in 1978 he noted that only about 17 percent of the land had been cleared for farms. Concluding on a worrisome note, he stated, "lately the wind break curtains are being taken down to be turned to agriculture."[128] With the incentive provided by soy, colonists intensified the process of deforestation. Colonist Johan Fehr, operating a bulldozer in the colony in those years, remembers that by 1985 the entire colony, once dense bush, had been cleared. The newly opened landscape may have looked like home to many Mennonites that had grown up in the arid valleys of northern Mexico. "Chihuahua was open," acknowledges

Johan Boldt, "we didn't know bush like that." Johan Fehr also emphasizes their misreading of the landscape. "The wind was very strong," he claims, "and we didn't know the climate very well. We didn't know what would happen if we cut down everything and did not leave a single tree."[129]

Riva Palacio's layout only exacerbated the effects of wind-based erosion.[130] For expediency, the positioning of villages had followed the original oil company routes traversing the colony from east to west that Peter Bergen, Peter Fehr, and Abraham Peters had explored back in 1967. The majority of farmers' long, narrow fields were oriented north to south along these arterial roads. The rectangular design effectively minimized distance between neighbors replicating a concentrated street-village (*Strassendorf*) system that Mennonites had successfully transplanted from Prussia, to Russia, Canada, and then Mexico. Unfortunately, these fields also aligned with perpetually strong north–south prevailing winds in Santa Cruz, effectively turning newly cleared lands into wind tunnels. Even newly planted windbreaks were less effective than they would have been on east–west fields. The consequences were painfully apparent to Abe Enns, who had lived in Mexico, Canada, and Belize before belatedly arriving in Riva Palacio in 1988. Unlike the dense bush that greeted original Mennonite settlers, Enns looked out on a vast treeless plain with neighboring Mennonite villages (often two or more kilometers away) clearly visible across the fields of soy. Above all he remembers the relentless wind that swept over these recently opened plains and precisely counted 178 days of intense wind in his first year in the colony. The following season, when Knelsen's production dropped to 0.3 tons per hectare, Enns also describes as "dry, very dry . . . and not one man got enough of a harvest to live off."[131] His entire crop failed that year, and he had to plead to purchase food on credit from an owner of one of the colony's small stores.

As soybeans failed to deliver on their promises of wealth, Mennonites might have turned back to the dairying practices they had helped to establish in Santa Cruz with imported Holstein cattle. In the depths of the drought, as even pasture failed, colonists purchased sugarcane to feed their herds. But while they continued to sell cheese and butter to Santa Cruz, the rapid growth of the Mennonite population in Bolivia, which by 1990 was nearing 20,000, saturated the regional and national market resulting in an abysmally low price for queso menonita. As aggressive creditors began to repossess machinery and have indebted Mennonites thrown in jail, the Mennonite Central Committee, long viewed by Bolivians as an unofficial Mennonite Embassy, attempted to intervene. As early as 1988, MCC was studying the situation with the help of a local Mennonite businessman, Jack Doerksen. Doerksen discovered that some individuals owed creditors between $20,000 and $150,000—most of it in debt servicing and estimated that, in total, 250 indebted Mennonite families owed approximately US$5 million dollars on unpaid equipment and planting loans.[132] Bernhard Sawatsky, for instance, had signed a blank form later used to charge him 9–10 percent interest per month suffered consecutive crop failures in 1987 and 1988. After auctioning off his possessions he still owed creditors $130,000 and spent the next two years and five months in jail.[133] Across Latin America in the 1980s and early 1990s, national governments negotiated with international financers over debt relief. Bolivia had just finished negotiating a debt cancellation agreement of its own with several financers in 1987. In Santa Cruz from 1988–93 this struggle played out in a miniaturized form as the MCC worked with Mennonite debtors to resolve the colonists' debts and restore their image as model producers.

MCC representative Tim Penner joined forces with the soy-producer organization ANAPO to pull both the Ministry of Agriculture and colony leadership into action in support of indebted colonists. With ANAPO's statistics on Mennonite soy production to support him, he warned the Ministry of the impact on the region and the country resulting from a crisis of Mennonite production if predatory creditors were not held at bay.[134] As a result of MCC and ANAPO interventions, the national government offered affected colonists access to a $2.5 million low-interest loan. Yet, colonists were unable to receive funds because banks processing the loans demanded property titles as collateral. Mennonites individually owned their property, but the title was held by the colony and considered inalienable. Colony leadership meanwhile, was ambivalent toward their debt-ridden fellow colonists whom "they criticized for not listening to warnings" even if they acknowledged that many of the indebted "did not act recklessly or with dishonorable motives."[135] Pushed forward by negative coverage in the local press, these leaders eventually took their own path to consolidate loans while they "resisted pressure . . . to pay unreasonable interest rates and inflated principal claims," and instead renegotiated several debts.[136] As with cattle importer Johan Guenther two decades earlier, Mennonite vorsteher auctioned off debtors' land while allowing individuals to continue living on the property and "retain several cows for food and the sale of milk." After 1990, conditions improved somewhat and colonists typically donated a hectare or two of soybeans to help their neighbors cover remaining debts.

Other debtors negotiated directly with their creditors. Jaime Duranović was working for Silvio Marinkovic's Aceite Rico oil factory in those years. Marinkovic had loaned substantial amounts to Mennonite producers and in

1992, Duranović was put in charge of a group of clients with $5 million in debt.[137] In a Bolivian economy that was increasingly turning to the informal sector, Aceite Rico and other processing companies renegotiated debts while accepting Mennonite cattle, products fashioned in the colony, and in-kind services (such as land clearing) in lieu of cash payments. When they faced economic hardship amid the long midcentury drought in northern Mexico, Mennonites had engaged in seasonal labor migration to Canada before coming to Bolivia. Those transnational routes remained open and intensified during the Mennonite debt crisis as a new generation of Mennonite colonists took part in this long-standing tradition of labor migration. While some Mennonite colonists sent home remittances to pay off their debts, buy new lands, and purchase new machinery to replace that which had been auctioned off, others fled to Canada with no intention of returning or paying back outstanding money. Duranović remembers receiving regular calls from colonists picking vegetables or working in factories in Ontario to cover their debts. Through this combination of transnational labor, auction, and in-kind services Duranović reduced his clients' debt-load to $1 million over the following six years.[138]

Mennonites were not the only lowland colonists who looked to transnational migration to escape debt and bleak economic prospects in Bolivia. Their Okinawan and Japanese neighbors were also returning home in those same years and for many of the same reasons. Facing a critical labor shortage in the late 1980s, the Japanese government altered its immigration law in 1990 to allow Japanese descendants (up to the third generation) to return as guest workers in unskilled factory labor. Along with a large-scale migration of Japanese Brazilians, many Japanese and Okinawan Bolivians took advantage of this program. Katoshi Higa of Colony Okinawa had just finished high school in the late 1980s and remembers that most of his close friends left for Japan.[139] The result was that by 1995 an estimated 200,000 Latin American *dekasegi* (migrant workers), lived in Japan including some remitting income to support family in Okinawa colony and San Juan.[140] Andean settlers also returned to prior migratory strategies and forged new ones in neoliberal Bolivia. With the Argentine peso fixed in relation to the U.S. dollar in the 1990s, large numbers of Bolivians migrated to Argentina, expanding the ranks of Bolivian braceros that had preoccupied the MNR in the 1950s. Although earlier migrations had centered on the harvest in northern provinces, these new migrants often traveled to Buenos Aires to work in construction and textiles, rapidly converting the former Jewish neighborhood of "Once" into a center of this new community of Bolivian *porteños* (residents of Buenos Aires).[141] In the industrial heartlands of Japan, Canada, and Argentina, the local effects of Bolivia's economic

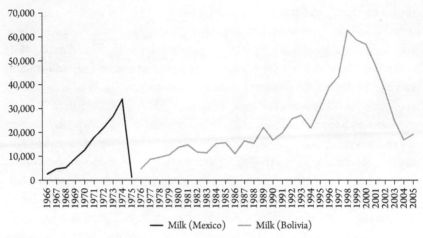

Jakob Knelsen's milk production in Mexico and Bolivia (in kilos). Produced by author from Knelsen's account book.

crisis on the Mennonite, Okinawan, and Andean diaspora were apparent in this resurgent transnationalism with roots in mid- and early twentieth century migrations.

Transnational migration emerged as one survival strategy in the face of drought and the debt crisis, but most Mennonites remained in Bolivia and began to shift the way they farmed. It is telling that colony officials allowed indebted Mennonites to maintain a few milk cows even when their agricultural equipment was auctioned off.[142] "Milk money" continued to provide colonists with a consistent source of income that stood in contrast to the risky, speculative nature of cash cropping in soybeans. Having lost everything in the late 1980s or in subsequent drought years, some colonists would turn from soy entirely and instead plant pasture and drought-resistant sorghum to serve as fodder for their expanding dairy herds. Once again colonist Jakob Knelsen's accounting is illustrative. Knelsen never planted more than thirty hectares of soy after the disastrous 1990 harvest while quadrupling his annual production of milk in those same years.

Overproduction kept cheese prices low, and for most of the decade Mennonite farmers remained trapped between unpredictable, but potentially lucrative, harvests and steady, but low-value, dairy. Some farmers turned to drought-tolerant cash crops. Wilhelm Buhler and Juan Fehr, whose father-in-law Peter Friesen had introduced them to the transnational tractor trade in the 1980s, would pioneer sesame and peanut production in Riva Palacio in the 1990s. It was at the end of the twentieth century, after two consecutive crop failures in which soy yields plummeted from 1.7 to 0.2 tons per hectare, that Mennonites made a critical decision. Negotiating with PIL Andina, S.A., the sole large milk processor and distributor in Bolivia, colonists had refrigerated holding tanks placed in colony villages and established daily milk collection. Internal colony tax records from the 2000s indicate the scope of this shift in which income from milk and cheese often exceeded that of soy and other grains.[143]

The intertwined economic and environmental crises that Mennonites experienced in the 1980s were simultaneously aberrant (attributable to extreme weather) and systemic (a product of the land-clearing practices of Mennonites that had created a vast treeless plain of the dense forest that the first Mennonite delegates had surveyed back in 1968). They shook the previous image of Mexican Mennonites as modern, prosperous farmers even as the eventual response—a negotiation of Mennonite debt by government officials, ANAPO, oil companies, and the MCC— suggested just how critical Mennonite production had become for Santa Cruz and the nation. Back in 1978 when barely over 10,000 Mennonites possessed more tractors than the 300,000 rural residents of the Department of Santa Cruz, the *modernity* of Mennonite agricultural practices was unquestionable. It did not matter that Mexican Mennonites drove tractors with steel wheels, although their rejection of rubber tires had been the internal conflict that had driven their emigration from Mexico. In Santa Cruz, it only mattered that they farmed with tractors at all. By the 1990s, the situation had changed dramatically. Japanese and Okinawan migrants continued to modernize their operations into large agroindustrial cooperatives. A wave of Brazilian expatriate soy farmers, priced out of the Brazilian soy boom, were also purchasing land across Santa Cruz and farming on an industrial scale. In this context, ANAPO officials and other agricultural extensionists increasingly critiqued deep plowing among Old Colony Mennonites.[144] They also claimed that their steel wheels increased soil compaction and had aggravated the environmental crisis of the late 1980s. Similarly, Bolivia's Forestry Service became involved in regulating the usage of windbreaks in Mennonite colonies after the passage of a 1996 Forestry Law. As new Mennonite colonies to the east of the Río Grande adopted no-till technology for farming soybeans, Bolivian officials saw the original Mexican Mennonite colonies of Riva Palacio, Santa Rita, and Swift Current, not as progressive farmers, but as backward, technological holdouts.[145]

If Mennonite production techniques lagged those of new more modern farmers, this was made up for by the continued expansion of Mennonite settlement across the Bolivian lowlands in those same years. Reflecting on the debt crisis in 1988, the MCC had attributed one cause of Mennonite indebtedness to the high cost of purchasing land in Riva Palacio colony. As in Mexico, high birth rates among colony Mennonites and the need to farm and reside in contiguous settlements created an extreme market juxtaposition in which uncleared land in the colony sold for $100–$200 per hectare while neighboring off-colony land was worth less than one tenth that amount.[146] In the late 1980s, Abe Enns remembers that there were nearly seven thousand people living in Riva Palacio, a colony originally designed to hold five thousand.[147] In the wake of the debt crisis, colony leadership actively responded to that pressure. Over the following years Bolivian Mennonites established daughter colonies across the lowlands.[148] Riva Palacio purchased land far to the south of Santa Cruz de la Sierra in the late 1980s forming the colony of Pinondi and in the following two decades purchased new lands to form the colonies of Manitoba, El Dorado, and New Mexico. Other Mexican and Paraguayan Mennonite colonies did the same. The pace of this expansion would have likely even surprised INC director Roberto Lemaitre who in 1969 had written that "this is only the beginning of an immigration which we have reasons to believe will gain momentum day by day."[149]

Conclusion: Two Guerrilla Visions

In late 1995, a retired Bolivian general revealed that the body of Che Guevara lay buried near the small Santa Cruz town of Vallegrande. Over the following two years, a joint Cuban-Argentine forensic team scoured the site without finding a trace of the remains. Finally, in July of 1997 investigators discovered several skeletons in a mass grave, one of which was subsequently confirmed as Guevara's. A few months later, Che's bones were ceremonially repatriated to Cuba where they were interned in a mausoleum constructed for the occasion. It had been exactly three decades since the two parties—Mexican Mennonites and Cuban revolutionaries—surveyed the densely forested frontier a short distance apart from one another. After that moment of unknowing proximity, the legacy of those two transnational delegations diverged sharply. Even as Guevara's revolutionary legacy in Latin America remained central at the close of the twentieth century—evident in popular culture and claimed by a spectrum of revolutionary organizations from the Revolutionary Armed Forces of Colombia (FARC) to the Zapatistas in Southern Mexico—the protracted search for his remains is testament to the negligible physical transformation engendered by his abortive revolutionary movement in Santa Cruz in the 1960s.

In contrast, while Mennonites continued to reside at the physical and metaphorical margins of Latin America, their environmental and agrarian impact was evident to anyone traveling through the Bolivian lowlands at the close of the century. Colony Mennonites, as the MCC had predicted in 1975, had demonstrated their paradoxical ability to both pursue physical and cultural isolation while simultaneously "revolutionizing native economies." By the end of the first decade of the 2000s there were approximately seventy Mennonite colonies in Bolivia with a combined population of seventy thousand spread out across Santa Cruz and the neighboring department of the Beni. The expansion of Mennonite settlement pushed the agricultural frontier in Santa Cruz eastward. Cruceño agriculture had long centered on the "integrated north"—the area to the west of the Río Grande—but in the 1990s and beyond, Mennonites and other farmers looked to the east side of the river when purchasing new land. Extending across the plains of the Chiquitania toward the Brazilian border, this region was the object of a World Bank–sponsored initiative from 1991–97. Known as the "Eastern Lowlands Project," it aimed to improve infrastructure along the frontier and help secure credit for small farmers, and World Bank officials identified Mennonites as key recipients of program funds. By the first years of the twenty-first century, the number of hectares under soybean cultivation in this eastern expansion zone was double that being farmed in the integrated north where many Mexican Mennonites in older colonies had transitioned to dairy and drought-resistant crops like peanuts.[150]

Mennonites of Mexican descent still represented the bulk of Bolivian Mennonites, but those new colonies also included Mennonites with Canadian, Paraguayan, and Belizean passports and a growing number of Bolivian-born Mennonites. Colonists ranged from horse-and-buggy Mennonites that used steel-wheeled tractors (the majority) to a small, but increasing percentage of Mennonites that owned pickup trucks, used the latest no-till technology and GPS-guided tractors, lived in airconditioned homes, and had internet access. Although they faced increasing competition from Brazilian expatriate soybean farmers, Mennonites continued to produce approximately 40 percent of Bolivia's soybeans, which in the mid-1990s accounted for half of all Bolivia's export income.[151] As their impact on the landscape of Santa Cruz, became ever more apparent in the twenty-first century, Mennonites continued to employ the range of transnational, legal, and agrarian strategies, described in this chapter, that legitimated and supported their conspicuous but increasingly commonplace presence.

Conclusion
Past and Present in the Bolivian Lowlands

As I prepared to board my flight to Bolivia in May of 2011, I waited in a queue behind several Mennonite men and women at Miami International Airport (MIA). As one of the main airline hubs connecting Bolivia with North and Central America, MIA is a key node for a constellation of transnational Mennonite farmers living across the frontiers of the Western Hemisphere. My fellow passengers might have been returning from any number of colonies in Canada, the United States, Mexico, Paraguay, or Belize, representing one of the diverse migrant streams that constitute Bolivia's March to the East. In their trademark cowboy hats, overalls, bonnets, and long dresses they were as easily identifiable to me in 2011 as they had been to U.S. diplomat William Dietrich, who had encountered them while stationed in Santa Cruz de la Sierra in the 1970s. As our plane touched down in that city, after a brief stopover in La Paz, I noticed the marked contrast between the airport of Bolivia's capital city and the decidedly modern infrastructure of Santa Cruz's Viru Viru International Airport. Constructed in the 1970s with a large loan from the Japanese government, the airport is one of a multiplicity of JICA-supported development projects in the lowlands, reflecting the indirect impact of another of Bolivia's migrant streams. Exiting the airport and heading into the city, on a highway busy with trucks transporting agricultural materials, the large-scale migration of Andeans to Santa Cruz over the last half century was unmistakable. By the year 2000, these transplanted highlanders constituted a full quarter of the department's population of two million.[1]

Leaving my rented apartment where Calle Beni meets Santa Cruz's multi-lane "Second Ring," the three migrant streams condensed to a single intersection. Once marking the urban periphery of Santa Cruz, where city gave way to bush, the thoroughfare is now one of twelve concentric ring roads encircling the rapidly expanding metropolis. If I hailed a cab at that corner, my driver was very likely to originate from Potosí or Oruro, having purchased his vehicle after a particularly successful harvest season. I could also cross the street for lunch at the Japanese Social Center, which included a dormitory for students from San Juan Yapacaní and Okinawa colonies that come to study in Santa Cruz as well as a Japanese restaurant, its windows plastered with post-

Mennonite colonist Abram Wiebe sells jam at a busy intersection.
Photo by author.

ers inviting cruceños to Okinawa Colony's upcoming sixtieth anniversary. As the traffic lights turned red, a range of predominantly Andean informal workers leapt into action. Newspaper and candy vendors competed for the attention of waiting commuters alongside young children who put on impressive displays of juggling and break dancing for spare change. They were joined by Mennonite families from drought-stricken and impoverished Durango colony in the south of the department who sold cheese, peanut butter, cookies, and jam to passing motorists. This distinctly cruceño street scene developed every Wednesday, an unofficial Mennonite market day when colonists from the seventy Mennonite colonies in lowland Bolivia stream into Santa Cruz to conduct business, obtain paperwork, make long-distance phone calls, shop, and socialize.

Considering the immediacy and visibility of this history of migration, my research felt highly relevant and surprisingly challenging. Whether I was taking the old Santa Cruz-Corumbá train to San José de Chiquitos, that had by turns excited and frustrated cruceños in the 1950s, or riding in a bus on the brand new, IDB-financed Interoceanic Corridor (connecting Bolivia to the soybean zones and export infrastructure of Brazil), which paralleled the railway, my personal movements seemed everywhere enabled and entangled in the past and present of this ongoing state project of eastern expansion that had created a landscape of migration out of this former frontier. The point was driven home on a visit to Canadiense II (Canadian II), a branch colony of Canadiense I founded by Paraguayan Mennonites in 1957.[2] As we sat drinking Paraguayan-style *tereré* (iced yerba mate), I explained to Peter Fehr, who was seven when his family entered the country, and his neighbor Abram Falk, manager of a colony store and dairy cooperative, my intention to write a history of lowland colonization from the 1952 revolution to debt crisis of the 1980s, roughly covering the period of state-led development in Bolivia. Fehr and Falk looked puzzled. "Why stop there?" they asked.[3]

I did not have an answer. For Mennonite, Okinawan, and Andean colonists who continue to expand across the eastern plains it seems that such a periodization makes little sense. When I last visited Bolivia in April of 2015, the Mexican Mennonite colony of Riva Palacio was contemplating a new settlement venture in Peru that would get underway that year (while a Mexican Mennonite colonization program in Colombia began in 2017). Within Bolivia, they were also in the process of purchasing two 30,000-hectare parcels to join the four daughter colonies they had opened since their arrival in 1968. In the face of a growing population and five hectare per year restrictions on new land clearing under Bolivia's Forestry Law, Riva Palacio's vorsteher had proposed donating one of these blocks as a forestry reserve in exchange for permission to more rapidly clear the second. In San Julián, the spatially unique settlement developed in cooperation with the CIU in the 1970s, one can visually track the expansion of the rapidly growing colony in the impromptu stores created by tractor trailers that simply back to the edge of the congested highway and open their doors for business.

Although framed in different terms, Bolivia's plurinational socialist government also seemed to be asking a variant of Fehr's "Why stop there?" In 2013, I visited Sagrado Corazón, a colony created in 1966 a short distance to the north of La Guabirá (Santa Cruz's largest sugar refinery), to attend the National Day of Corn. After wandering through the experimental plots and machinery demonstrations, I listened as Bolivia's Vice President Álvaro García Linera (a former guerrilla fighter) repeatedly proclaimed to a large group of farmers that "without corn, there is no country (*sin maíz, no hay pais*)." With that poetic evocation of agrarian nationalism and food security, two of the original tenets of the March to the East, he launched into an explanation of his administration's plans to triple agricultural production and frontier expansion over the next five years.[4] As promised, the years after García Linera's speech saw rapid growth in the eastern expansion zone several hundred kilometers east, across the Río Grande, from Sagrado Corazón. In August of 2019, as this book nears publication, fires raged across the Chiquitano dry forest. But rather than corn, it has been cattle, along with soy, sunflower, and other cash crops that have accounted for much of that change. Indeed, the same month the fires exploded onto the international news circuit, Bolivia sent its first shipment of 48 tons of beef to China as part of a new export deal with a forecasted $800 million dollar value by the end of the next decade.[5]

The past history of the March to the East was undeniably present in the preceding scenes. in Santa Cruz. Yet paradoxically, the continuing growth of the Santa Cruz region tends to obscure as much as highlight that earlier history. Since the debt crisis of the 1980s, migration to the lowlands has increased exponentially, and the urban and rural regions of Santa Cruz have grown apace. At times, this growth seems poised to devour earlier histories of settlement. When Peter Fehr was colony administrator of Canadiense I in the early 2000s he had decided to commemorate their fiftieth anniversary with a modest monument. When he joined many of his neighbors in Canadiense II (a *daughter* colony on the east side of the Río Grande) a few years later, he elected to take the monument with him. The original colony was on the verge of being swallowed by the expanding city reflecting the increasing urban trend of lowland migration to Santa Cruz.[6] It now sits in relative obscurity (in his backyard) signifying a personal, rather than public, claim to the past. Mexican Mennonites, like those in Riva Palacio who were on the verge of their own fifty-year anniversary in Santa Cruz as I conducted fieldwork, repeatedly explained to me they had no intention of officially commemorating the date. They had little need to. That history has been inscribed on the landscape of Santa Cruz, evident in dozens of second and third generation colonies scattered throughout the lowlands.

Other colonies mobilize the past to a much greater extent. Yet in this process of official remembering, certain aspects of settler history fade from view. As with the Japanese colony of San Juan Yapacaní, Okinawa Colony contains a historical museum and frequently publishes official colony histories. In 2014, Okinawa celebrated its sixtieth anniversary by inviting Bolivian and Japanese officials as well as the general public to a grand multiday public ceremony. The event was equally a performance of their integration in national society and their continued transnationalism. In this scripted spectacle, the colony carefully eluded the overt racism they experienced and actively contested in their first years in Bolivia. In rapidly growing San Julián colony, I attended the forty-sixth anniversary of the colony's founding at which speakers made frequent references to the original pioneers' struggles against a hostile nature, the utter lack of support from government, and the foundational 1984 roadblock.[7] Missing was any reference to the history of the interfaith coalition of Methodists, Mennonites, and Maryknolls that administered the project throughout the 1970s. Likely, this would have pleased members of the United Church Committee who frequently emphasized that, in their quest to reduce settler dependency, they wanted to leave "no monuments."[8] However, this absence from the colony narrative also points to the tenuousness of historical memory in a region that continued to receive massive waves of new migrants well into the twenty-first century. Indeed, the majority of San Julián's 38,000 residents—many of whom maintain homes in the colony's expanding urban center—had arrived long after Harry Peacock, Marty Miller, and other members of the CIU departed.

Atlanta (the place where this book began) famously promoted itself as a "city too busy to hate," amid the intense and often violent civil rights struggles of the 1960s.[9] Santa Cruz de la Sierra, a city that has grown from 40,000 to 2 million in a little over half a century, and in which an estimated quarter of the population was born in other departments of Bolivia, currently carries the inclusive migrant-friendly motto of "we are all Santa Cruz."[10] It might just as easily describe itself as a city and region "too busy to remember," situated in opposition to the nation's Andean core, where a much more explicit relationship to the nation's history of colonialism and injustice is evident in official government policies to decolonize Bolivian society. This conclusion and epilogue seek to respond to the impermanence of the past and inescapability of the present in Santa Cruz's development. Drawing together the past half century of settlement and regional development I return to the diverse cast of mobile actors that people the preceding chapters. Secondly, I highlight this book's contributions to local, national, and transnational history and our periodization of the second half of the twentieth century in Bolivia. In the epilogue I address Fehr and Falk's question (Why stop there?), by returning to the ways that migration and development continue to reshape twenty-first century Bolivia.

A transnational history of migration (or mobility in the broadest sense), *Landscape of Migration* employs these five distinct case studies to answer a common question. What happens when people, ideas, and technologies are transplanted from one location to the next? More specifically, how are the meanings of that mobility interpreted, conscribed, and enacted in a frontier landscape? Santa Cruz provides an exceptionally rich terrain for addressing this question. Okinawan migrants moved from the rocky islands of the Pacific to the tropical lowlands. Andeans have not only traveled from the Andes to Amazonia. They have also crossed the border to work as sugarcane harvesters in northern Argentina before settling in Santa Cruz. Mennonites carried village structures, language, and social institutions from Russia and Canada through Mexico, Belize, and Paraguay before arriving in Bolivia. Human actors are not the only migrants in this narrative. It is a story that equally depended on the movement of nonhuman migrants from flora and fauna to farming technology. In chapter 5, I explained how aging U.S. machinery (nearly obsolete in the North American context) was repurposed by transnational Mennonite famers to jumpstart a modern agricultural revolution in Santa Cruz. I might have just as easily turned to Gir, Kankrej, and Ongole cattle brought from India to Brazil in the nineteenth and twentieth centuries and crossbred to form the hardy tropical Zebu breed of beef cattle. Taken across the border to Bolivia in the 1950s and 1960s, introduction of this Indian breed initially received a hostile response from local ranchers (and was the subject of lengthy congressional debate) but soon replaced creole stock. Today, Zebu and Nelore cattle (another Indo-Brazilian transplant) dominate an increasingly export-driven cattle industry in Santa Cruz and across much of Amazonia, part of what historian Robert Wilcox describes as a "zebu revolution."[11]

Agroindustrial innovation also brought a different sort of migrant—in the form of the humble soybean—from Asia to the United States and then to Brazil. There it was adapted for the semitropical climate of the cerrado in the 1960s and 1970s before crossing the border into Bolivia. In the process the meaning of soy changed. A food crop in Asia became a versatile agroindustrial cash crop in South America circulating in animal feedlots, as a food preservative and, most recently, as biodiesel. Along with its derivatives (oil, soy cake), soy accounts for the largest share of Bolivia's export earnings after natural gas, resulting in a profound impact on the economy and ecology of the Bolivian lowlands. Its dramatic expansion in Santa Cruz is part of a much broader transnational story that needs to be told. As agrochemical company Syngenta proudly proclaimed in the 2003 advertisement that opens this book, a "United Soy Republic" has arisen at the heart of South America with Brazil, Bolivia, Argentina and Paraguay now producing most of the world's soybeans. Just as miracle rice and dwarf wheat have been central to earlier histories of the Green Revolution in south and Southeast Asia, soy is at the center of that narrative for South America. In Bolivia, the history of migrant soybeans was directly tied to the success of another temperate transplant—Mennonite migrants, who pioneered this novel agrarian transformation.

This is also a narrative of transplanted knowledge, beliefs, and ideas—of development, modernization, faith, and ethnicity rearticulated out of context. Jorge Ruiz's films, shown in La Paz, Santa Cruz, at the 1961 Punta del Este conference and beyond, provide a suggestive response. Distinct audiences embraced certain elements of Ruiz's narrative and aesthetic repertoire while bringing their own meanings to his films. This was as true of regional audiences in Santa Cruz as it was for Che Guevara who criticized the U.S. vision of rural development displayed in Ruiz's *Los Ximul*, at the inaugural conference for the Alliance for Progress in 1961. U.S. authorities even dubbed, *La Vertiente*, a Ruiz film about a community water project in the small town of Rurrenabaque, into multiple languages and distributed copies across the developing world. We do not know if the film was ever shown or how it might have been received across the global south, but we do know that Ruiz transferred his aesthetic repertoire quite smoothly as he left Bolivia to film similar rural modernization projects across the Americas. U.S. informal empire provided the physical and financial infrastructure that made Ruiz's impressive mobility possible. However, it was the pervasiveness of the broader concept of development that made his vision appealing to a variety of regimes in Guatemala, Ecuador, Peru, and elsewhere.

We can see a similar desire to transplant ideas and practices of rural modernization in the 1970s, 1980s, and 1990s as a stream of rural sociologists, anthropologists, and NGOs flocked to Bolivia's largest (and purportedly most successful) colonization project, San Julián, before moving to other projects across the Americas, Africa, and Southeast Asia. In tracing the religious origins of San Julián, I have also shown how U.S. Methodists many hailing from Texas (a former frontier that, like Santa Cruz, became a center of ranching and oil production) understood the spiritual and social challenges of Bolivia's lowland colonists in an era characterized by liberation theology and authoritarian rule. Conversely, the travels of Bolivian Methodist Jaime Bravo during the Banzer dictatorship flip that north-south perspective and suggestively demonstrate how his Aymara background and experience organizing migrant farmworkers in Bolivia carried over to politicized civil rights discourses of red power and brown power during his exile in North America.[12]

A national project of internal colonization, the March to the East involved a compelling mixture of national and foreign actors standing in as proxies for an absent state. I began by exploring the way that Jorge Ruiz and his collaborators produced a seductive narrative and aesthetic of lowland colonization.

State projects, as James Scott has famously argued, need to be visualized in order to be enacted.[13] This was certainly the case in Bolivia. Before large-scale colonization got underway, the state worked extensively to represent its outcome. Through the work of individuals like Ruiz and the National Cinematography Institute, the revolutionary government consolidated a frontier imaginary that was no less enduring for being wildly optimistic. The family unit played a privileged role in this territorial process of integration.[14] Although lesser known than the classics of Latin American literature (referred to by Doris Sommer as "foundational fictions"), the fictional documentary style of Ruiz sought to expose the national and regional tensions of lowland settlement in a conscribed and manageable space where they could be swiftly reconciled through a gendered union that was both racial and environmental. To cite the Ruiz film titled *A Little Bit of Economic Diversification*, the resulting new mestizo of Bolivia, would be a fusion of "kolla blood and camba land."

Ruiz provided both the state, viewers, and international financers with easily digestible narratives about what settlement should look like. "The jungle disappears and progress advances," assures a suggestively titled pamphlet, "A Human Transplant," about the challenges of an unfamiliar tropical environment.[15] "They will soon be good friends," that same pamphlet continued, in which the image of a young child and a tropical bird, stood in for good relationships between new transplants and established farmers in the settlement zones. The settlers, "are now totally adjusted to their new environment and they don't get homesick for their mountains since they can visit them any weekend," the author insists while remaining silent on high rates of settler abandonment that worried officials in the colonization zones.[16]

The happy union of camba (lowlander) and kolla (highlander) was a frequent trope for Ruiz, but lowlanders—and particularly the elite of Santa Cruz—were not always receptive to the large-scale migration of indigenous Bolivians even as they welcomed the promising infrastructure and development projects he portrayed. Cruceño elites had long nurtured a discourse that their region had been abandoned by the state, but they quickly shifted this rhetoric to one of invasion when confronted with a wave of incoming Andean migrants. This regionalist response intensified over the course of the March to the East and is critical to understanding a central paradox. Although Okinawan and Mennonite migrants were highly visible foreigners in Santa Cruz and met some initial opposition, they were ultimately embraced by cruceños as market-driven, desirable producers, a position they consolidated over the following decades. In a surprising contrast that betrays the racial underpinnings of regional identity, cruceños treated their fellow indigenous

Bolivians from the nation's highlands as an undesirable, and even foreign, presence.

While I began by considering the March to the East as a way of "seeing like a state," I ultimately privilege the voices and practices of migrants—Mennonite, Okinawan, Andean—who answered the MNR's call to colonize the east. In chapter 2, I troubled the idea that eastern expansion was solely a national project by placing both Okinawans and Paraguayan Mennonites in the streets of La Paz and the fields of Santa Cruz in the radical early days of the 1952 Revolution. Traveling to U.S.-administered postwar Okinawa and drought-stricken midcentury Mexico strains the boundaries of Bolivian historiography. Yet the histories of repatriated Okinawans and their struggles against U.S. military base construction or the travels of Mennonite braceros to harvest sugar beets and tomatoes in Ontario are not simply transnational curiosities. The experiences of Mennonite braceros and Okinawan evictees structured the way that those new migrants established themselves in Santa Cruz.

These transnational histories offer surprising parallels with Andean migrants who also experienced landlessness, overcrowding, and undocumented bracero labor before settling in Bolivia's lowlands. Their hopes and experiences emerge in a wealth of petitions that Andeans addressed to the state both before and after migrating to the tropics. Uncovering those voices inverts the limited perspective of high modernism as a process by which elites and planners have attempted to implement their visions of modernity. These subaltern actors made their own meanings—sometimes complimentary, sometimes confrontational—of internal colonization projects. On one hand, this approach draws on a rich history of petitioning paternalistic state and crown authorities in modern and colonial Latin America. On the other hand, this genre takes on a unique meaning within the context of mid-twentieth-century development. Initially, the letters of Andean settlers containing bitter complaints of missing services and absent officials seem to resonate with the larger narrative of failure in the historiography of agrarian reform and colonization in Latin America at large.[17] Yet for letter writers, failure was not an end point in the discussion but a call for new rounds of state intervention.

Surprisingly, those subsequent initiatives were often delivered, not by government officials, but rather by religious organizations acting as state proxies in Bolivian colonization. The activities of missionaries and other faith-based workers encompassed a broad spectrum of work that moved from traditional religious activities (ministering) to decidedly secular ones (administering). Even as evangelical organizations like the World Gospel Mission offered to convert displaced lowland indigenous groups and manage their incorporation into Bolivian society, Methodists, Mennonites, and Maryknolls worked with the settlers themselves in public health programs, education, and agricultural extension that overlapped with the labor performed by national and international agencies. In doing so, they came to see themselves as buffers between a heavy-handed state and an abandoned, radicalized colonist. Critically, they became conduits for North American and European funding in a decade when financing for development was waning. By turning furloughs into degrees and actively writing about their work in academic formats, faith-based workers also moved between the field and the university and used church-based work as a springboard for careers with international development agencies and nongovernmental organizations. As a form of flexible capacity that often worked in the absence of the state, these organizations were favorites of Bolivia's revolutionary leaders and expanded their role during the dictatorship of Hugo Banzer. The authoritarianism that characterized Bolivia, and much of Latin America in the 1970s, foreclosed many of the possibilities for radical peasant organizing. In San Julián, under the apolitical language of capacitation, faith-based workers encouraged forms of settler organizing that quietly continued this tradition. With the return to democracy in 1982 and the beginnings of Bolivia's debt crisis, which saw reductions in state funding, the radical organizing characteristics of the pre-Banzer era reemerged with force.[18]

Returning to the experiences of Mexican Mennonites in the final chapter of this book presents several compelling and interrelated ironies that drive home the paradox of lowland development in Bolivia. A revolutionary nation-state that sought to transform traditional indigenous subjects into citizens welcomed foreign horse-and-buggy Mennonites and granted them special exemptions that explicitly exempted them from the central domains of citizenship. Seeking to develop modern small farmers *and* agribusiness on its eastern frontier, the MNR invited a communitarian, traditionalist agricultural community that shunned a wide-range of technological innovations. Yet, horse-and-buggy Mexican Mennonites emerged over fifty years as exactly the sort of model, mechanized and market-oriented, the Bolivian state hoped to create of its own citizenry. By the end of the twentieth century they were the largest producers of Bolivia's most important new cash crop—soybeans. Additionally, they dominated milk and cheese production in Santa Cruz and cultivated new foodways among the Bolivian population. They did so while still predominantly farming the characteristic fifty-hectare plots designated as small producers by the National Institute of Colonization.[19] These twin production cultures—soy and cheese—reflected the tension between a new logic of lowland development based on export agriculture and an older and enduring rhetoric of food security. For Mennonite farmers they also represented twin logics—of risky but lucrative farming and secure but subsistence-level dairying. Freed from participation in the military, school, and politics, Mennonites consistently defined their sense of belonging in agrarian rather than legal terms. As such they were welcomed by cruceños who favored this Mennonite brand of agrarian citizenship based on their status as model producers for, rather than participants in, the nation. In contrast to the image of pacifist, productive Mennonites, cruceños cast Andeans as violent, politicized, and disorderly invaders. By the late twentieth century, this comparison emerged most clearly in relation to Bolivia's largest agricultural exports— with Andeans imagined as aggressive coca farmers (cocaleros) and Mennonites as model soybean farmers (soyeros).

This book extends over several distinct political periods from the revolutionary era of the MNR (1952–64); the transition to authoritarianism and military rule under Barrientos, Hugo Banzer, and others (1964–82); and the return to democracy with its accompanying neoliberal reforms (1982–2000). These new political modalities carried distinct consequences for forms of political organizing. Peasant syndicates flourished under the MNR and were suppressed, often violently, by Barrientos and Banzer only to reemerge in neoliberal Bolivia. Yet, as environmental historians have argued, when viewed in terms of nature, broad similarities are also apparent across disparate political regimes.[20] Although Bolivia's postrevolutionary governments may have broken with many of the progressive policies of the MNR, they each maintained a consensus around the policy of continuous frontier expansion. Over half a century, a series of politically divergent regimes—from the revolutionary and authoritarian to the neoliberal—insisted that the future of national development lay in the east.

In the twenty-first century, as twenty years of neoliberal hegemony collapsed in the face of popular protests, Bolivia experienced yet another political transition that also confirmed the role of the frontier in the national imaginary. In 2005, the election of Evo Morales marked a new era of pluriethnic socialism that resonated with the victories of left-leaning governments across Latin America. Along with prominent leaders of the pink tide like Rafael Correa, Hugo Chávez, and Lula da Silva, "Evo"—as he is typically referred to in Bolivia—was outspoken in his criticism of U.S. involvement in the region (twice expelling USAID, which had supported earlier coca eradication programs). He also nationalized key resources and implemented a series of broad social reforms and antiracism campaigns framed in the powerful language of decolonization. Although its extent is debated among Bolivian historians, Morales's party, the Movement to Socialism (MAS), revived much of the state socialism of the MNR's 1952 revolution from an explicitly indigenous perspective.[21] But whereas Morales's election marked a clear break from late twentieth century politics of neoliberalism, in terms of the March to the East a narrative of political rupture is misplaced. Despite Evo's strong rhetoric of environmentalism, employing concepts like *buen vivir* as a model for alternative development predicated on respect for mother earth (*Pachamama*), the infrastructure programs, resource development, and colonization in the nation's lowlands have continued apace under his government. As such, his vision of the east evinces clear continuities with revolutionary *and* neoliberal regimes. To those familiar with Morales's biography, this is hardly surprising. His indigeneity and his socialism are foundational elements of his political formulation and agenda, but his continued support for frontier expansion is linked to another aspect of his history. Like Okinawans, Mennonites and many of his fellow indigenous Bolivians, the President is a settler.

Throughout this book I explore the ways in which personal trajectories—like those of Methodists Harry Peacock and Jaime Bravo, filmmaker Jorge Ruiz, sociologist James Tigner, and Mennonite leader Johan Wiebe—intersected with the broader dynamics of the March to the East. In the following epilogue I return to Morales's own history. While useful for understanding his policies, Morales's biography also provides us with another convenient device for weaving together the diverse strands of this book. As a colonist his personal trajectory and political development offer a narrative of transnational and internal migration. Intertwined with the long history of the March to the East, Morales's movements parallel, or directly intersect with, many of the central subjects and actors I have discussed. That past is particularly revealing in light of the challenges he faced from the nation's lowlands in the early and latter years of his presidency, a period in which emergent discourses changed the terms, if not the direction, of eastern expansion. Ideas of autonomy have replaced conceptions of abandonment as a mobilizing metaphor for a surprising range of lowland actors. First, I explore the simultaneous demands for autonomy put forward by regional elites and lowland indigenous groups. I then consider how these conceptions of autonomy might be provocatively compared with the conspicuous privileges and forms of cultural and economic autonomy practiced by Bolivian Mennonites.

Epilogue
From Abandonment to Autonomy

In January of 2006 an Aymara settler became the first indigenous president of Bolivia. Juan "Evo" Morales Ayma was born in rural Orinoca canton in the highland department of Oruro in 1959.[1] In the immediate postrevolutionary period, the region appeared frequently in desperate petitions sent in to state authorities like those analyzed in chapter 3. Five years before Morales's birth, the corregidor of Santiago de Andamarca (fifty kilometers to the north of Orinoca along the shores of Lake Poopó) wrote to the director of colonization requesting lands in the new settlement zones. Describing the arid landscape of Oruro as little more than "shifting sands . . . with no benefit for the life of man," he worried that many members of his rural region were leaving their communities to seek work abroad.[2] In 1964, Morales and his family joined the hundreds of thousands of Bolivians that became what politician and writer Fernando Antezana would refer to as a "human river" and "nomads without a country."[3] They traveled south across the altiplano and into neighboring Argentina heading as far as Calilegua in the province of Jujuy to work the sugar harvest. Antezana lamented the fate of these Bolivian braceros whose "modest illusions of investment projects or supporting their family during the harvest, to buy a little house or a cow or start a new stage of life," were rarely realized.[4] Like these Bolivian emigrants, the Morales family found opportunity scarce in Argentina, and they were soon back in Oruro where little had changed for those who eked out a living through small-scale farming and herding. In the late 1960s, a group of farmers from Corque (100 kilometers to the north of Orinoca) wrote to the president demanding help in the form of agronomists, engineers, seeds, and technical assistance. "We want to develop," wrote Maximo Luna and Felix Machaca, in a letter (which opens chapter 2 of this book) "so that one day our sons can be good citizens and good Bolivians."[5]

The Morales family remained in Oruro until the early 1980s when an environmental disaster struck the highlands. El Niño conditions, which had decimated fish stocks in Peru in 1972 and spurred the nascent Mennonite-dominated soy industry in Santa Cruz, repeatedly devastated the Bolivian altiplano in those years. The accompanying severe weather wiped out 70 percent of agricultural production and killed nearly half of the region's herding animals.

In the face of the disaster, Morales's father once again decided to relocate the family. Their first move to the Alto Beni colonization zone near La Paz—the focus of Alliance for Progress–era projects documented by Jorge Ruiz two decades earlier—was unsuccessful. Ultimately, the Morales family chose the Chapare in the tropical lowlands of Cochabamba department. Settlement had expanded along a new highway linking Cochabamba and Santa Cruz whose heroic construction filmmaker Jorge Ruiz had also documented when he returned to Bolivia from Peru in the late 1960s. While the Chapare and neighboring Chimoré river systems were one of the principal colonization zones of the MNR's March to the East, the El Niño disaster and economic crisis of the early 1980s brought new transformations to the region. Following the neoliberal model proposed by economist Jeffrey Sachs, Victor Paz Estenssoro's government laid off the majority of the state-run mining workforce in 1985. In a real-life enactment of Ruiz's 1955 documentary, *A Little Bit of Economic Diversification*, many of these ex-miners, "moved to the Chapare and reinvented themselves."[6] From 1981–86, the population of the region went from 40,000 to 215,000.[7]

The Morales family found land in a tiny settlement of thirty houses, Villa 14 de Septiembre. Coca production was booming throughout the Chapare fueled by strong international demand for cocaine. In response, the U.S. government made the region a central focus for its Andean drug policy, pressuring Bolivia to curb coca production and encouraging small farmers, like the Morales family, to switch to other crops. USAID officials searched for experienced administrators to run the project and, in a compelling twist, turned to some of the faith-based development workers that had operated in San Julián in the 1970s. Former Methodist missionary Harry Peacock joined the program along with MCC volunteer Milton Whitaker.[8] Peacock had spent the last eight years (the height of the Contra War) working with a USAID development project along the Nicaragua–Costa Rica border and was familiar with the bipolar logic of U.S. foreign policy, which in the Bolivian case provided funding for both a militarized policy of coca eradication and a community development initiative to support alternative crops.[9] The credibility that officials like Peacock garnered from previous colonization administration in the Bolivian lowlands helped them in their interactions with coca farmers. Peacock remembers that a particularly bad meeting with "some of Evo's people who were angry with us," was salvaged when one farmer spoke up to let his fellow colonists know that, "before I came to the Chapare, I was in San Julián, I knew this man . . . he never lied to us."[10] Yet neither U.S. funding nor Peacock's reputation could overcome the logistical and economic challenges of the region. It was a near-Sisyphean endeavor—an apt metaphor considering that the main logistical challenge to marketing was transporting produce up the steep, rough mountain roads separating the Chapare from the markets of Cochabamba. As Morales remembers, in the late 1980s 100 pounds of coca leaves could fetch the same as 15,000 oranges.[11] Additionally coca could be harvested four times a year and transported out on a donkey rather than by truck. Milton Whitaker remembers an exasperated U.S. embassy official insisting that, "surely there is some crop that can replace coca." "Maybe opium?" Whitaker responded in jest, acknowledging that "my star kind of fell within the organization after that."[12] Peacock also resigned after six years in the Chapare. USAID had spent approximately $300 million in the alternative cropping program with few gains. In response, the U.S. Drug Enforcement Agency stepped up its aggressive eradication campaign.

Like earlier generations of radical settlers in Yapacaní and San Julián, cocaleros led marches to La Paz, blockaded roads, and engaged in direct confrontations with eradication agents. Over those years Morales moved from a lower-level representative within the union to executive secretary of the Federation of the Tropics responsible for thousands of new settlers. He was briefly imprisoned in 1995 but elected to Congress two years later and continued his political rise in opposition to neoliberalism while actively supporting the Water War in Cochabamba in 2000—a conflict over the privatization of the city's water system—and protests against President Gonzalo Sánchez de Lozada's plans to export gas to Chile in 2003. In 2005, the shepherd-turned-bracero-turned-lowland migrant won Bolivia's presidential elections.

In the 1950s the MNR had hoped that sending impoverished migrants to settle and cultivate Bolivia's neglected frontiers could transform the nation. On a more cynical level colonization also involved a spatialization of dissent in which landless Bolivians and radicalized ex-miners were sent off to the periphery. With Morales's surprising victory, one such migrant returned to the seat of national political power. If Morales's election seemed to offer a convenient narrative resolution to Bolivia's Amazonian-Andean identity, this was not to be the case. After half a century of state-driven integration in the half-moon of the Bolivian lowlands, the departments of Pando, Beni, Santa Cruz, and Tarija met Morales's election with calls for regional autonomy. Their resistance reflected the shift of economic power eastward in Bolivia along with the racialized regional identity that cast highland migrants as foreign invaders. Despite his indigenous background, Morales also faced a very different set of demands for autonomy from lowland indigenous groups who had mobilized politically over the past two decades. Long silenced in official plans for the March to the East, a range of indigenous lowlanders found themselves under increasing pressure as settler frontiers and agribusiness expanded well beyond the integrated north near Santa Cruz de la Sierra. Along with the cruceño elite, they often viewed highland indigenous settlers like Morales as outside invaders of their traditional lands. These simultaneous demands for autonomy, elite and indigenous, have been well documented.[13] In this epilogue I take a different approach, exploring their strange intersection with a very different form of autonomy exercised by Mennonite colonists in Santa Cruz.

Bulls, Beauty Queens, and Buggies

Morales entered the presidency with a commitment to rewriting the nation's constitution and holding a national referendum on departmental autonomy. In mid-2006, Bolivians voted on convening a constituent assembly to address both of those issues. The first motion passed. The second vote was split with the four lowland departments voting in favor of autonomy, while the nation's five highland departments were opposed. As cruceños celebrated, the national government pointed out that because the majority of departments had voted against autonomy, the decision held for the entire nation. The conflicting interpretations spurred protests in Santa Cruz led by the Committee Pro-Santa Cruz, with strong parallels to the civic struggles of the late 1950s. In December of 2007, an estimated one million people rallied for autonomy at the large statue of Christ the Redeemer on the city's second ring. In 2008, autonomy conflicts often pitted cambas against kollas and other people deemed to be foreign.[14] These were led by organizations like Camba Nation and the Cruceño Youth Brigade, an armed wing of the Committee Pro-Santa Cruz. The latter occupied government buildings and marched through Santa Cruz neighborhoods that were home to many Andean migrants leading to violent confrontations in the streets. In one of the most notorious moments of the movement, an autonomy supporter attacked an indigenous Andean woman in Santa Cruz's public plaza before being pulled away by several bystanders.

At the time, foreign commentators and U.S. officials were warning of a civil war or race war in Bolivia, but such dire predictions proved unfounded.[15] A new Constitution passed with a 67 percent majority in 2009, guaranteeing expanded forms of autonomy for Bolivian departments. Although the 2007–9 autonomy movement did not lead to sovereignty for Santa Cruz or transform Bolivia into a failed state, it did provide fertile ground for a range of political scientists, anthropologists, and other academics to reflect on ideas of regional and national identity.[16] Conducting fieldwork in Santa Cruz in the mid-2000s, Bret Gustafson observed a common theme in impromptu gatherings, like the 2007 mass rally as well as in recurring spectacles such as the annual celebrations of Carnival and Santa Cruz's massive business and agricultural fair, EXPOCRUZ. At these events, elites expressed ideas of autonomy through a gendered discourse of hypermasculinity and femininity in which local competitions over prize bulls and beauty queens figured as key sites for regional self-fashioning. In holding up their cattle and their women, cruceños rehearsed their region's defining environmental and racial tropes—natural fecundity and a supposed non-Andean ethnicity.[17] They also paired nostalgic pastoralism with a capitalist cosmopolitanism, thus moving directly from resolute localism to ardent transnationalism while skipping the nation entirely. Although it may seem a surprising place to hedge regional difference, it was a cultural move with notable parallels to other regional contexts in Latin America, especially southern Brazil and northern Mexico.[18]

For Gustafson, few men epitomized the intersection of beauty, bovines, and cruceño autonomy better than Branko Marinkovic. President of the Committee Pro-Santa Cruz and a vocal supporter of departmental autonomy, Marinkovic was the owner of Santa Cruz's largest soybean processing company IOL, S.A., holds extensive ranching properties, and naturally is married to a Bolivian beauty queen. The public face of lowland autonomy in the 2000s, Marinkovic fled Bolivia in 2010 after being implicated in a plot to assassinate Morales at which point his sister Tatiana took over leadership of the family business. With immigrant parents and a degree from a U.S. university, Branko embodies an added paradox of cruceño identity that "revolves around claims of deeply rooted historical particularity [but] also thrives on accommodations with transnational sources of wealth and power."[19] Yet the dramatic regional growth that underpinned his politics—and a good part of Marinkovic's own wealth—was the product of another transnational source. Branko may have courted a Santa Cruz beauty queen of German descent, but his father Silvo Marinkovic—who arrived in Bolivia from Croatia in 1954, the very year the first Mennonite Privilegium was signed by Víctor Paz Estenssoro—had built the family fortune by courting Low-German speaking Mennonite soy producers in the 1970s and 1980s. By the time of the autonomy movements there were approximately 70,000 Mennonites in nearly seventy colonies on the plains of Santa Cruz and neighboring lowland departments, producing nearly half of the raw material for Marinkovic's oil plant. The majority, like Branko, were Bolivian-born children of immigrant parents. For understandable reasons, Mennonites did not become symbols of regional autonomy in Santa Cruz. Yet, it is worth exploring their continuing economic and anecdotal relationship, not only to Marinkovic, but also to bulls and beauty queens that emerged as key symbols in this national-regional conflict.[20]

The distinctive humped white Zebu cattle are seen to symbolize a deep-seated rural identity, but they are in fact recently imported breeds of Indo-Brazilian stock that have displaced traditional creole herds. In a further irony, Mennonite ranchers have become primary producers of this cruceño symbol just as they continue to produce significant quantities of soybeans, another defining cruceño export that is everywhere on display in agricultural fairs. Mennonite cattlemen might not compete in the yearly best-of-breed competitions at EXPOCRUZ, but they crowd the numerous cattle auction houses that dot the eastern outskirts of Santa Cruz where immature cattle are brought in from the Beni to be sold and fattened on local farms. Mennonite farmers like Abram Hamm quietly stroll the raised walkways above the outdoor pens, inspecting the new arrivals and planning out their bids. In the face of volatile commodity prices and unpredictable weather, cattle-raising has emerged in the early 2000s as a more stable, if less lucrative, business than cash cropping for many Mennonites. Of the distinctive white Zebu (the border-crossing Indo-Brazilian cattle that have come to define Santa Cruz), one border-crossing Mexican Mennonite says, "You sell the white ones once per year, you have to see if the money [lasts.]"[21] With the opening of direct beef exports to China in 2019 that market seems poised to surge further, just as soybeans had with the first exports to Argentina in the mid-1980s.

In contrast to rural cattle culture, Santa Cruz's beauty queens seem unbelievably distant from the conservatively dressed Mennonite population of the department. Yet their transnational lives and conspicuous visibility can occasionally bring both Mennonites and cruceña models together in unexpected ways. One freelance journalist captured this literal proximity in 2014, writing "got off the plane today in Santa Cruz, the line to board the next flight was composed of beauty queens and Mennonites."[22] As the diary of Johan Wiebe discussed in chapter 5 indicates, Mennonites exercise a surprising degree of mobility whether they are traveling to Sucre for medical attention, to La Paz for paperwork, or back to Canada or Mexico to work or visit family. With Santa Cruz, the center of a regional fashion industry, cruceña models flying between national and international destinations often find themselves queuing behind globetrotting colony Mennonites.

If the airplane represents the unexpected transnationalism of Mennonite settlers, the distinctive horse-and-buggy stands at the opposite end of this spectrum of mobility and seems to symbolize the rootedness, simplicity, and deliberate antimodernity of Mennonite communities.[23] Yet in an ironic twist, Mennonite buggies and Santa Cruz beauty queens have also come together in strange ways. In May of 2012, the Santa Cruz newspaper *El Día* carried an article entitled "Villamontes Shines with the Most Beautiful of Bolivia."[24] The author reported on the visit of aspiring contestants for Miss Bolivia 2012 to the city several hours south of Santa Cruz and on the edge of the Gran Chaco in the neighboring department of Tarija. In contrast to the autonomy movements, here regional feminine beauty was mobilized, albeit competitively, in the service of national unity. Miss Santa Cruz, Alexia Viruez, would go on to win the competition in June. Yet, on that day, the entire delegation of twenty-one contestants was paraded through town atop horse-drawn buggies provided by the neighboring Mennonite colony of El Palmar. The strange image of Bolivian beauty queens circulating in Mennonite buggies is, again, less surprising than it may initially appear. Mennonite horse-drawn carts are well-known in rural areas of Santa Cruz. Fashioned in thriving colony machine shops, they contain modern suspension and travel easily across the region's dirt roads that alternate between sand dunes in the dry season and deep mud in the rainy season. As such they are prized possessions not just for Mennonites but also for neighboring Andean settlers and sell for upward of $5,000 or $6,000. In Villamontes, the reporter from *El Día* bore witness to one literal enactment of a broader reality, the Mennonite production that undergirds Santa Cruz's regional economy and, in part, drives prosperity that has pushed forward calls for regional autonomy.

Guaraní and Mennonite Autonomies

As cruceños protested the central government in 2007, other ideas of autonomy, building on three decades of political organizing, were also being voiced by the indigenous communities of the lowlands. The MNR, regional elites, and Andean settlers had been silent on the land's indigenous inhabitants, when they described much of Santa Cruz and other colonization zones as vacant or empty, in the 1950s and 1960s. Yet, Guaraní, Chiquitano, Guarayo, Mojeño, and Ayoreo peoples were present throughout the settlement regions even if their extensive, low density land use led the tropical lowlands "to be considered abandoned by the western logic of land use."[25] As the accounts of World Gospel missionaries Geyer and Tamplin make clear, the MNR often welcomed evangelical organizations to incorporate seminomadic indigenous communities like the Ayoreo and the Sirionó into the nation. More populous lowland indigenous groups like the Guaraní and the Guarayo had a different and much longer history of laboring on mission stations. When sugar and cotton boomed in the 1960s and 1970s, many were pressured by regional elites into coerced and poorly paid labor alongside seasonal migrants from the Andes.[26]

In 1982, facing this legacy of exclusion and displacement that accompanied the March to the East, the Guaraní, Chiquitano, Guarayo, Mojeño, Ayoreo, and other lowland communities organized a Confederation of Indigenous Peoples of the Bolivian Oriente (CIDOB). In 1990, while Morales and Andean settlers were resisting coca eradication, CIDOB held a March for Territory and Dignity that brought the silenced struggles of lowland indigenous communities into the national political debate. In a symbolic inversion of the March to the East, indigenous delegates from the lowland Beni department marched to La Paz and ultimately secured the creation of four protected indigenous territories that year. Subsequent marches were held throughout the late 1990s and early 2000s, resulting in additional Communal Lands of Origen (TCOs) some of which have received, or are in the process of obtaining, title. In 2007, as delegates were meeting to rewrite the Bolivian Constitution and lowland elites were demanding regional autonomy, indigenous lowlanders also marched on Sucre to demand their own form of autonomy—focusing on land and control over local resources. These marches led to commitments to the extension of indigenous autonomy and the preservation of indigenous language and culture in the 2009 Constitution alongside a degree of autonomy for Bolivia's departmental governments. Some lowland indigenous autonomy movements were framed in environmental terms *and* in opposition to the perceived settler-friendly policies of the Morales government.[27] In the 2010s, new marches, some met with violence, took place in opposition to Morales's planned construction of a highway through the Isiboro-Sécure Indigenous Territory (TIPNIS), one of the first TCOs created in 1990.

In mid-2011, Guaraní leaders in the small town of Charagua, a few hours north of Villamontes where Mennonite buggies would parade Bolivian beauty queens the following year, were also actively pursuing forms of municipal autonomy that had been guaranteed by the 2009 constitution. While seeking autonomy as part of a national political process, in conversations with volunteers of the Mennonite Central Committee, which was engaged in community development projects in the region, the Guaraní also actively framed their ideal conception of autonomy in relation to their neighbors in Charagua—Old Colony Mennonites.[28] That surprising conversation emerged from some compelling parallels. Both the Guaraní and the Mennonites are culturally distinct minorities in Santa Cruz yet, in the rural area around Charagua, they form the two largest ethnic groups.[29] Furthermore, the central aims of indigenous autonomy movements are precisely those legal rights enjoyed by Mennonite colonists in Bolivia under Decree 06030. As the vice president of CIDOB, Nelly Romero explains, "our dream is to consolidate and expand collectively managed indigenous territories. There we can exercise self-government."[30] The Guaraní maintain an alternative landholding system to preserve the traditional lands that they have won back from the government. Mennonites also hold land in an alternative system that allows for the preservation of large, homogenous blocks blending individual ownership and inalienable communal title. Within the colonies, individuals own and transfer property, but titles are held collectively. This system functioned to restrict land ownership to members of the community, notably in the face of the debt crisis of the late 1980s, when urban lenders unsuccessfully demanded that individual Mennonite lands be placed as collateral for renegotiated loans.

Lowland indigenous groups like the Guaraní also seek to preserve their language by gaining control over local education. It was this issue that drove earlier Mennonite migrations from Canada to Mexico and Bolivia and the right to maintain separate Low German schooling is one of the central elements in the 1962 set of privileges signed by Bolivian President Victor Paz Estenssoro that also included freedom from military service and the right to maintain independent social systems such as a widow and orphans fund. The common objection to Mennonite privileges in Mexico and Bolivia was that they constituted a "state within a state." In a further irony, Bolivia's new constitution attempted to give exactly those rights "a distinctive space within the state," to indigenous groups. As Linda Farthing elaborates, the idea of plurinationality as defined in the new constitution evokes, not just "cultural diversity but also an acceptance of varying values, cultural organizations, forms, and worldviews. Depicting indigenous groups as separate nations within the broader state each with substantive rights to consultation, autonomy, and self-determination turns the idea of the nation on its head, upsetting longstanding notions of race, identity, and territory."[31]

If Mennonite autonomy proved appealing for some indigenous autonomy seekers, it also stood as a difficult system to imagine replicating. Back in 1978, Jesus Bolívar concluded his exhaustive study of Mennonite contributions to regional development with a telling paradox. The very cultural, social, and religious systems that had enabled Mennonites' rapid success in converting the forests of Santa Cruz into productive fields also insulated them from neighboring farmers. He puzzled over how these model farmers might serve as actual models for other lowland communities. Despite the parallels between Guaraní and Mennonite conceptions of autonomy as seen through control over land, language, culture, and ethnic institution-building, it would be disingenuous, at best, to suggest the Mennonite model might be simply transplanted from one community to the other while ignoring the racial privilege, economic power, transnational connections, and legal sanction that have produced autonomous Mennonite communities on the plains of Santa Cruz. Furthermore, while relations between Mennonites and Guaraní in Charagua were largely amicable in the early 2010s, a significant conflict over land existed between urban Guaraní (grouped under the Asamblea de la Nación Guaraní Urbana) and a Mennonite colony (Santa Rita), a short distance from Santa Cruz.[32] Finally, the political route to autonomy opened by the 2009 constitution is one expressly rejected by Mennonite communities who refuse to participate in national politics, even if they voice their acquiescence to whatever form of political organization emerges.[33]

In a further irony, even as indigenous autonomy was being enacted through the new constitution and the Guaraní were looking at Mennonites as model form of autonomy, Mennonite autonomy appeared an increasingly tenuous privilege. In 2010, Morales's government had passed a resolution ending a special permanent resident card for Mennonites (created in 1998) who would now be subject to a renewal process and fee every two years. Mennonites that were non-Bolivian citizens would also need to obtain foreign passports. For the next several years many Mennonites had no valid form of identification as they contested the resolution.[34] In March 2014, as I was completing research for this book, a more worrying draft congressional bill was revealed that planned to annul the 1962 special decree. Although the Guaraní in Charagua had spoken of Mennonites as a model for their autonomy in 2011, national legislators made explicit references to the land rights of lowland indigenous communities as justification for challenging the privileges held by Bolivian Mennonites.[35]

The bill was not introduced at the time, and has not moved forward since, but it revealed some of the new challenges that Mennonites would face in maintaining their conspicuous ethnoreligious autonomy in plurinational Bolivia. On a deeper level, it pointed to an enduring paradox in the broader history of Mennonite settlement in Russia, Canada, Mexico, and Bolivia. Mennonites' eventual success in transforming a former frontier into an agricultural breadbasket, a condition for the initial extension of special exemptions, had consistently eroded the basis for those privileges. As the frontier disappeared, long-absent state authorities and institutions stepped in. Conflicts with the Russian imperial state in the 1880s, the Canadian federation in the 1920s, and the Mexican republic in the 1950s had spurred new migrations. By the early twenty-first century, Santa Cruz had also undergone a similarly dramatic transformation—from a frontier to the center of economic power in Bolivia. It remained unclear if this growth would eventually erode Mennonite privileges in Bolivia. Yet the history of other Latin American Mennonite communities challenges that declensionist narrative. Canada had forced Mennonites to take part in national schooling in the twenties, but Mexican Mennonites still maintained the set of privileges granted by President Álvaro Obregón in 1921. Paraguayan Mennonites also continued to exercise the privileges they had received in exchange for settling the Chaco in 1926. For their part, Bolivian Mennonite leaders responded to these challenges in new and familiar ways that highlighted their diplomatic acumen and transnational ties. In the case of the 2010 identity card issue, colony Mennonites enlisted the assistance of the Mexican, Canadian, and Paraguayan embassies along with a host of local legal counsel and, by early 2015, had normalized their residency. In the case of the 2014 bill, Mennonites reacted just as Johan Wiebe and others had when a 1976 decree by Hugo Banzer threatened their privileges and special exemptions, by boarding planes to La Paz to negotiate directly with government officials. While they continued to brandish their credentials as agricultural citizens or producers for the nation, Mennonites could also claim a legal citizenship when negotiating with the state. By the twenty-first century nearly 80–85 percent of the country's 70,000 Mennonites were Bolivian citizens, many without any secondary or dual citizenship.[36] ANAPO, the regional soy-producing organization, recognized this fact in their annual reports when they finally stopped separating Mennonite and national producers as they had done throughout the 1980s, 1990s, and early 2000s.[37] Most tellingly, when Mennonites in California Colony—a granddaughter colony of the first Mexican Mennonite colony in Bolivia (Riva Palacios)—attempted to legalize their title in the late 2000s, they invoked their rights as campesino communities under the Agrarian Reform Law rather than as Mennonite colonies defined by the 1962 set of privileges.[38]

Mennonites were not the only lowland settlers to find both challenges and opportunities in plurinational Bolivia. Okinawan and Japanese colonists had also made a significant impact on the region over the previous half-century and consecrated their role as producers for the nation. Yet unlike Mennonites who typically let their production speak for itself or negotiated with government ministers behind closed doors, Okinawan and Japanese settlers, and their children, actively performed their integration in the nation. This was apparent in carefully scripted annual celebrations which government officials and the Bolivian public were invited to attend. Colonists displayed both their transnational ties—evident in the Japanese General Hospital in Santa Cruz, visits from members of the Japanese imperial family, Japanese cultural centers (promoting everything from sushi to sumo), and projects supported by the Japanese International Cooperation Agency. They also branded their local production. Wandering in markets of Santa Cruz, one can purchase rice from unmarked bags—the product of Andean settlers—next to packaged superior quality rice, which proudly proclaimed its origins in the Japanese colony of San Juan Yapacaní as well as Okinawa Noodles, products of the agro-industrial cooperatives of San Juan (CAISY) and Okinawa Colony (CAICO). A sign at the entrance to the latter cooperative boasts that Okinawa is also the center of wheat production in Bolivia thus laying claim to a crop that Bolivian officials have unsuccessfully tried to increase production of over the past decades among Mennonite and Andean farmers. These two organizations, CAICO and CAISY, also hold permanent seats on the board of directors of the regional soy producers' organization (ANAPO) and actively participate in other cruceño business organizations like the Chamber of Eastern Agriculture (CAO) and the Bolivian Institute for Foreign Commerce (IBCE). When Jaime Duranović, who renegotiated the loans of Mennonite debtors in the 1990s before opening a daily bus service to the Mennonite colony of Riva Palacio, had confidently proclaimed that someday there would be a "Mennonite President" in Bolivia, we had both laughed. Despite his sincerity, the idea seemed a distant prospect. In contrast Michiaki Nagatani, the son of Japanese migrants to San Juan Yapacaní in 1955, had already been elected to Congress nearly a decade earlier, as a member of the MNR, in 2005.

Andean migrants have also refashioned themselves over the last decade. Colonies like San Juan Yapacaní and San Julián were islands of support for Evo Morales in a region largely committed to autonomy. His election promised continued representation for highland migrants at the national level, but in Santa Cruz, the autonomy movements of indigenous and elite lowlanders (while entirely distinct from one another) both framed these Andean colonists as outside invaders. In response, Andeans and their descendants have turned to new discursive strategies in the twenty-first century. Many settlers have rejected the title of colonist altogether. Arguing that it is impossible to colonize one's own nation, they have rebranded themselves as *interculturales*, who, like Morales, have united a fractured nation with their personal migrations. In doing so they have also challenged racialized regional distinctions. "The *camba* has become *kolla*, and the *kolla* has become *camba*," one man, whose father had migrated to one of the first government supported colonies in Santa Cruz in 1955, told me.[39] The Methodist Jaime Bravo—who blurred the line between development practitioner, missionary, and migrant over his long career in the lowlands—made a similar point. "Look I am an Indian [*indio*], of Aymara origin, from the altiplano," Bravo explained. "I came to Santa Cruz and found a woman of the east, a camba who lived in Mineros, there close to a zone of colonization. I got married to her. And she taught me her culture, and I taught her mine. And the two of us had children, and now my children are not cambas nor kollas, my children are Bolivians.[40]

Bravo's narrative eloquently ties together romance, intergenerationality, and regional and national identity. It is a migrant narrative, like those offered by many in this book—from Evo Morales to filmmaker Jorge Ruiz. Ironically, Bravo told me this story as we sat at the entrance to Plan 3000, the poor and largely Andean neighborhood of Santa Cruz de la Sierra that was the target of violent attacks pitting supporters of departmental autonomy like Camba Nation against kolla residents during the 2008 autonomy conflict. The Japanese Bolivian congressman, Michiaki Nagatani, offered a different version of this narrative. He claimed that his decision to enter politics had been inspired by earlier conflicts between cambas and kollas in Santa Cruz, noting that "I was not one or the other and I could work with rural people, as such I could serve as cushion [between the two]."[41] It is a narrative also suggested by Germán Bravo (Jaime's brother) who lives in, and is married to a member of, Okinawa colony. For Germán, annual colony celebrations such as Okinawa's harvest festival, as well as larger anniversaries were opportunities not only for renewed transnationalisms (the colony was celebrating its sixtieth anniversary in 2014 precipitating a flood of returning Okinawan–Bolivian *dekasegi* workers from Japan), but also multiple forms of integration. In addition to broadcasting Okinawan agrarian citizenship to Bolivian officials these spectacles also provided a neutral site for *cambas* and *kollas* to come together.[42] Peter Wieler from the Mexican Mennonite colony of Swift Current that borders Riva Palacio, offered a more playful and palatable attempt at resolving transnational and regional dynamics, in branding the popular tortilla chip company that he founded (with assistance from the Mexican embassy in Bolivia) in the same years as the Camba Nation–led autonomy movement in Santa Cruz was gaining strength. Wieler's "Nachos Mexicambas" produced with corn grown on Mennonite farms and sold in stores across Bolivia offers an easily digestible packaging of Mennonites' Mexican origins and their new lives as camba pro-ducers, or agrarian citizens, in Santa Cruz.[43] These three examples of narra-tive and naming—Andean, Okinawan-Japanese, and Mennonite—certainly conceal as much as they reveal. But they also return us to the central themes of this book, demonstrating the ongoing work of migrants to lay claims to a contested landscape produced through the long and ongoing history of mobility in Bolivia's March to the East.

Notes

Abbreviations in the Notes

AGN	Archivo General de la Nación, Mexico City
AHD	Archivo Histórico Departamental Hermanos Vásquez Machicado, Santa Cruz
AHINM	Archivo Histórico del Instituto Nacional de Migración
AHUAGRM	Archivo Histórico de la Universidad Autónoma Gabriel René Moreno
ALP	Archivo La Paz
ANAPO	Asociacíon Nacional de Productores de Oleaginosas
ANB	Archivo Nacional de Bolivia
BAHCN	Biblioteca y Archivo Histórico del Congreso Nacional
BIICA	Biblioteca del Instituto Interamericano de Cooperación para la Agricultura
BMLT	Biblioteca Miguel Lerdo de Tejada, Mexico City
BNB	Biblioteca Nacional de Bolivia
CB	Cineteca Boliviana, La Paz
CBF	Bolivian Development Corporation
CBF-BID	Corporación Boliviana de Fomento-Banco Interamericano de Desarrollo
CONACINE	Consejo Nacional del Cine
IC	Instituto de Colonización
INC	Instituto Nacional de Colonización
INE	Instituto Nacional de Estadística
INRASC	Instituto Nacional de Reforma Agraria, Santa Cruz de la Sierra
IWGIA	International Work Group for Indigenous Affairs
JICA-CIAT	Japanese International Cooperation Agency-Centro de Investigación de Agrícola Tropical
MCCA	Mennonite Central Committee Archives
MDRT	Ministerio de Desarrollo Rural y Tierras, Depository
PIEB	Fundación Para la Investigación Estratégica en Bolivia
PR	Presidencia de la República
SEGOB	Secretaría de Gobernación (Mexico)
UN	United Nations
UNESCO	United Nations Educational, Scientific and Cultural Organization
USCAR	United States Civil Administration of the Ryukyu Islands
USNA	U.S. National Archives—College Park
WGM	World Gospel Mission

Introduction

1. Jaime Bravo, interview with author, Santa Cruz de la Sierra, August 2014.

2. This controversial advertisement was discussed in "The United Republic of Soybeans" and Lapenga, *Soybeans and Power*, 5.

3. Dinani, "En-gendering the Postcolony."

4. Malaysian Prime Minister Tunku Abdul Rahman's creation of a Federal Land Development Authority in the late 1950s to resettle impoverished Malaysians on small cooperative farms.

5. Tsing, *In the Realm of the Diamond Queen*.

6. Garfield, *In Search of the Amazon*, 22.

7. Skidmore, *The Politics of Military Rule*, 145.

8. As Robert Wilcox points out, the view held that both "the Cerrado and the Amazon Rain forest were underexploited and simply needed modern science to render them economically viable and thereby attractive regions for settlement." Wilcox, *Cattle in the Backlands*, 231.

9. Latta and Wittman, *Environment and Citizenship*.

10. Sharon, "Inscribed in the Margins."

11. Cullather, *The Hungry World*.

12. Field, "Ideology as Strategy," 147–83, esp. 149.

13. Scudder, *The Development Potential of New Lands Settlement*.

14. Sandoval Arenas, *Santa Cruz, economía y poder*, 53. The estimate for Santa Cruz up to 1980 was 28,712 families. These figures are contested and likely understate the numbers primarily because many families migrated spontaneously without government support, and others came down as seasonal laborers before settling permanently.

15. Sivak, *Evo Morales*, 160.

16. "Conclusiones y resoluciones," BNB.

17. "The World's Fastest Growing Cities."

18. INE, *Censo agropecuario*, 12.

19. Thomson, *The Bolivia Reader*.

20. Estenssoro, "Decreto Supremo 4192, Oct. 6, 1955."

21. "Charles Stuart Kennedy interview with William Dietrich Director, Cultural Center, USIS Santa Cruz (1970–72)."

22. Loewen, *Village among Nations*; Suzuki, *Embodying Belonging*; Gill, *Peasants, Entrepreneurs, and Social Change*; Stearman, *Camba and Kolla*.

23. Lim, *Porous Borders*; Putnam, *Radical Moves*; Young, *Alien Nation*.

24. See, for example, in Tejada and Tatar, eds., *Transnational Frontiers*.

25. Buchenau, "Small Numbers, Great Impact," 23–49.

26. Lesser, *Negotiating National Identity*.

27. Whiteford, *Workers from the North*.

28. Anthropologist Lund Skar in, *Lives Together-Worlds Apart* explores similar indigenous migration patterns in Peru that were both rural–urban (to Lima) and rural–rural (to the Amazon basin).

29. Soliz, "Fields of Revolution"; Gildner, "Indomestizo Modernism"; Pacino, "Prescription for a Nation."

30. Tinsman, *Partners in Conflict*.

31. To cite one of many examples, settlers in San Julián colony only received full title to their land in the early 2000s, thirty years after the colonization project began. Harry Peacock conversation with the author, San Julián, July 2012.

32. Armiero and Tucker, *Environmental History of Modern Migrations*, 5.

33. Identified as a major subfield by Sutter, "The World With Us," 94–119.

34. MacLennan, "Waves of Migration: Settlement and Creation of the Hawaiian Environment," in Armiero and Tucker, *Environmental History of Modern Migrations*, 38; Nodari, "Immigration and Transformation of Landscapes in Misiones Province, Argentina and Southern Brazil," in Blanc and Freitas, eds., *Big Water*; Nash, *Inescapable Ecologies*.

35. Nodari, "Immigration and Transformation of Landscapes."

36. A role not unlike that of Syrian-Lebanese migrants in Brazil. See Karam, *Another Arabesque*.

37. Grandin, *Fordlandia*; Gordillo, *Rubble*. Also see Hetherington, *Guerrilla Auditors*.

38. A gap pointed to in a recent issue of the *Journal of Peasant Studies* dedicated entirely to soybeans. See introductory essay by Oliveira and Hecht, "Sacred Groves, Sacrifice Zones."

39. McKay and Colque, "Bolivia's Soy Complex."

40. INE, *Censo agropecuario*, 18.

41. Fabricant and Gustafson, eds., *Remapping Bolivia*.

42. Cornelio Peters, interview with author, Riva Palacio Colony, Bolivia, May 2014; Jaime Bravo, interview. This critical idea of farming as "mining" routinely appeared in discussion of agricultural expansion related to soy and other speculative booms during my fieldwork.

43. Dimitrina Mihaylova also identifies the strategic frontier discourse of abandonment in a radically different context, the post-socialist Bulgarian frontier with Greece. See "Reopened and Renegotiated Borders."

44. The relationship between foreign actors and state-building is particularly well developed in relation to missionaries and NGOs in Latin America. For an example of the former see, Hartch, *Missionaries of the State*. For the latter, see Bebbington and Thiele, eds., *Non-Governmental Organizations*.

45. Minchin, "Eastern Bolivia and the Gran Chaco," 404.

46. The perspectives of Minchin, Orton, and other observers regarding transport, territoriality, and sovereignty in Bolivia are also discussed in chapter 3 of Fifer, *Bolivia: Land, Location, and Politics*.

47. Minchin, "Eastern Bolivia and the Gran Chaco," 417.

48. Truett, *Fugitive Landscapes*.

49. Minchin, "Eastern Bolivia and the Gran Chaco," 417. For an explanation of the concept see Piper, *Cartographic Fictions*.

50. Duguid, *Green Hell*, 222.

51. Duguid, 259.

52. Duguid, 200.

53. Nobbs-Thiessen, "Channeling Modernity: Nature, Patriotic Engineering, and the Chaco War," in Chesterton, *The Chaco War*, 67–89.

54. Alcaldía Municipal de Santa Cruz de la Sierra, "Tarija y el tratado con la Argentina— Argentina y Bolivia y su vincculación ferroviaria efectiva en el oriente," vol. 23, Folletos Históricos, AHUAGRM, 4.

55. Hoyos, *Territorios ignorados.*

56. See Garfield, *In Search of the Amazon.* U.S. wartime concern over strategic resources also led to a renewed development of rubber supplies in the Brazilian Amazon.

57. White, *Railroaded.*

58. Hajdarpasic, *Whose Bosnia?* Hajdarpasic invokes the concept of "nationalist self-fashioning" and its intellectual trajectory in a primarily ethnographic sense.

59. Raffles, *In Amazonia.*

60. Yu, "Los Angeles and American Studies."

61. Tinsman, *Buying into the Regime*, 9.

62. For a discussion of the way these new Brazilian immigrants have been viewed in Santa Cruz see Mackey, "Legitimating Foreignization."

63. See McNeill, *Something New Under the Sun.*

64. In this chapter I concentrate on Mennonite and Methodist roles in colonization to engage a broader narrative about the far-reaching effects of Latin America's "Protestant Boom" while devoting less attention to the Maryknolls that worked alongside them because that history has been well documented. See, for example, Fitzpatrick-Behrens, *The Maryknoll Catholic Mission in Peru*; Fitzpatrick-Behrens, "From Symbols of the Sacred"; Fitzpatrick-Behrens and LeGrand, "Canadian and U.S. Catholic Promotion of Co-operatives."

Chapter One

1. Valdivia, *Testigo de la realidad*, 22.

2. See similar approaches to U.S. and Brazilian frontier history in Limerick, *The Legacy of Conquest*, and Garfield, *Indigenous Struggle at the Heart of Brazil.*

3. Silva, "A construção simbólica do oeste brasileiro (1930–1940)," in Silva et al., eds., *Vastos sertões: História e natureza na ciência e na literatura* (Rio de Janeiro: Mauad, 2015).

4. Brandellero, *Brazilian Road Movie.*

5. Guevara Arze, *Plan inmediato*, 12.

6. Guevara Arze, 77.

7. Guevara Arze, 33.

8. For a contemporaneous discussion of some of the problems arising from the early Agrarian Reform see, Carter, *Aymara Communities.*

9. Guevara Arze, *Plan inmediato*, 83.

10. Guevara Arze.

11. Valdivia, *Testigo de la realidad*, 70.

12. Sommer, *Foundational Fictions*, ix.

13. Malitsky, *Post-Revolution Nonfiction Film.*

14. Rankin, *¡México, la patria!*, 2.

15. Garitano, *African Video Movies*, 46.

16. Malitsky, *Post-Revolution Nonfiction Film*, 61.

17. Kohl and Farthing, *Impasse in Bolivia*, 95.

18. Burton, *The Social Documentary*, 17.

19. Smith, "Film as Instrument of Modernization," 73.

20. Valdivia, *Testigo de la realidad*, 94.

21. Sánchez-H., *The Art and Politics of Bolivian Cinema*, 30–31. Initially created as the Ministry of Press and Propaganda in 1952 and subsequently renamed.

22. Instituto de capacitación política, *Lecciones de propaganda, organización y agitación* (La Paz: SPIC, 1953), BNB, 6, 10.

23. For a discussion of the use of images in religious indoctrination in colonial Latin America, see Gruzinski and MacLean, *Images at War*.

24. Instituto de capacitación política, *Lecciones de propaganda*, 24.

25. Instituto de capacitación política, 21.

26. Paz Estenssoro, "Decreto Supremo 3342, March 20, 1953."

27. Gandarilla, *Cine y televisión en santa cruz*, 186. From 1953–56, 42 of 136 news bulletins deal with lowland themes. From 1958–59, 20 of 47 shorts related directly to Santa Cruz.

28. Shaw, "Vargas on Film," 212.

29. Cerruto, *Viajando por nuestra tierra*, CB.

30. Cerruto.

31. Cerruto.

32. Cerruto.

33. Diputados, *Redactor: Actas públicas* (La Paz: Hemeroteca,1956), 73, BAHCN.

34. Malitsky, *Post-Revolution Nonfiction Film*, 85.

35. Diputados, *Redactor: Actas públicas* (La Paz: Hemeroteca, 1956), 73, BAHCN.

36. Li, *The Will to Improve*.

37. Ruiz, *La Vertiente*.

38. Ruiz, *Los Primeros*.

39. Sommer, *Foundational Fictions*, 24.

40. Li, *The Will to Improve*.

41. Earlier regional movements also took place during the First World War as part of the "Railway or Nothing" movement. See Peña, *La permanente construcción de lo cruceño*.

42. Harvey, "The Materiality of State Effects," 135.

43. Comité pro-defensa intereses cruceños, *Argentina y bolivia y su vinculación ferroviária efectiva en el oriente* (La Paz: Artistica, 1942), 4, AHUAGRM.

44. "Inauguración del ferrocarríl y entrevista de los Presidentes," *El Deber*, January 9, 1955.

45. For a discussion of the incompleteness and spectacle of railway inauguration in Guatemala, see Grandin, *The Blood of Guatemala*, 161, 174.

46. "El tráfico en el Río Grande," *El Deber*, January 1955.

47. "Son cinco las personas ahogadas en el Río Grande," *El Deber*, February 9,1955; "A propósito de la tragedía última en el Río Grande," *El Deber*, February 9, 1955.

48. "Noticias del F. C. Corumbá-Santa Cruz–el servicio ferroviario es penoso," *El Deber*, January 30, 1955.

49. "Punibles irregularidades en el ferrocarríl Corumbá-Santa Cruz," *El Deber*, July 2, 1956. Only two days later, *El Deber* covered another accident, product of the "caprice and disorder" of the train authorities in which a young merchant Claudina Vaca Flor had been killed when traveling back from Puerto Suarez on the Brazilian border. "El último accidente ferroviario," *El Deber*, July 4, 1956.

50. White, *Railroaded*, xxix.

51. Clark, *The Redemptive Work*.

52. P. C., "Marcada importancia economica del ferrocarríl Corumbá-Santa Cruz," *El Deber*, March 11, 1955.

53. Indiano, "Vasto y prometedor mercado extranjero para la producción frutícola de Santa Cruz," *El Deber*, March 14, 1955.

54. "Es una tragedía el servicio ferroviario Corumbá-Santa Cruz," *El Deber*, December 19, 1956.

55. "Adquiere proporciones inusitadas el contrabando de Santa Cruz a Brasil," *El Diario* (La Paz), reprinted in *El Deber*, May 1, 1956.

56. For a Brazilian perspective on the railway, see Wilcox, "Ranching Modernization in Tropical Brazil."

57. P. C., "Santa Cruz, epicentro de la economía nacional tiene derecho a sus beneficios," *El Deber*, May 4, 1955.

58. P. C., "Santa Cruz, epicentro de la economía nacional," *El Deber*, May 4, 1955.

59. Clouzet, "Por la nueva carretera," *El Deber*, December 14, 1955.

60. Ruiz, *Juanito sabe leer*.

61. For a critique of the "closed corporate community" model of village life, see Langer, "Bringing the Economic Back in."

62. Sanabría, "Nuestro indio," *El Deber*, August 2, 1956.

63. For a further discussion of the emergence of regional identity in relation to local historical and anthropological descriptions emphasizing the uniqueness of Santa Cruz's indigenous population, see Pruden, *Cruceños into cambas*.

64. Siles Zuazo, "Decreto Supremo 4740. Sept. 23, 1957."

65. Diputados, *Redactor: Actas publicas*, 31st Sesion Ordinaria, September 24, 1957 (La Paz: Hemeroteca), 331–33, BAHCN.

66. Diputados, *Redactor: Actas publicas*, 31st Sesion Ordinaria, September 24, 1957 (La Paz: Hemeroteca), 331, BAHCN.

67. "Santa Cruz: Crucificado," *El Diario* (La Paz), reprinted in *El Deber*, September 24, 1957.

68. Siles Zuazo, "Supreme Decree 4758. Oct. 29, 1957."

69. "Reacción extraordinaria del pueblo cruceño," *El Deber*, November 1, 1957.

70. *El Deber* printed full copies of the speeches the following day. Pinto, "Mensaje al pueblo boliviano," *El Deber*, November 1, 1957.

71. Gandarilla, *Cine y televisión en santa cruz*, 178.

72. Maldonado, "Crítica cinematográfica," *El Deber*, April 26, 1958.

73. Maldonado.

74. Gandarilla, *Cine y Televisión en Santa Cruz*, 179.

75. Gandarilla, 180.

76. Gandarilla, 208–9.

77. Gandarilla, 210.

78. "El cine en puerto pailas," *El Deber*, 1955.

79. Gandarilla, *Cine y Televisión en Santa Cruz*, 152.

80. Gandarilla, 153.

81. Parker, *Hearts, Minds, Voices*, 36.

82. Nacify, *A Social History of Iranian Cinema*, xxxvii.

83. Parker, *Hearts, Minds, Voices*, 118.

84. Valdivia, *Testigo de la realidad*, 81.

85. Zhuoyi Wang, *Revolutionary Cycles*.

86. Valdivia, *Testigo de la realidad*, 81.

87. Ernesto Guevara, *Our America and Theirs*.

88. Douglas Dillon to President Kennedy, "Telegram From the Embassy in Uruguay to the Department of State," August 9, 1961.

89. Guevara, *Our America and Theirs*, 38.

90. Guevara, 44.

91. Guevara, 50.

92. Field, *From Development to Dictatorship*.

93. Field.

94. Parker, *Hearts, Minds, Voices*, 71.

95. Valdivia, *Testigo de la realidad*, 135.

96. Valdivia, 131–32.

97. Ruiz, *Los Ximul*.

98. Ruiz.

99. Valdivia, *Testigo de la realidad*, 136.

100. Beltrán, "Comunicación para el desarrollo."

101. Shaw, "Vargas on Film," 208.

102. Valdivia, *Testigo de la realidad*, 137–40.

103. Guevara, *Our America and Theirs*, 86.

104. King, *Magical Reels*, 157.

105. Valdivia, *Testigo de la realidad*, 136.

106. Field, "Ideology as Strategy," 149. From 1953–64, Bolivia had received more absolute funding from the United States than from any other Latin American nation.

107. Ruiz, *Las montañas no cambian*.

108. Beltrán, "Comunicación para el desarrollo."

109. Soria and Sanjínes, *Qué es el plan decenal*.

110. Urquidi et al., "Principios de cooperativismo," BIICA. The SAI, with its base in Montero was already producing visual materials for colonists by the late 1950s.

111. "Como viviré y trabajaré," BIICA.

112. Welch, *The Seed Was Planted*, 297. For an analysis of the Chilean case, see Tinsman, *Partners in Conflict*.

113. Mesa Gisbert, *La Aventura del cine boliviano*, 65.

114. "Como viviré y trabajaré."

115. "La ruta del desarrollo," BIICA.

116. Salmon and Ruiz, "Un transplante humano."

117. Salmon and Ruiz.

118. Díaz, *Autonomías Departamentales*.

119. Wright and Wolford, *To Inherit the Earth*, 39. For discussion of images of Brasilia, see Sadlier, *Brazil Imagined*, 257.

120. Beltrán, "Comunicación para el desarrollo."

121. Smith, "Film as instrument of modernization."

122. Field, *From Development to Dictatorship*.

Chapter Two

1. A standard account of the Revolution is Malloy, *Bolivia: The Uncompleted Revolution.*

2. Regier, "Bolivien-Reise," *Mennoblatt,* January 1953.

3. Regier.

4. Expediente "Uruma Society," ANB, IC, 547.

5. Expediente "Uruma Society," ANB, IC, 547.

6. See Putnam, *Radical Moves*; Guridy, *Forging Diaspora.*

7. See McKeown, "Conceptualizing Chinese Diasporas."

8. The literature on Okinawan identity in relation to Japanese annexation and U.S. administration is vast and often divisive. For a condensed history of Japanese assimilationist policies, see Christy, "The Making of Imperial Subjects in Okinawa."

9. Amemiya, "The Bolivian Connection."

10. See Lai "Southern Encounters: Ethnic Nationalism and Okinawa Military Migration to Latin America," paper presented at the American Historical Association Conference, Washington, D.C., January 2018.

11. Sawatzky, *They Sought a Country,* 25–26.

12. Tigner, "The Okinawans in Latin America," ii.

13. Tigner, 520.

14. Also see Iacobelli, "James Tigner and the Okinawan Emigration Program" in Tejada and Tatar, eds., *Transnational Frontiers of Asia and Latin America since 1800.*

15. Iacobelli, "James Tigner and the Okinawan Emigration Program," 520.

16. Amemiya, "The Bolivian Connection."

17. Iacobelli, *Postwar Emigration to South America* estimates Okinawa's population at 500,000 in 1946 growing to more than 787,000 the year Tigner's report was published in 1954.

18. The contentious issue has been explored from a variety of diplomatic, social, and cultural angles. For an approach that draws on Okinawan perspectives, see Hein and Selden, *Islands of Discontent.*

19. Translated news clipping, *Okinawa Shimbun,* March 12, 1955, Record Group (hereafter RG) 260, Box 228, Land Issues, USNA. The *Okinawa Shimbun* is a conservative newspaper but one that frequently criticized U.S. forces during the land conflicts. It ceased publication in 1958.

20. Petition posted at a U.S. military base, November 11, 1955, RG 260, Box 281, Land Issues, USNA.

21. Tigner, "The Okinawans in Latin America," 522.

22. Toguchi and Oroku Village Incident Reports, September and December 1953, RG 260, Box 104, Internal Political Activities, USNA. Amemiya, "The Bolivian Connection."

23. Miyume Tanji, *Myth Protest and Struggle in Okinawa,* 54.

24. Petition to USCAR Deputy Governor from residents of Maja District, September 30, 1955, RG 260, Box 281, Land Issue, USNA.

25. Petition to USCAR Deputy Governor from residents of Maja District..

26. Memo regarding the Land Evacuation Issue of Yagihara-Ku, March 24, 1955, RG 260, Box 280, Land Issues, USNA.

27. Toguchi Village Incident Report, 1953, USNA.

28. Memo from James Ross, USCAR Director of Economic Development, March 30, 1959, RG 260, Box 257, Economic Development Files, USNA.

29. Report on Fiscal Year 1960 on Financial Assistance to Overseas Migration, Department of Social Welfare, Naha, Japan, RG 260, Box 258, Visitors-Emigration, USNA.

30. Amemiya, "The Bolivan Connection."

31. Amemiya.

32. Amemiya.

33. Masterson and Funada-Classen, *The Japanese in Latin America*, 179.

34. Higa, "Investigation of the Okinawan Colonies in Bolivia," 1961, RG 260, Box 257, Economic Development Files, USNA.

35. Article summary in internal memo. Southern Area Fellow Countrymen Assistance Society, "Kuroshio," *Okinawa and Ogasawara*, May 15, 1958, RG 260, Box 220, Entry and Exit Control Files, 1955–70, USNA.

36. Internal Memo from General Lemnitzer, January 1956, RG 260, Box 220, Entry and Exit Control Files, 1955–70, USNA.

37. Correspondence with General Moore, January 1956, RG 260, Box 220, Entry and Exit Control Files, 1955–70, USNA.

38. Correspondence, Donald Booth to Chief of Civil Affairs, July 28, 1958, RG 260, Box 220, Entry and Exit Control Files, 1955–70, USNA.

39. Minutes from Conference on USCAR Certificate of Identity, October 17, 1960, RG 260, Box 220, Entry and Exit Control Files, 1955–70, USNA.

40. Minutes from Conference on USCAR Certificate of Identity.

41. Ryukyuan Overseas Emigration Corporation, "Proposed Plan for Use of ICA Funds," July 29, 1961, RG 260, Box 257, Economic Development Files, USNA. The ROEC had replaced the Uruma Society as the entity responsible for Okinawan emigration.

42. Memo, "Preliminary Meetings of Defense-State-ICA Working Group on Emigration from the Ryukyu Islands," August 8, 1958, RG 260, Box 220, Entry and Exit Control Files, 1955–70, USNA.

43. Letter from USOM Lieutenant Colonel Isaac Cundiff to Chief of Civil Affairs in Okinawa, January 14, 1957, RG 260, Box 220, Entry and Exit Control Files, 1955–70, USNA.

44. Letter from USOM Lieutenant Colonel Isaac Cundiff to Chief of Civil Affairs in Okinawa.

45. Report from U.S. Embassy La Paz, "Staff Study on Okinawan Immigration," May 29, 1959, RG 260, Box 220, Entry and Exit Control Files, 1955–70, USNA.

46. Report from U.S. Embassy La Paz.

47. "La inmigración a Bolivia debe ser cuidadosamente seleccionada," *El Deber*, May 6, 1955.

48. Although the meaning of "whitening" could differ widely in regional and national contexts, Peter Wade provides the most commonly cited account in *Race and Ethnicity in Latin America*. Also see Stepan, *The Hour of Eugenics*.

49. "Del altiplano al oriente: Dezplazimiento de poblaciones altiplánicas al oriente," *El Deber*, May 6, 1955.

50. Another article joined in the nativist posturing. "Santa Cruz is living a historic moment in its production," it began. The benefits were clearly meant for its own "natives." "La producción agraria cruceña va cumpliendo su hora histórica," *El Deber*, May 1955.

51. "Capital cruceño dentro del desarrollo agrario departamental," *El Deber*, May 15, 1955.

52. "En torno a la inmigración japonesa," *El Deber*, June 1955.

53. "Santa Cruz ausente en la solución de sus propios problemas," *El Deber*, September 5, 1956.

54. The announcement of Bonoli's arrival appeared in an article titled "Otra colonia italiana." In the same issue, the paper elaborated on the proposed migration in the article "Colonización italiana en bolivia," *El Deber*, December 14, 1955.

55. Luis Simón García, "Mi saludo a Santa Cruz," *El Deber*, June 7, 1956.

56. "La última novedad: 'Colonizadores' que 'descolonizar,'" *El Deber*, August 1, 1956.

57. "La última novedad: 'Colonizadores' que 'descolonizar.'"

58. "Santa Cruz ausente en la solución de sus propios problemas," *El Deber*; "Marcha al oriente," *El Deber*, July 15, 1957.

59. Higa, "Investigation of the Okinawan Colonies in Bolivia."

60. Romero, *The Chinese in Mexico*. As historian Raanan Rein points out for the latter group, "anti-Semitic discourses, even when emerging from powerful centers of political power, do not always translate into absolute oppression." Rein, *Argentine Jews Or Jewish Argentines?*, 17.

61. Suzuki, *Embodying Belonging*, 56.

62. Dionicio Condori (mayor of Colonia Okinawa), interview with author, Colonia Okinawa, Bolivia, August 2014.

63. Suzuki, *Embodying Belonging*, 27, 204.

64. Correspondence Ministry of Agriculture to Alcalde of Santa Cruz de la Sierra, February 15, 1954, Folder: Historia, Primeros Japoneses en Santa Cruz, Provincias Santistevan Ichilo, AHD.

65. Correspondence from Toshimichi Nishikawa to Alcalde, August 25, 1955, Folder: Historia, Primeros Japoneses en Santa Cruz, AHD.

66. "Sólo el boliviano puede resolver el problema de la producción agrarian," *El Deber*, August 1957.

67. "¿Qué beneficios trae la inmigración japonesa a Santa Cruz?," *El Deber*, March 26, 1958.

68. "Sobre la inmigración japonesa," *El Deber*, 1958.

69. "Excursión a Okinawa," *El Deber*, 1958.

70. "Los inmigrantes japoneses de Okinawa," *El Deber*, 1958.

71. Untitled article, *El Deber*, July 27, 1958.

72. Putnam, *The Company They Kept*; Balderrama and Rodriguez, *Decade of Betrayal*.

73. Letter from Sieryo Nagamine to Alcalde of Santa Cruz, December 17, 1959, Folder: Japoneses en Bolivia, Palometillas Uruma y El Pailon, Colonia Okinawa, AHD.

74. Letter from Sieryo Nagamine to Alcalde of Santa Cruz.

75. Higa, "Investigation of the Okinawan Colonies in Bolivia."

76. Ruiz, *Las montañas no cambian*.

77. Field, *From Development to Dictatorship*, 6.

78. Ozawa, *Multinationalism, Japanese Style*, 13.

79. Correspondence, Adolfo Linares to Victor Andrade, November 4, 1959, CBF, Box 12, "Varios," Folder: "Japón," ALP.

80. Victor Andrade to CBF, "Planificación económica de Villamontes–agricultura y colonización," April 7, 1960, ALP-CBF, Box 88, Proyecto de Riego Villamontes, ALP.

81. Victor Andrade to Adolfo Linares Correspondence, June 4, 1960, CBF, Box 12, Folder: "Japón," ALP.

82. Technical Supervisor, Ricardo Urquidi to Adolfo Linares, President of CBF, May 4, 1960, "Inmigración japonesa a Villamontes," CBF, Box 85, ALP.

83. Irquidi, "Inmigración," ALP.

84. Internal Report, Augusto Valdivia, October 1960, CBF, Box 85, ALP.

85. Internal Report.

86. Correspondence, Linares to Andrade, La Paz, June 4, 1960, CBF-ALP, Box 12, "Varios," Folder: "Japón," ALP.

87. "Proposed Plan for Use of ICA funds," July 29, 1961, RG 260, Box 257, Economic Development Files, USNA.

88. Expediente "Primera colonia menonitas, denominada, 24 de Septiembre, Adjudicación de 1200ha," 1954, AHD.

89. Expediente "Primera colonia menonitas," AHD.

90. Expediente "Primera colonia menonitas," AHD.

91. Paz Estenssoro, "Decreto Supremo 4192, Oct. 6, 1955."

92. Paz Estenssoro.

93. Paz Estenssoro.

94. Hollweg, *Alemanes en el oriente boliviano*.

95. Notary Public Luis Landivar Sala, "Poder especial que confieren Peter B. Fehr Abram Doerksen en favor del Dr. J. Felix Pérez Baldivieso," Expediente "de colonia Canadiense," IC, Vol. 2, p. 628, ANB.

96. Letter from Felix Pérez, Expediente "de colonia Canadiense 063643-42," October 10, 1957, IC, Vol. 2, p. 628, ANB.

97. Hurtado et al., Registration of Sale in "Expediente Canadiense," December 5, 1957, IC, Vol. 2, p. 628, ANB.

98. Fretz, "A Visit to the Mennonites in Bolivia," *Mennonite Life* 15, no. 1 (January 1960): 14.

99. Peter Fehr (son of Peter Fehr) in conversation with author, Canadiense Colony, June 2011.

100. For a discussion of letter-writing within the Low-German Mennonite diaspora and in the *Mennonitische Post* of Steinbach, Manitoba, see Loewen, *Village Among Nations*. For the reference to money orders within the Okinawan diaspora, see Kerr, *Okinawa: History of an Island People*, 439.

101. Paz Estenssoro, "Decreto Supremo 06030, March 16, 1962."

102. Buchenau, "Small Numbers, Great Impact."

103. Will, "The Old Colony Mennonite Colonization of Chihuahua."

104. Chang, "Racial Alterity in the Mestizo Nation"; Knight, "Racism Revolution and Indigenismo"; Renique, "Race, Region, and Nation."

105. Navarro, *Xenofobia y xenofilia*.

106. Loewen, *Village Among Nations*.

107. "Tierras ingratas que se tornan en vergeles: Brillante éxito de los colonos mennonitas establecidos en chihuahua," *El Nacional* (Mexico City), May 9, 1931 (taken from Subject Press Clippings, "Menonitas," BMLT).

108. "Extranjeros que deben venir al país libremente," *La Prensa* (San Antonio, Texas), November 24, 1931, BMLT.

109. "Nos convienen Mennonitas y no 'Douhobors,'" *El Nacional*, March 12, 1933, BMLT.

110. "Vanse los Mennonitas al estado de Coahuila," *El Nacional*, March 6, 1944, BMLT.

111. Fretz, *Mennonite Colonization in Mexico*.

112. This current historiographic "flood" extends from colonial Latin America, Walker, *Shaky Colonialism*, to the twentieth century Healey, *The Ruins of the New Argentina*.

113. Joseph et al., *Fragments of a Golden Age*.

114. Expediente "4-352-1951-133," AHINM.

115. Wolfe, *Watering the Revolution*, 182.

116. Letter from Daniel Salas López, August 2, 1954, Expediente "4-352-1951-137," AHINM.

117. Letter from Gabino Aguilar, April 9, 1957, Ruiz Cortines, Presidential Papers, "Menonitas," AGN.

118. Letter from Gabino Aguilar.

119. Letter from Gabino Aguilar, April 13, 1963, Díaz Ordaz, Presidential Papers, "Menonitas," AGN.

120. Schmill, "Los Menonitas," *Novedades*, May 16, 1957, BMLT.

121. Letter from Daniel Salas López, July 7, 1955, Ruiz Cortines, Presidential Papers, "Menonitas," AGN.

122. Wolfe, *Watering the Revolution*, 182.

123. Mize and Swords, *Consuming Mexican Labor*, 221.

124. Basok, *Tortillas and Tomatoes*, 25.

125. Expediente "4-355-1-1930-132922," AHINM.

126. Expediente "4-355-1-1922-87099," AHINM.

127. Expediente "4-352-1951-138," AHINM.

128. Expediente "4-350-7-1962-3," AHINM.

129. Expediente "4-350-7-1962-3," AHINM.

130. Expediente "4-350-7-1962-3," AHINM.

131. See Cohen, *Braceros*, 201, and Cohen and Sirkeci, *Cultures of Migration*.

132. Letter from Daniel Salas López, May 12, 1954, Ruiz Cortines, Presidential Papers, "Menonitas," AGN.

133. Andrés Ortega Estrada, "Letter to the Editor," *Novedades*, July 19, 1965, BMLT.

134. See Kniss, *Disquiet in the Land*, for the growth of evangelical movements.

135. Medina, "Los Menonitas constituyen un 'estado,' dentro de México," *El Heraldo de Mexico*, August 23, 1967, BMLT.

136. Letter from Daniel Salas López, May 12, 1954, Ruiz Cortines, Presidential Papers, "Menonitas," AGN.

137. Expediente "4-351-0-1963-36254," AHINM.

138. Expedient "4-351-0-1963-36254," AHINM.

139. Expediente "4-351-0-1963-36254," AHINM.

140. Expediente "4-351-0-1963-36254," AHINM. Mennonite colonists had sent victims "forty-five tons of beans which represents a considerable cost."

141. Expedient "4-351-0-1963-36254," AHINM.

142. Expediente "4-351-0-1964-36444," and Expediente "4-351-0-1964-42834," AHINM.

143. Randolph, "A People Abandon Us," *Excelsior*, June 4, 1968, BMLT.

144. "Ryukyus Emigration Project Reports," Box 258, Visitors, Emigration, USCAR, USNA.

145. "Ryukyus Emigration Project Reports," RG 260, Box 257, Economic development re. immigration, USCAR, USNA.

146. Reynolds story was published in the Japanese magazine, *Shukan Asahi*, issue no. 49, September 27, 1963. It was clipped and translated into English by U.S. authorities who had closely monitored Reynolds' tour through Okinawa, Box 103, Local Activity Files, USCAR, USNA.

147. "Ryukyus Emigration Project Reports," RG 260, Box 257, Economic Development re. immigration, USCAR, USNA.

Chapter Three

1. Lomnitz, *Deep Mexico*, 11.

2. Scott, *Seeing Like a State*.

3. Hines, "The Power and Ethics of Vernacular Modernism," 252.

4. Owensby, *Empire of Law*.

5. Méndez, *The Plebeian Republic*.

6. Belmessous, *Native Claims*.

7. "Are you not Equal? Manuel Isidoro Belzu," trans. Seemin Qayum, in Thomson et al., eds., *The Bolivia Reader*.

8. James, *Resistance and Integration*; Fischer, *A Poverty of Rights*.

9. See Larson, "Capturing Indians Bodies," in Grindle, Serrill, and Domingo, eds., *Proclaiming Revolution*, 186. Also, nineteenth-century indigenous state interactions throughout the Andes in Larson, *Trials of Nation-Making*.

10. Gotkowitz, *A Revolution for Our Rights*, 85.

11. Gotkowitz, 270.

12. Rivera, *Oprimidos pero no vencidos*, 166.

13. Soliz, "Land to the Original Owners."

14. The proliferation of institutions in postrevolutionary Bolivia and their changing titles could be a source of confusion for petitioners. In the period under consideration, the director of colonization was a part of the Ministry of Agriculture. The CBF was an independent crown corporation like the state mining company (COMIBOL). In 1965 a separate organization, the National Institute of Colonization (INC), was created within the Ministry of Agriculture.

15. The phrase "incorporate into national life" is repeated in numerous pamphlets produced by the MNR. One book commemorating "10 years of Revolution" suggested that colonization and the agrarian reform had transformed the Bolivian campesino from "a thing" into a "full person." Libermann, *Bolivia: 10 años de revolución*, 143.

16. For a discussion of the evolution of the term "indio" in Bolivia, see Gotkowitz, *A Revolution for Our Rights*, 14.

17. Boyer, *Becoming Campesinos*; Mayer, *Ugly Stories of the Peruvian Agrarian Reform*.

18. Stearman, *Camba and kolla*.

19. Letter from Corregidor of Andamarca to Ministerio de Colonización, October 26, 1954, IC 628, ANB.

20. Letter from Corregidor of Andamarca to Ministerio de Colonización.

21. Letter from members of Cooperative "Phasa" to Ministerio of Colonization, November 29, 1955, IC 216, ANB.

22. Letter from members of Cooperative "Phasa" to Ministerio of Colonization. Others claimed that in the arid environment of the Andes campesino labor was entirely "in vain," letter from Cooperative Ninoca de Pacollo, October 13, 1955, IC 592, ANB.

23. Platt, "Calendars and Market Interventions."

24. Letter from Moises Colque and Casimiro Flores, January 26, 1955, IC 628, ANB.

25. Letter from Mamani, Presidente Junta Rural Agraria, April 4, 1955, IC 628, ANB.

26. Letter from Toribio Tarqui Calderón of Sora Cooperative, February 21, 1957, IC 628, ANB.

27. Schweng, "Quarterly Report on the Pillapi Project, March–June 1956," UNESCO, http://unesdoc.unesco.org/Ulis/cgibin/ulis.pl?catno=158849&set=0058009235_3 _24&gp=0&lin=1&ll=2.

28. Letter from Agrarian Judge A. Goitia to Ministry of Asuntos Campesinos, March 11, 1955, IC 628, ANB. Caupolicán has since been renamed Franz Tamayo.

29. Letter from Mamani, Presidente Junta Rural Agraria, April 4, 1955, IC 628, ANB.

30. Letter from Moises Colque and Casimiro Flores, January 26, 1955, IC 628, ANB.

31. Letter from Evorista Mayorga to director of colonization, April 14, 1958, IC 592, ANB.

32. Letter from Evorista Mayorga to director of colonization.

33. Letter from Felix Oroza to Ministry of Agriculture, February 5, 1954, IC 592, ANB.

34. Letter from Julio Menses to director of colonization, January 12, 1956, IC 592, ANB.

35. Letter from Julio Menses to director of colonization.

36. Letter from Angel Cossio to Minister of Agriculture, April 14, 1958, IC 592, ANB.

37. Coronil, *The Magical State.*

38. Letter from Cochabamba (signatories unclear), March 12, 1955, IC 228, ANB.

39. Letter from ex-miners of "la Chojilla" to Minister of Agriculture, February 20, 1958, IC 592, ANB.

40. Letter from Ángel Vargas, December 3, 1959, IC 216, ANB.

41. Letter from Eduardo Pérez, president of Co-op, Mario Torres to Minister of Agriculture, Ranching and Colonization, May 19, 1955, IC 228, ANB.

42. Letter from Victor Hugo Zelada to director of colonization, February 27, 1956, IC 592, ANB.

43. Guevara Arze, *Plan inmediato,* 86.

44. Segundio Gómez to Ministry of Agriculture, April 10, 1958, IC 592, ANB.

45. Whiteford, *Workers from the North,* 28.

46. See, for example, *Redactores del Congreso,* December 5, 1956, 368, AHDC.

47. Antezana, *Los braceros bolivianos.*

48. Antezana, 16.

49. Antenaza, 41.

50. *Presencia* was owned by the Catholic Church and, unlike other conservative La Paz newspapers, frequently published articles addressing the nation's social issues.

51. Undated clipping from *Presencia,* Wálter Guevara Arze Collection, ANB. Likely from the mid-1960s.

52. Letter from Rogelio Miranda to Roberto Lemaitre, March 16, 1966, Unlabeled folder, Depository, MDRT.

53. Letter from Rogelio Miranda to Roberto Lemaitre, March 16, 1966. It is likely Miranda was responding to a meeting of church leaders that had taken place in Cochabamba in March where participants had spoken at length about the bracero problem.

54. Letter from Mamani, Presidente Junta Rural Agraria, April 10, 1955, IC 628, ANB.

55. Letter from Evorista Mayorga to director of colonization, April 14, 1958, IC 592, ANB.

56. Letter from Ministry of Asuntos Campesinos to president's office, January 16, 1965, PR 1143, ANB.

57. Letter from Miguel Riveros to Ministry of Agriculture, May 30, 1962, PR 991, ANB.

58. Francisco Condori, interview with author. Colonia Yapacaní, August 2014.

59. Francisco Condori, interview.

60. Cavagnaro and Borquez, *Estudio sociológico*.

61. Dios, "La frontera invisible: Bolivianización," *El Mundo*, November 24, 1965. Clipping in PR1140, ANB.

62. Letter from Co-op Agropecuaria "Exodo" to Ministerio Of Agriculture, February 23, 1966, Unlabeled Folder, MDRT.

63. Evident in newspaper articles like "Capital cruceño dentro del desarrollo agrario departamental," *El Deber*, May 15, 1955.

64. Many of these reports and proceedings are found in the library of the Inter-American Institute for Cooperation on Agriculture in Chasquipampa, La Paz. Cappelletti, *Informe al gobierno de Bolivia sobre colonización*; Linares, *Problemas de colonización*.

65. Signed statement from Martín Quispe, September 29, 1956, CBF, Box 88, Migraciones Internas, Folder 105, ALP.

66. Signed statement from Eustaquio Ayaviri Villca, February 22, 1957, CBF, Box 88, Migraciones Internas, Folder 105, ALP.

67. Signed statement from Eustaquio Ayaviri Villca, February 22, 1957.

68. Published as Alfred Métraux, *Croyances et pratiques magiques*.

69. Jeanne Sylvain to Richardot, "Progress Report," September 20, 1955, https://unesdoc.unesco.org/ark:/48223/pf0000181184?posInSet=1&queryId=5b48bc29-6a60-4079-93f8-5ea724a29271, 2, 5.

70. Jeanne Sylvain to Richardot, 2.

71. Jeanne Sylvain to Richardot, 3.

72. Jeanne Sylvain to Richardot, 3.

73. Letter Jeanne Sylvain to Byron Hollinshead, Director Technical Assistance, "Rapport Trimestriel," January to March 1957.

74. Sylvain, "Rapport Trimestriel," 8.

75. Sylvain, 3.

76. Gunder Frank, *Chile; el desarrollo del subdesarrollo*.

77. See, for example, INC, *Proyecto de las zonas de colonización*; Epp, "Establishing New Agricultural Communities in the Tropical Lowlands"; Cappelletti, *Informe al gobierno de bolivia*.

78. Sylvain "Rapport Trimestriel," 7.

79. Letter from colonists at Campanero to René Barrientos, December 5, 1965, PR 1143, ANB.

80. Letter from colonists at Campanero to René Barrientos.

81. Letter from colonists at Campanero to René Barrientos.

82. Letter from colonists at Campanero to René Barrientos.

83. Sylvain, "Rapport Trimestriel," 4.

84. Letter from colonists of Jorochito to René Barrientos, March 1, 1966, PR 1143, ANB.

85. Letter from colonists of Jorochito to René Barrientos.

86. Letter from colonists of Jorochito to René Barrientos.

87. Response to colonists of Jorochito from Min. Asuntos Campesinos, November 4, 1966, PR 1143, ANB.

88. Letter to president from residents of Communidades Humacha y Saladillo, September 14, 1966, PR 1143, ANB.

89. Letter to president from residents of Communidades Humacha y Saladillo, September 14, 1966, PR 1143, ANB.

90. Letter from Eustaquio Chuquimia to director of colonization, October 7, 1956, IC 592, ANB. Letter from Max Molina to director of colonization, November 12, 1956, IC 592, ANB.

91. Letter Colonia Juan Lechín to Paz Estenssoro, December 19, 1962, PR 991, ANB.

92. Letter from Colonias Campesinas of Provincia Larecaja to Barrientos October 3, 1967, PR 1097, ANB.

93. Letter from District Supervisor of Education Roboré to Barrientos, August 24, 1966, PR 1143, ANB.

94. Letter from Mayor of San José Elio Montenegro Banegas to Barrientos, March 6, 1966, PR 1097, ANB.

95. "La ruta del desarrollo."

96. Letter from Colmena Cooperative to Barrientos, June 30, 1966, PR 1143, ANB.

97. Letter from Colmena Co-op to INC, October 22, 1968, PR 1231, ANB.

98. Salmon and Ruiz, *Un transplante humano*.

99. "La ruta del desarrollo."

100. Other Latin American historians have taken this perspective on colonization and resettlement. Turits, *Foundations of Despotism*; Tinsman, *Partners in Conflict*.

101. Hetherington, *Guerrilla Auditors*.

102. Internal Correspondence INC, April 29, 1959, IC 216, ANB.

103. Max Molina to director of colonization, November 27, 1956, IC 592, ANB.

104. Pablo Gutiérrez to Director of Colonización, "Denuncia venta de lote agrícola," November 9, 1962. IC 139, ANB.

105. Various correspondence Gutiérrez to director of colonization, 1961–62, IC 139, ANB.

106. Letter from José Vino to Ministry of Agriculture, July 8, 1959, IC 592, ANB.

107. Letter from Max Molina, to director of colonization, July 31, 1957, IC 592, ANB.

108. Letter from Max Molina.

109. Letter from Arturo Cuenca to Ministry of Agriculture, June 1956, IC 628. Letter from Angel Gómez to Ministry of Agriculture, December 17, 1958. IC 592, ANB.

110. Celestino Roque Chuquimia to Min of Agriculture, June 28, 1960, IC 216, ANB.

111. Excombatant in the Chaco War to director of colonization denouncing Luisa Valdés, IC 216, ANB.

112. Excombatant in the Chaco War to director of colonization denouncing Luisa Valdés, IC 216, ANB.

113. Turits, *Foundations of Despotism*, 107.

114. Li, *The Will to Improve*, 19.

115. Scott, *Seeing Like a State*, 301.

116. Unlabeled reel, CB.

117. "La ruta del desarrollo," BIICA.

118. Letter to Barrientos from Circuata, Inquivisi Province, January 24, 1965, PR 1097, ANB.

119. Letter from El Palmar Cooperative to Paz Estenssoro, April 26, 1961, PR 1143, ANB.

120. Letters from Colonia Huaytú to Paz Estenssoro, May 10, 1963 and May 5, 1964, PR 1143, ANB.

121. Letter from Virgilio Zabalaga to Paz Estenssoro, May 5, 1964, PR 987, ANB.

122. Letter from Colonia Huaytú to Ministerio de Asuntos Campesinos, July 12, 1964, PR 1143, ANB.

123. Letter from Cuatro Ojitos and Aroma colonies to Paz Estenssoro, April 13, 1964, PR 987, ANB.

124. Letter from Cuatro Ojitos and Aroma colonies to Paz Estenssoro.

125. Letter from residents of General Saavedra town, May 14, 1964, PR 987, ANB.

126. Letter from Victor Paz colony to Victor Paz Estenssoro, June 29, 1963, PR 987, ANB.

127. Thayer Scudder, pioneer of comparative approaches to this phenomenon, trained and encouraged a variety of students to study the issue in the 1970s and 1980s, see Scudder, *The Development Potential of New Lands Settlement*.

128. See Field, *From Development to Dictatorship*.

129. From a methodological perspective, the disastrous state of state archives for the Banzer years also makes it increasingly difficult to read for a subaltern voice in government archives related to colonization.

Chapter Four

1. "The Pax Program in Cuatro Ojitos," MCC Orientation Documents for New Volunteers, MCC-Bolivia Files, 1969–72, MCCA.

2. Fitzpatrick-Behren, "The Maya Catholic Cooperative Spirit," 291.

3. Cullather, *The Hungry World*, 160.

4. I also devote less attention to Maryknolls because several authors have already explored their role, and that of other Catholic organizations, in frontier development in Latin America.

5. Bornstein, *The Spirit of Development*, 1.

6. O'Donnell, *Modernization and Bureaucratic Authoritarianism*.

7. Barbieri, "That Strange Land Called Bolivia," in Methodist Church Joint Section of Education and Cultivation, *Lands of Witness and Decision* (New York, Methodist Church, 1957).

8. See Gill, "Religious Mobility and the Many Words of God."

9. Geyer, *Death Trails in Bolivia*, 59.

10. Geyer, 60.

11. Letter from Carroll Tamplin, "Señor Ministro del Estado en el despacho de colonización," April 14, 1950, IC 237, ANB.

12. For a discussion of the escalating tensions that led to secularization of the missions in 1949, see Langer, *Expecting Pears from an Elm Tree*.

13. Hartch, *Missionaries of the State.*

14. Garrard-Burnett, *Protestantism in Guatemala*, xi.

15. Letter from Tamplin, April 14, 1950, ANB.

16. Langer, *Expecting Pears*, 251.

17. Geyer, *Death Trails in Bolivia*, 72.

18. Letter from Dirección General de Colonización y Tierras to Carrol Tamplin, November 15, 1955, IC 237, ANB.

19. Geyer, *Death Trails in Bolivia*, 87.

20. Milton Whitaker, interview with author, Santiago de Chiquitos, November 2013.

21. Geyer, *Death Trails in Bolivia*, 91.

22. The young man returned, "to win for Christ those who had killed Don Ángel, and who now had almost taken his life!" Geyer, *Death Trails in Bolivia*, 91.

23. Palmer, *Red Poncho and Big Boots*, 159.

24. Methodist Church, *Lands of Witness and Decision*, 7.

25. "Bolivia: A Land of Decision."

26. Barbieri, "That Strange Land," 40.

27. Virginia Burnett notes a similar process at work among Protestant missionaries in Bolivia.

28. Palmer, *Red Poncho*, 7.

29. Harry Peacock, interview with author.

30. They soon recruited a Japanese missionary couple to minister to the Okinawans.

31. Palmer, *Red Poncho*, 115. Also see Quispe, *Historia de la iglesia evangélica metodista en Bolivia*, 203.

32. Harry Peacock, interview with author.

33. Harry Peacock, interview with author.

34. Ana Fajardo, interview with author, Colonia Yapacaní, July 2014.

35. Jaime Bravo, interview with author.

36. Jaime Bravo, interview with author.

37. Jaime Bravo, "The Alto Beni—Another Hope for Bolivia," *Highland Echoes*, February 1966, 4.

38. Hickman, "The Alto Beni Team and Community Development," *Highland Echoes*, February 1966, 5.

39. Ana Fajardo, interview.

40. Palmer, *Red Poncho*, 119.

41. Palmer, 119.

42. Fretz, "A Visit to the Mennonites in Bolivia."

43. Janzen, "Bolivia Monthly Report: Tres Palmas," June 1966, MCC-Bolivia Office Files 1966, Akron, PA, MCCA.

44. Fretz, "A Visit to the Mennonites in Bolivia," 13.

45. Kniss, *Disquiet in the Land.*

46. "Other Workers in Bolivia," *Highland Echoes*, October 1966, MCCA, 7.

47. Stauffer, *The Grindstone: Voice of MCC-Bolivia*, November 1972, MCCA.

48. See Mumaw, "The Bolivia Mystique," *MCC-Bolivia*, 1984, MCCA. A collection of unpublished stories from MCC workers in Bolivia during the sixties and seventies. Collected for MCC-Bolivia's twenty-fifth anniversary.

49. Gerald Mumaw, "In the Spirit of the Slaughterhouse," *Grindstone*, July 1974, MCCA.

50. Memo from Arthur Driediger in MCC-Bolivia Field Office Files, 1968, MCCA.

51. "8 personas habrían perecido ahogadas en turbión Río Grande," *El Deber*, February 13, 1968.

52. "Inundacíones llevan la desolación a las zonas urbanas y rurales," *El Deber*, February 20, 1968.

53. "Accesso terrestre a Santa Cruz se halla obstruido," *El Deber*, February 21, 1968.

54. "Emergencia nacional: Se estima en un millón de dólares las pérdidas por las inundacíones," *El Diario*, February 23, 1968.

55. "Emergencia nacional," *El Diario*, February 23, 1968.

56. Shrock, "Bolivian Flood Victims," April 10, 1968, MCC-Bolivia News Releases, 1963–75, MCCA.

57. Shrock, "Bolivian Flood Victims."

58. See Agier, *On the Margins of the World*, and Epp, *Women Without Men*.

59. "Deprevados operan en Montero alarmando a todo el vecindario," *El Deber*, March 10, 1968. In this case, the fear of radicalized flood victims intersected with President Barrientos portrayal of Bolivia as threatened from all sides by Communist infiltrators after the capture of Che Guevara.

60. "Hasta el 25 durarán los recursos para víctimas de las inundacíones aquí," *El Deber*, April 6, 1968.

61. Mercado, "Rehabilitación de zona inundada," editorial in *El Deber*, April 18, 1968.

62. Barrientos speech printed in *El Diario*, February 28, 1968.

63. Letter to Juan José Torres, "Informe Preliminar," April 30, 1968, PR 1231.

64. CIU, "Hardeman Colony Resettlement Project," July 16, 1968, MCC-Bolivia Files, 1969–72, MCCA.

65. Harry Peacock, interview.

66. "Progress Report: Hardeman," July 30, 1968, MCC-Bolivia Files, 1969–72, MCCA.

67. "Progress Report: Hardeman."

68. Methodists and Mennonites differed from Maryknoll missionaries on this issue. The latter preferred to remain as long-term members of the community, whereas the former two groups pushed for rapid disengagement.

69. "Progress Report: Hardeman," July 30, 1968, MCC-Bolivia Files, 1969–72, MCCA.

70. "Progress Report Hardeman."

71. "Progress Report: Hardeman."

72. "Orientation Documents for Volunteers: The United Methodist Church in Montero," 1970, MCC-Bolivia Files, 1951–71, MCCA.

73. "Orientation Documents for Volunteers."

74. Barrientos speech printed in *El Diario*, February 28, 1968.

75. Harry Peacock, Mary de Porres Pereyra et al., "Orientation Center for Colonists, Santa Cruz," August 20, 1971, MCC-Bolivia Files, 1971–74, MCCA.

76. Peacock, Mary de Porres Pereyra et al., "Orientation Center for Colonists."

77. "Report of the First Phase of Orientation, Piray," 1971, MCC-Bolivia Files, 1971–74, MCCA.

78. "Report of the First Phase of Orientation, Piray."

79. Harry Peacock, interview.

80. Harry Peacock, interview.

81. Harry Peacock, interview.

82. Harry Peacock, interview.

83. Harry Peacock, interview.

84. "Report of the First Phase of Orientation, Piray," 1971, MCC-Bolivia Files, 1971–74, MCCA.

85. "Report of the First Phase of Orientation, Piray."

86. "Report of the First Phase of Orientation, Piray."

87. "Report of the First Phase of Orientation, Piray."

88. Memorandum of Understanding between MCC and INC, re, San Julián, September 21, 1970, MCC-Bolivia Files, 1971.

89. Scott, *Speaking to a State.*

90. Jaime Bravo, interview.

91. Bonino, *Doing Theology in a Revolutionary Situation.*

92. Jaime Bravo, interview.

93. Gill, *Peasants, Entrepreneurs, and Social Change,* 97.

94. "Open letter from UAGRM," July 25, 1969.

95. Jaime Bravo, interview.

96. Ticoña, *Historia de Yapacaní,* 39.

97. Ticoña, 62.

98. Ticoña, 96.

99. "Colonos invaden Buena Vista domingo," *El Deber,* April 10, 1969.

100. Mercado, "Clima que hay que aquietar," *El Deber,* April 11, 1969.

101. Espejo, *Historia de Yapacaní,* 90.

102. "Open letter from UAGRM," July 25, 1969, PR 1277, ANB.

103. "Conclusiones y resoluciones del primer congreso nacional de colonizadores realizado del 17 al 20 de febrero de 1971," CP-GLE-FOL/3619, BNB.

104. These meetings also took place at the regional level including a National Congress of Bolivian Colonists in Montero in June of 1971.

105. Espejo, *Historia de Yapacaní,* 103.

106. Espejo, 103.

107. A common activity for Methodist high school students from congregations across the U.S. south.

108. Espejo, *Historia de Yapacaní,* 108.

109. Espejo, 109.

110. Harry Peacock, interview.

111. Harry Peacock, interview.

112. Espejo, *Historia de Yapacaní,* 115–20.

113. Goff and Goff, "Set Back in Bolivia," April 27, 1972, MCC-Bolivia Files, 1969–72, MCCA.

114. Dale Linsenmeyer memo to Akron office, September 9, 1971, MCC-Bolivia Files, 1971, MCCA.

115. Harry Peacock, interview.

116. Goff and Goff, "Set Back in Bolivia."

117. Jaime Bravo, interview.

118. Harry Peacock, interview.

119. Ana Fajardo, interview.

120. Ana Fajardo, interview.

121. Dale Linsenmeyer to Akron office, August 23, 1971, MCC-Bolivia Files, 1971, MCCA.

122. Dale Linsenmeyer to Akron office.

123. Latin America Director Edgar Stoesz, Trip Report, January 30 to February 9, 1973, MCC-Bolivia-Files, 1973, MCCA.

124. Latin America Director Edgar Stoesz, Trip Report.

125. Loucks, "Report on TAP-Bolivia," June 1974, MCC-Bolivia Files, 1974, MCCA. The Teacher Abroad Program (TAP), developed by MCCer Lynn Locuks in 1962 was one of the largest MCC initiatives, placing North American volunteers in rural villages without schools across the globe and was particularly active in Bolivia.

126. Amatutz, "Three Month Report," February 28, 1975, MCC-Bolivia Files, 1975.

127. Miller, "Three Month Report," July 30, 1974, MCC-Bolivia Files, 1974, MCCA.

128. Reimer, "Three Month Report," May 1974, MCCA.

129. Murray Luft, "Slaughterhouse," *Grindstone Newsletter*, 1972, MCC-Bolivia Files, 1972, MCCA.

130. Harry Peacock, interview.

131. "Seminario de ideas," 1. The book contains the published conference proceedings, including transcriptions.

132. "Seminario de ideas," 2.

133. "Seminario de ideas," 3.

134. "Seminario de ideas," 423.

135. "Seminario de ideas," 428–33.

136. "Seminario de ideas," 434.

137. "Seminario de ideas," 434. This amount included rations for the entirety of the orientation period.

138. "Seminario de ideas," 434.

139. INC, *Proyecto de las zonas de colonización*, 1.

140. INC, 1.

141. INC, 39.

142. INC, 35.

143. INC, 358.

144. Epp, "Establishing New Agricultural Communities," 135.

145. Epp, 87.

146. Harry Peacock, interview.

147. Luft, "Slaughterhouse," *Grindstone Newsletter*, 1972, MCC-Bolivia Files, 1972, MCCA.

148. Harry Peacock, interview.

149. Jaime Bravo, interview.

150. Harry Peacock, interview.

151. Fifer, "A Series of Small Successes."

152. Dozier, *Land Development and Colonization*.

153. Harry Peacock, interview. Nelson, *The Development of Tropical Lands*.

154. Stearman describes her participation in the project in Stearman, *Camba and Kolla*, 183.

155. As Stearman points out, there was little change in membership as the CIU transitioned into FIDES. Stearman, *Camba and Kolla*, 183.

156. Jaime Bravo, interview.

157. Harry Peacock, interview.

158. Stearman, *Camba and Kolla*, 183.

159. Flora Gómez, interview with author, San Julián, June 2014.

160. Harry Peacock, interview.

161. Jaime Bravo, interview.

162. Solem, "Bolivia," 8.

163. Manshard and Morgan, eds., *Agricultural Expansion and Pioneer Settlements*.

164. *San Julián: Bloqueos campesinos*, 37–42.

165. See Taboada, *San Julián, la colonizacion y su fortaleza*. Just as it had for Yapacaní's Juan Espejo.

166. Jaime Bravo, interview.

167. Wachtel, *Gods & Vampires*.

168. Ayers, *Conflicts or Complementarities?*

169. David Hess, interviewed by Stephen Ladek on Aidpreneur "TOR064: Consulting after a USAID Career with David Hess," http://aidpreneur.com/.

170. Jaime Bravo, interview.

171. Jaime Bravo, interview.

Chapter Five

1. Guevara, *The Bolivian Diary*, 69.

2. Letter to General Néstor Valenzuela Jefe de la Division de Colonización from Oscar Rivera Prado, January 31, 1967, Expediente "Riva Palacio," 1957, INRASC.

3. Guevara, *The Bolivian Diary*, 73.

4. Letter to General Valenzuela, January 31, 1967.

5. Craib, *Cartographic Mexico*.

6. Comité Pro-Defensa Intereses Cruceños, *Argentina y Bolivia*, 4, AHUAGRM.

7. Redekop, *The Old Colony Mennonites*, vii.

8. Yu, *Thinking Orientals*.

9. Wiebe, "First Draft Bolivia Report," March 20, 1975, MCC-Bolivia Files, 1975, MCCA.

10. Colonel Roberto Lemaitre, "What Benefits does the Mennonite Immigration Bring?" from *La Crónica* (Santa Cruz) and reprinted in *Los Tiempos* (Cochabamba), March 14, 1968. Translated into English by Elwood Shrock, MCC-Bolivia Files, 1961–68, MCCA.

11. Recounted in the Diary of Johan Wiebe, typed copy given to author by Jakob Giesbrecht, Riva Palacio Colony.

12. "Campesinos Nemonitas [sic]," *El Diario*, February 28, 1968.

13. "Campesinos Nemonitas [sic]," *El Diario*, February 28, 1968.

14. "Campesinos Nemonitas [sic]," *El Diario*, February 28, 1968.

15. Colonel Roberto Lemaitre, "What Benefits does the Mennonite Immigration Bring?" from *La Crónica* (Santa Cruz) and reprinted in *Los Tiempos* (Cochabamba), March 14, 1968. Translated into English by Elwood Shrock, MCC-Bolivia Files, 1961–68, MCCA.

16. Lemaitre, "What Benefits?"

17. Lemaitre.

18. Arguedas, *Pueblo enfermo.*

19. Lemaitre, "What Benefits?"

20. Lemaitre.

21. Lemaitre.

22. Shesko, "Constructing Roads," 6–28.

23. Lemaitre, "What Benefits?"

24. Discussed in Spitzer, *Hotel Bolivia;* Suzuki, *Embodying Belonging.*

25. Lemaitre, "What Benefits?"

26. Peters et al., "Minimum Plan of Work," February 10, 1967, Expediente "Riva Palacio," INRASC.

27. Peters, "Minimum Plan," Expediente "Riva Palacio."

28. Peters.

29. Taffet, *Foreign Aid as Foreign Policy.*

30. Letter to Colonel Roberto Lemaitre from Oscar Ribero Prada, "Referente. Expediente solicitud de Colonia Mennonite Riva Palacio," September 2, 1967, Expediente "Riva Palacio," INRASC.

31. Letter to the Director of the INC from General, Valenzuela, March 7, 1967, Expediente "Riva Palacio," INRASC.

32. Peters et al., "Minimum Plan of Work," Letter to the Director of the INC from Epifanio Rios, n.d., Expediente "Riva Palacio," INRASC.

33. Letter to the Director of the INC from Epifanio Rios, n.d., Expediente "Riva Palacio," INRASC.

34. Letter to the Director of the INC from Epifanio Rios, n.d., Expediente "Riva Palacio," INRASC.

35. Letter to the Director of the INC from Epifanio Rios, n.d., Expediente "Riva Palacio," INRASC.

36. Letter to Roberto Lemaitre from Carlos Zambrana, March 29, 1967, Expediente "Riva Palacio," INRASC.

37. Contract signed by Peter Fehr and Abram Peters represented by Carlos Zambrana, July 10, 1967, Expediente "Riva Palacio," INRASC.

38. Letter to Edgar Stoesz MCC-Latin America director from Arthur Driediger, August 22, 1967, MCC-Bolivia Files, 1967, MCCA.

39. Kopp, "Report to MCC Office," n.d. (ca. August 1968), MCC-Bolivia Files, 1968, MCCA.

40. Letter to Arthur Driediger from MCC-Latin America director Edgar Stoesz, September 18, 1968, MCC-Bolivia Files, 1968, MCCA.

41. Letter to Arthur Driediger from MCC-Latin America director Edgar Stoesz, MCCA.

42. Letter to MCC Executive Committee from Menno Wiebe and Edgar Stoesz, March 21, 1975, MCC-Bolivia Files, 1975, MCCA.

43. Menno Wiebe to Minister of Agriculture Alberto Natusch Busch, December 10, 1974, MCC-Bolivia Files, 1974, MCCA.

44. Schumacher, *Small is beautiful.*

45. Letter to MCC Executive Committee from Menno Wiebe and Edgar Stoesz, March 21, 1975, MCCA.

46. Peters et al., "Minimum Plan," Expediente "Riva Palacio," INRA, INRASC.

47. Ingeniero Oswaldo Quevedo, "Informe sobre el cumplimiento del Plan de Trabajo," October 15, 1969, Expediente "Riva Palacio," INRASC.

48. See Lesser, *Welcoming the Undesirables*, and Spitzer, *Hotel Bolivia*.

49. Report John Friesen, May 1, 1971, MCC-Bolivia Files, 1971, MCCA.

50. Wiebe, "First Draft Bolivia Report," March 20, 1975, MCC-Bolivia Files, 1975, MCCA.

51. Letter to Gerald Shenk from George Hildebrandt of Evangelical Mennonite Mission Conference, April 29, 1976, MCC-Bolivia Files, 1976, MCCA.

52. Wiebe, "First Draft Bolivia Report," March 20, 1975, MCC-Bolivia Files, 1975, MCCA.

53. McKeown, "Conceptualizing Chinese Diasporas," 322.

54. Diary Johan Wiebe, 1966–83, Copy obtained from Jakob Giesbrecht, Riva Palacio Colony, Bolivia.

55. Correspondence from November 11, 1968 to July 21, 1969, Expediente "Riva Palacio," INRASC. The confusion also meant that a fellow group of Mennonites who had planned to settle the land were instead forced to purchase a separate 10,000-hectare property to the northeast near the town of Paurito.

56. Letter to José Luis Roca from Roberto Lemaitre, March 25, 1970, MDRT.

57. Letter from Bernard Peters, Abram Schmitt, and Bernard Penner from Riva Palacio, Swift Current, and Santa Rita colonies to Ministry of Agriculture, April 17, 1970, Ministerial Resolutions, 1970, MDRT.

58. Letter from Bernard Peters, Abram Schmitt, and Bernard Penner, MDRT.

59. Alredo Ovando Candía, "Decreto Supremo 09238," June 4, 1970, Listado de Decretos, Gaceta Oficial del Estado Plurinacional de Bolivia, http://www.gacetaoficialdebolivia.gob.bo/index.php/normas/lista/11.

60. Diary, Johan Wiebe.

61. Oscar Arze, "Situación lechera," *El Mundo* (Santa Cruz de la Sierra), September 26, 1979. Contains references to the original 1969 study.

62. Resolución Ministerial 326/68, Resoluciones ministeriales, 1969, MDRT.

63. Resolución Ministerial 253/69, July 28, 1969, Resoluciones ministeriales, 1969, MDRT.

64. Enrique Siemens, interview with author, Riva Palacio Colony, May 20, 2014.

65. Resolución Ministerial 154/69, June 13, 1969, Resoluciones ministeriales, 1969.

66. Telegram from Ignacio Kinn, Regional Adminstrator in Santa Cruz to Licenciado Raúl Vega, July 7, 1969, Resoluciones ministeriales, 1969, MDRT.

67. Redekop, *The Old Colony Mennonites*, 90. This vibrant cross-border trade is also evident in the AHINM, Mexico City.

68. Loewen, *Diaspora in the Countryside* brings Shover's concept in dialogue with the experience of North American Mennonites.

69. Johan Fehr, interview with author, Riva Palacio Colony, May 7, 2014.

70. Peter Wall, interview with author, Riva Palacio Colony, March 27, 2014.

71. Wiebe, "First Draft Bolivia Report," March 20, 1975, MCC-Bolivia Files, 1975, MCCA.

72. Bolívar, "Las colonias menonitas," 45.

73. Bolívar, 11.

74. Johan Fehr, interview with author.

75. Kopp, "Report to MCC Office," n.d. (ca. August 1968), MCC-Bolivia Files, 1968, MCCA.

76. Letter from Luis Barrón, August 30, 1967, Box 60, CBF, ALP.

77. Wuhl, "Superioridad de la leche como alimento," *El Deber*, August 14, 1968.

78. Arze, "Producción lechera en Santa Cruz," *El Mundo*, September 18, 1979.

79. Bolívar, "Las colonias menonitas," 73.

80. Bolívar, 1.

81. Bolívar, 141.

82. Abram Falk and Peter Fehr, interview with author, Canadiense Colony, October 22, 2013.

83. Maldonado, *Algodón*, BIICA.

84. Candía, "Evaluación del cultivo del algodon," BIICA.

85. Bolívar, "Las colonias menonitas," 73.

86. "Soya! Recetas para toda la familia" (La Paz: Banco Nacional de Bolivia, 1984), BNB.

87. A mixture of ground corn and cheese typically wrapped in a corn husk and boiled.

88. "Soya!"

89. Bolívar, "Las colonias menonitas," 128.

90. Hecht and Mann, "How Brazil Outfarmed the Americans," 92.

91. Oliveira and Hecht, "Sacred Groves," 272.

92. Johan Boldt, interview with author, Riva Palacio, April 5, 2014.

93. Johan Fehr, interview with author.

94. Jaime Duranovic, interview with author, Santa Cruz de la Sierra, May 15, 2014. Duranovic is a former employee of Silvo Marinkovic.

95. Hecht and Mann, "How Brazil Out-farmed the American Farmer," 98.

96. Heidi Tinsman, *Buying into the Regime*.

97. Hecht and Mann, "How Brazil Outfarmed the American Farmer," 98.

98. Alves et al., "The Success of BNF."

99. *Memoria Annual*, ANAPO, 1989–90.

100. Cushman, *Guano and the Opening of the Pacific World*, 309–10.

101. International Fishmeal and Fish Oil Organization, "The production of fishmeal."

102. Laws, *El Niño and the Peruvian Anchovy Fishery*.

103. Guillermo Ribera, interview with author, Santa Cruz de la Sierra, April 22, 2014.

104. Johan Buhler, interview with author, Pinondi Colony, Charagua, April 19, 2015.

105. Johan Boldt, interview with author, Santa Cruz de la Sierra, April 5, 2014.

106. Johan Boldt, interview with author.

107. Letter to INRA officials from Jakob Klassen, November 30, 1984, Expediente "Del Norte," 458-SC, INRASC.

108. Letter to INRA officials from Jakob Klassen.

109. Letter to INRA officials from Jakob Klassen, October 15, 1982, Expediente "Del Norte."

110. Letter to INRA officials from Jakob Klassen.

111. See Kohl and Farthing, *Impasse in Bolivia*.

112. L. Jemio, *Debt, Crisis, Reform Bolivia*, 216.

113. Guillermo Ribera, interview with author.

114. Healy, "The Boom Within the Crisis."

115. Healy, 101.

116. "Protección a las colonias," *El Deber*, April 15, 1987; "Santa Cruz no se rinde," *El Deber*, April 15, 1987.

117. "Protección a las colonias," *El Deber*.

118. Knelsen, Account Book, 1966–2014, Personal Papers, Riva Palacio Colony.

119. Imaizumi, Clementelli, and Terrazas, "Estudio sobre las situaciones reales."

120. Knelsen, Account Book, 1966–2014.

121. *Memoria Annual*, ANAPO, 1989–90.

122. "Proyecto agrícola," *El Deber*, March 16, 1990.

123. Interviews in Riva Palacio, San Juan Yapacaní, and Okinawa Colony.

124. Ediger, "Report on the Debt Crisis," n.d., MCC-Bolivia Files, Centro Menno Debts, 1979–99, MCCA.

125. Alongside that 1990 article were several about the ongoing drought in Santa Cruz in which "small producers have lost everything" which provide some indication of the factors that might have placed Mennonites on the wrong side of the law.

126. Knelsen, Account Book, 1966–2014.

127. "Santa Cruz no se rinde," *El Deber*, April 15, 1987.

128. Bolívar Menacho, "Las colonias menonitas," 141.

129. Johan Fehr, interview with author.

130. Jakob Buhler, interview with author, Riva Palacio, July 21, 2014.

131. Abe Enns, interview with author, Riva Palacio, August 11, 2014.

132. Ediger, "Report on the Debt Crisis," n.d., MCC-Bolivia Files, Centro Menno Debts, 1979–99, MCCA.

133. Warkentin, "A Visit to some Mennonite debtors in a Bolivian prison, February 1991," Kanadier Concerns staff report, April 12, 1991, Kanadier Mennonite Concerns Committee 1989–91, XIV-3, Mennonite Archives of Ontario, Conrad Grebel University College.

134. Tim Penner report to Akron, October 20, 1988, MCC-Bolivia Files, Centro Menno Debts, 1979–99, MCCA.

135. William Janzen, report to Akron on January 9–16 Bolivia trip, January 20, 1988, Centro Menno Debts, 1979–99, MCCA.

136. Dick Plett, report to Akron, n.d., MCC-Bolivia Files, Centro Menno Debts, 1978–99, MCCA.

137. Jaime Duranovic, interview with author.

138. Jaime Duranovic, interview with author.

139. Katoshi Higa, interview with author, Colonia Okinawa, July 3, 2014.

140. Avila-Tàpies and Domínguez-Mujica, "Postcolonial Migrations and Diasporic Linkages."

141. Guano, "The Denial of Citizenship." With the Argentine economic collapse of 2001, Bolivians shifted their labor migration toward Brazil and Spain.

142. Dick Plett, report to Akron, n.d, MCC-Bolivia Files, Centro Menno Debts, 1978–99, MCCA.

143. Financial records Riva Palacio colony, Personal Documents of Vorsteher Abram Klassen. Viewed and discussed April 21, 2015, Riva Palacio.

144. Barber and Navarro, "Perfil del Ensayo"; Morales Coca, "Efectos de labranza profunda."

145. David Cortez Vargas (ANAPO technician), conversation with author, Santa Cruz de la Sierra, April 15, 2014.

146. Dick Plett, report to Akron, n.d., MCC-Bolivia Files, Centro Menno Debts, 1978–99, MCCA.

147. Abe Enns, interview with author, Riva Palacio, August 11, 2014.

148. The gendered term "daughter" colony is widely used among scholars working on Mennonites and other similar communal religious communities such as Hutterites. It is likely derived from biology (cell division by mitosis) or entomology (from the colony practices of ants). For an example see, Regehr, *Mennonites in Canada.*

149. Lemaitre, "What Benefits."

150. *Memoria Annual,* ANAPO, 2002, 81.

151. *Memoria Annual,* ANAPO, 1997–98, 78.

Conclusion

1. Gustafson, "Spectacles of Autonomy," 355.

2. Expediente "de colonia Canadiense," IC 628, ANB.

3. Falk and Fehr, interview with author.

4. Álvaro García Linares (Vice President of Bolivia), Speech, Dia del Maíz exhibition, Sagrado Corazón, Santa Cruz, August 30, 2013.

5. Miriam Telma Jemio, "Bolivia's Forest Fires and the Rise of Beef Exports," *Diálogo Chino,* accessed October 25, 2019, https://dialogochino.net/29896-bolivias-forest-fires-and -the-rise-of-beef-exports/.

6. Falk and Fehr, interview with author.

7. Official Speeches, San Julián forty-sixth anniversary, San Julián, June 23, 2014.

8. Hoey, "No Monuments, No Heroes."

9. Kruse, *White Flight.*

10. Gobierno Autónomo Departamental, Santa Cruz, http://www.santacruz.gob.bo/.

11. Wilcox, "Zebu's Elbows"; Wilcox, *Cattle in the Backlands.*

12. Jaime Bravo, interview with author.

13. Scott, *Seeing Like a State.*

14. Sommer, *Foundational Fictions.*

15. Salmon and Ruiz, "Un transplante humano."

16. Salmon and Ruiz.

17. See, for example, Mayer, *Ugly Stories of the Peruvian Agrarian Reform.*

18. Condori, "En San Julián sigue el bloqueo de la ruta," *La Razón* (La Paz), July 18, 2013, http://www.la-razon.com/sociedad/San-Julián-sigue-bloqueo-ruta_0_1871812826.html.

19. Larger farmers in Riva Palacio might operate 100- to 150-hectare parcels, but these were typically farmed and owned jointly by parents and children (or in-laws). Notable exceptions include some Mennonites that had left traditional colonies, and the small rebel colony of Chihuahua, where farm sizes could exceed 1,000 hectares.

20. McNeill, *Something New Under the Sun.*

21. Dunkerley, "Evo Morales"; Morales, "From Revolution to Revolution."

Epilogue

1. Biographical notes about Evo Morales drawn from Sivak, *Evo Morales.*

2. Letter from Corregidor of Andamarca to Ministerio de Colonización, October 26, 1954, IC 628, ANB.

3. Antezana, *Los braceros bolivianos*, 16.

4. Antezana, 7.

5. Letter to President's office from Maximo Luna and Felix Machaca, December 15, 1968, PR 1097, ANB.

6. Sivak, *Evo Morales*, 41.

7. Sivak, 39.

8. Anthropologist Harry Sanabría, who had studied San Julián in the early 1980s also turned his attention to the coca boom.

9. Historian Lars Schoultz identifies a similarly bifurcated logic in the El Salvadorian context where in the 1980s, "the United States regularly filled two trucks, one with a Food for Peace shipment and development specialists to address the needs of hungry campesinos, and the other with U.S.-armed and trained Salvadoran soldiers to attack the Communist guerrillas, and sent both trucks down the road toward whatever region of the country happened to be unstable." Schoultz, *Beneath the United States*, 358.

10. Harry Peacock, interview with author.

11. Sivak, *Evo Morales*, 39.

12. Milton Whitaker, interview with author, Santiago de Chiquitos, November 2013.

13. Fabricant and Postero, "Sacrificing Indigenous Bodies and Lands."

14. Farthing and Kohl, *Evo's Bolivia*, 44–50.

15. Gustafson, "By Means Legal and Otherwise."

16. See essays in Fabricant and Gustafson, eds., *Remapping Bolivia*.

17. Gustafson, "Spectacles of Autonomy and Crisis."

18. Most notably in southern Brazil and northern Mexico, like Santa Cruz, ranching frontiers that also define themselves as white regions within mestizo and mulatto nations. See Oliven, *Tradition Matters*.

19. Gustafson, "By Means Legal or Otherwise."

20. Janzen, "Media Portrayals of Low German Mennonite 'Ghost Rapes,'" makes a similar argument over a very different issue. In a recent article on popular media's portrayal of the prosecution of a series of notorious rape cases in one Mennonite colony, Janzen advocates bringing together these neighboring conceptions of autonomy (regional, Mennonites).

21. Abe Enns, interview with author, Riva Palacio, August 11, 2014.

22. Sara Shahriari (freelance journalist), social media post, October 21, 2014.

23. See, for example, Kraybill and Hurd, *Horse-and-Buggy Mennonites*.

24. Chávez, "Villamontes brilla con las más bellas de Bolivia," *El Dia*, May 26, 2012, https://www.eldia.com.bo/index.php?cat=356&pla=3&id_articulo=91946.

25. Farthing and Kohl, *Evo's Bolivia*, 118. For a discussion of ongoing lowland indigenous responses to colonization in La Paz department, see Sturtevant, "Missions, Unions, and Indigenous Organization."

26. Simón et al., *I Sold Myself, I Was Bought*.

27. See Emily Achtenberg, "Why is Morales Reviving Bolivia's Controversial TIPNIS Road," *NACLA*.

28. Lucas Land, MCC Volunteer in Charagua, "Two Kingdoms: Low German Mennonites in Charagua, Bolivia," https://wwje.wordpress.com/2011/08/16/two-kingdoms-low-german-mennonites-in-charagua-bolivia/.

29. Faguet, *Decentralization and Popular Democracy*.

30. Farthing and Kohl, *Evo's Bolivia*, 119.

31. Farthing and Kohl, 42.

32. Kopp, *Las colonias menonitas en Bolivia: antecedentes, asentamientos y propuestas para un diálogo*, 123–25.

33. Lucas Land, MCC Volunteer in Charagua, "Two Kingdoms: Low German Mennonites in Charagua, Bolivia."

34. Wilhelm Buhler, interview with author, Riva Palacio, March 23, 2014.

35. "Menonitas son sometidos a leyes vigentes," *Cambio* (La Paz), March 13, 2014. The official newspaper of the Plurinational State of Bolivia published the text of the proposed decree.

36. "Población menonita en Bolivia: Distribución porecentual por colonia y pais de origen," October 1, 2010, Dirección General de Migración. Unpublished documents provided by Adalberto Kopp.

37. ANAPO officials, conversation with author, Santa Cruz de la Sierra, May 2014.

38. Attorney Waldemar Rojas, interview with author, Santa Cruz de la Sierra, June 20, 2014.

39. Prudencio, conversation with author, Santa Cruz de la Sierra, November 20, 2013.

40. Jaime Bravo, interview with author.

41. "De una granja a la política," *La Razón*, March 25, 2012.

42. Germán Bravo, interview with author, Colonia Okinawa, August 2014.

43. Peter Wieler, interview with author, Santa Cruz de la Sierra, April 27, 2015.

Bibliography

Primary Sources

ARCHIVES AND COLLECTIONS

Bolivia

Archivo Histórico Departamental Hermanos Vásquez Machicado, Santa Cruz de la Sierra

Archivo Histórico de la Universidad Autónoma Gabriel René Moreno, Santa Cruz de la Sierra

    Folletos Históricos

    Hemeroteca

Archivo La Paz, La Paz

    Corporación Boliviana de Fomento

Archivo de Ministerio de Relaciones Exteriores, La Paz

Archivo Nacional de Bolivia, Sucre

    Instituto de Colonización

    Presidencia de la República

Asociación Nacional de Productores de Oleaginosas, Santa Cruz de la Sierra

Biblioteca y Archivo Histórico del Congreso Nacional, La Paz

    Camara de Diputados, *Redactor: Actas Publicas*

    Hemeroteca

Biblioteca del Instituto Interamericano de Cooperación para la Agricultura, Chasquipampa

Biblioteca Nacional de Bolivia, Sucre

Centro Cultural Japonés-Boliviano, La Paz

Cineteca Boliviana, La Paz

Dirección General de Migración, La Paz

Instituto Nacional de Reforma Agraria, Santa Cruz de la Sierra

Instutito Nacional de Reforma Agraria, Pailón

Ministerio de Desarrollo Rural y Tierras, La Paz

    Depository

    Resoluciones Ministeriales

Muséo Histórico Okinawa, Colonia Okinawa, Santa Cruz

Canada

Mennonite Archives of Ontario, Conrad Grebel University College, Waterloo, Ontario

Mexico

Archivo General de la Nación, Mexico City

    Fondo de Presidentes

        Diaz Ordaz Papers

        Ruiz Cortines Papers

    Investigaciones Políticas y Sociales

Archivo Histórico del Instituto Nacional de Migración, Mexico City
  Expedientes
Biblioteca Miguel Lerdo de Tejada, Mexico City
Universidad Nacional Autónoma de México, Mexico City
  Hemeroteca
Secretaría de Relaciones Exteriores, Acervo Histórico Diplomático, Mexico City

United States
Mennonite Central Committee Archives, Akron, PA
  MCC-Bolivia Files
  MCC-Mexico Files
  MCC-Zaire Files
National Archives at College Park, College Park, MD
  Records of the U.S. Civil Administration of the Ryukyu Islands

Online Archives
Foreign Affairs Oral History Collection, Association for Diplomatic Studies and Training,
  Arlington, VA, accessed February 2, 2016, https://www.adst.org.
Gaceta Oficial del Estado Plurinacional de Bolivia, accessed November 2, 2014, http://www
  .gacetaoficialdebolivia.gob.bo.
  Listado de Decretos
UNESCO, accessed December 19, 2017, https://unesdoc.unesco.org.
  Project Documents

Personal Archives
Jakob Giesbrecht, Riva Palacio Colony
  Diary Johan Wiebe, 1966–83
Jakob Knelsen, Riva Palacio Colony
  Personal account book 1966–2014
Abram Klassen Riva Palacio Colony
  Colony tax records, Personal Documents of Vorsteher Abram Klassen

NEWSPAPERS AND MAGAZINES

Cambio, La Paz, Bolivia
La Crónica, Santa Cruz de la Sierra
El Deber, Santa Cruz de la Sierra
El Día, Santa Cruz de la Sierra
El Diario, La Paz
Fortune
The Grindstone, Voice of MCC Bolivia,
  Santa Cruz de la Sierra
El Heraldo de México, Mexico City
Highland Echoes, Bolivian Methodist
  Church, La Paz

Mennoblatt, Fernheim Colony, Paraguay
El Mundo, Buenos Aires, Argentina
El Mundo, Santa Cruz de la Sierra
El Nacional, Mexico City
The New York Times
Novedades, Mexico City
Okinawa Shimbun, Naha, Japan
Okinawa Times, Naha, Japan
La Prensa, San Antonio, Texas
Presencia, La Paz, Bolivia
La Razón, La Paz, Bolivia

AUDIOVISUAL SOURCES

Viajando por nuestra tierra. Various Episodes, 1953–56. Directed by Walter Cerruto.
Juanito sabe leer. 1954. Directed by Jorge Ruiz. DVD (2010).
Las montañas no cambian. 1962. Directed by Jorge Ruiz. DVD (2010).
La vertiente. 1958. Directed by Jorge Ruiz. DVD (2010).
Un poquito de diversificación económica. 1955. Directed by Jorge Ruiz. DVD (2010).
Los primeros. 1959. Directed by Jorge Ruiz. DVD (2010).
Los ximul. 1960. Directed by Jorge Ruiz. DVD (2010).

ORAL HISTORY INTERVIEWS

Boldt, Johan. April 5, 2014. Riva Palacio Colony, Santa Cruz, Bolivia.
Braun, Willy. March 19, 2014. Canadiense II Colony, Santa Cruz.
Bravo, Germán. August 2014. Okinawa Colony, Santa Cruz.
Bravo, Jaime. July 30, 2014. Santa Cruz de la Sierra.
Buhler, Jakob. July 21, 2014. Riva Palacio Colony, Santa Cruz.
Buhler, Johan. April 19, 2015. Pinondi Colony, Charagua, Bolivia.
Condori, Dionicio. August 1, 2014. Colonia Okinawa, Santa Cruz.
Condori, Francisco. August 7, 2014. Yapacaní, Santa Cruz.
Cortez Vargas, David (ANAPO technician). April 15, 2014. Santa Cruz de la Sierra.
Duranovich, Jaime. May 15, 2014. Santa Cruz de la Sierra.
Enns, Abe. August 11, 2014. Riva Palacio Colony, Santa Cruz.
Fajardo, Ana. July 1, 2014. Yapacaní, Santa Cruz.
Falk, Abram, and Peter Fehr. October 22, 2013. Canadiense Colony.
Fehr, Johan. May 7, 2014. Riva Palacio Colony, Santa Cruz.
Friesen, Peter. April 21, 2014, Riva Palacio Colony, Santa Cruz.
Gómez, Flora. June 23, 2014. San Julián Colony, Bolivia.
Hamm, Abram. April 1, 2014. Riva Palacio Colony, Santa Cruz.
Higa, Katoshi. July 3, 2014. Colonia Okinawa.
Klassen, Abram. April 24, 2015. Riva Palacio Colony, Santa Cruz.
Klassen, Cornelio. July 26, 2014. Riva Palacio Colony, Santa Cruz.
Klassen, Peter. July 29, 2014. Riva Palacio Colony, Santa Cruz.
Knelsen, Jakob. April 27, 2015. Riva Palacio.
Miller, Marty. June 20, 2013. Harrisonburg, VA.
Painter, Michael. May 1, 2012. Atlanta, GA (Skype).
Peters, Cornelio. May 11, 2014. Riva Palacio Colony, Santa Cruz.
Peters, Isaac. April 30, 2014. Riva Palacio Colony, Santa Cruz.
Peacock, Harry. August 13, 2014. Santa Cruz de la Sierra.
Prudencio. November 20, 2013. Santa Cruz de la Sierra.
Rempel, Abram. June 26, 2014. Campo Chihuahua Colony, Santa Cruz.
Rempel, Franz. June 26, 2014. Campo Chihuahua Colony, Santa Cruz.
Ribera, Guillermo. April 22, 2014. Santa Cruz de la Sierra.
Rojas, Waldemar (lawyer). June 20, 2014, Santa Cruz de la Sierra.
Siemens, Enrique. May 20, 2014. Riva Palacio Colony, Santa Cruz.
Wall, Peter. March 27, 2014. Riva Palacio Colony, Santa Cruz.

Wieler, Jakob. April 27, 2015. Santa Cruz de la Sierra.
Whitaker, Milton. November, 2013. Santiago de Chiquitos.

Published Sources

Achtenberg, Emily. "Why is Morales Reviving Bolivia's Controversial TIPNIS road." *NACLA*. Accessed October 26, 2019, https://nacla.org/blog/2017/08/22/why-evo -morales-reviving-bolivia%E2%80%99s-controversial-tipnis-road.
Agier, Michel. *On the Margins of the World: The Refugee Experience Today.* Cambridge: Polity, 2008.
"Alto Beni, Bolivia." La Paz: Special Projects Office, 1964.
Alves, Bruno J. R., Robert M. Boddey, and Segundo Urquiaga. "The Success of BNF in Soybean in Brazil." *Plant and Soil* 252, no. 1 (May 2003): 1–9.
Amemiya, Kozy K. "The Bolivian Connection: U.S. Bases and Okinawan Emigration." WPRI Working Paper No. 25. Oakland: Japan Policy Research Institute, 1996.
Andrews, George Reid. *Blacks and Whites in São Paulo, Brazil, 1888–1988.* Madison: University of Wisconsin Press, 1991.
Antezana, Fernando. *Los braceros Bolivianos: Drama humano y sangría nacional.* La Paz: Editorial Icthus, 1966.
Appelbaum, Nancy P., Anne S. Macpherson, and Karin A. Rosemblatt. *Race and Nation in Modern Latin America.* Chapel Hill: University of North Carolina Press, 2003.
Arguedas, Alcides. *Pueblo enfermo.* Santiago, Chile: Ediciones Ercilla, 1937.
Armiero, Marco, and Richard Tucker. *Environmental History of Modern Migrations.* New York: Routledge, 2017.
Avila-Tàpies, Rosalia, and Josefina Domínguez-Mujica. "Postcolonial Migrations and Diasporic Linkages between Latin America and Japan and Spain." *Asian and Pacific Migration Journal* 24, no. 4 (2015): 487–511.
Ayers, Alison. *Conflicts or Complementarities?: The State and Non-Government Organizations in the Colonisation Zones of San Julián and Berlin, Eastern Bolivia.* London: Overseas Development Institute, 1992.
Baily, Samuel L., and Eduardo José Míguez. *Mass Migration to Modern Latin America.* Wilmington, DE: Scholarly Resources, 2003.
Balderrama, Francisco E., and Raymond Rodriguez. *Decade of Betrayal: Mexican Repatriation in the 1930s.* Albuquerque: University of New Mexico Press, 1995.
Barber, Richard, and Felipe Navarro. "Perfil del Ensayo Sobre Sistemas de Labranza Conservacionista de Suelos Brecha 5.5, Campo 104, 1991–1997." Santa Cruz: Centro de Investigación Agrícola Tropical, 1991.
Basok, Tanya. *Tortillas and Tomatoes: Transmigrant Mexican Harvesters in Canada.* Ithaca, NY: McGill-Queens University Press, 2002.
Beattie, Peter M. *The Tribute of Blood: Army, Honor, Race, and Nation in Brazil, 1864–1945.* Durham, NC: Duke University Press, 2001.
Bebbington, Anthony, and Graham Thiele, eds. *Non-Governmental Organizations and the State in Latin America: Rethinking Roles in Sustainable Agricultural Development.* New York: Routledge, 2005.

Belmessous, Saliha. *Native Claims: Indigenous Law against Empire, 1500–1920*. Oxford: Oxford University Press, 2012.

Beltrán, Luis Ramiro. "Comunicación para el desarrollo en Latino America: Una evaluación suscinta al cabo de 40 años." IV Mesa Redonda sobre Comunicación. Lima, Perú: Instituto Peruano Alemán (IPAL), February 23–26, 1993.

Bender, Harold S., Martin W. Friesen, Menno Ediger, Isbrand Hiebert, and Gerald Mumaw. "Bolivia." *Global Anabaptist Mennonite Encyclopedia Online*. June 2013. Accessed June 1, 2016. http://gameo.org/index.php?title=Bolivia&oldid=122239.

Blanc, Jacob, and Frederico Freitas, eds. *Big Water: The Making of the Borderlands Between Brazil, Argentina, and Paraguay*. Tucson: University of Arizona Press, 2018.

Bolívar Menacho, Jesús. "Las Colonias Menonitas: Aporte y Participación en la Producción Agropecuaria Regional." MA Thesis, UAGRAM, 1978.

"Bolivia: Población por Censos Según Departamento, Area Geográfica y Sexo, Censos de 1950–1976–1992–2001." Instituto Nacional de Estadísticas. Accessed October 10, 2019. https://bolivia.unfpa.org/sites/default/files/pub-pdf/Caracteristicas_de_Poblacion_2012.pdf

Bornstein, Erica. *The Spirit of Development: Protestant NGOs, Morality, and Economics in Zimbabwe*. New York: Routledge, 2003.

Boyer, Christopher R. "Becoming Campesinos: Politics, Identity, and Agrarian Struggle." In *Post-Revolutionary Michoacán, 1920–1935*. Stanford: Stanford University Press, 2003.

Brandellero, Sara Lucia Amelia. *Brazilian Road Movie: Journeys of (Self) Discovery*. Cardiff: University of Wales Press, 2013.

Buchenau, Jürgen. "Small Numbers, Great Impact: Mexico and its Immigrants, 1821–1973." *Journal of American Ethnic History* 20 (Spring 2001): 23–49.

Buchenau, Jürgen, and Lyman L. Johnson. *Aftershocks: Earthquakes and Popular Politics in Latin America*. Albuquerque: University of New Mexico Press, 2009.

Burton, Julianne. *The Social Documentary in Latin America*. Pittsburgh: University of Pittsburgh Press, 2013.

Cañás Bottos, Lorenzo. *Old Colony Mennonites in Argentina and Bolivia: Nation Making, Religious Conflict and Imagination of the Future*. Leiden, NL: Brill, 2008.

Candia, J. D. "Evaluación del Cultivo del Algodon en Santa Cruz en la Campaña 1972–73." Asociación de Productores de Algodón. La Paz: ADEPA, 1973.

Cappelletti, Fausto. *Informe al gobierno de Bolivia sobre colonización*. Rome: Food and Agriculture Organization, 1965.

Carter, William E. *Aymara Communities and the Bolivian Agrarian Reform*. Gainesville: University of Florida Press, 1964.

Chang, Jason. "Racial Alterity in the Mestizo Nation." *Journal of Asian American Studies* 14, no. 3 (2011): 331–59.

Christy, Alan S. "The Making of Imperial Subjects in Okinawa." *Positions: East Asia Cultures Critique* 1, no. 3 (1993): 607–39.

Clark, Kim. *The Redemptive Work: Railway and Nation in Ecuador, 1895–1930*. Wilmington: SR Books, 1998.

Cohen, Deborah. *Braceros: Migrant Citizens and Transnational Subjects in the Postwar United States and Mexico*. Chapel Hill: University of North Carolina Press, 2011.

Cohen, Jeffrey H. and Ibrahim Sirkeci. *Cultures of Migration: The Global Nature of Contemporary Mobility*. Austin: University of Texas Press, 2011.

Comité Pro-Defensa Intereses Cruceños. *Argentina y Bolivia y su vinculación ferroviaria efectiva en el oriente*. La Paz: Artistica, 1942.

"Como Viviré y Trabajaré en mi Nueva Parcela." La Paz: CBF-BID, 1962.

Coronil, Fernando. *The Magical State: Nature, Modernity and Money in Venezuela*. Chicago: University of Chicago Press, 1997.

Cote, Stephen. *Oil and Nation: A History of Bolivia's Petroleum Sector*. Morgantown: West Virginia University Press, 2016.

Craib, Raymond. *Cartographic Mexico: A History of State Fixations and Fugitive Landscapes*. Durham: Duke University Press, 2004.

Cullather, Nick. *The Hungry World*. Cambridge: Harvard University Press, 2011.

Cushman, Gregory. *Guano and the Opening of the Pacific World: A Global Ecological History*. New York: Cambridge University Press, 2013.

Dinani, Husseina. "En-gendering the Postcolony: Women, Citizenship and Development in Tanzania, 1945–1985." PhD diss., Emory University, 2013.

Dozier, Craig L. *Land Development and Colonization in Latin America; Case Studies of Peru, Bolivia, and Mexico*. New York: Praeger, 1969.

Duguid, Julian. *Green Hell: Adventures in the Mysterious Jungles of Eastern Bolivia*. London: Century Co., 1931.

Dunkerley, James. "Evo Morales, the 'Two Bolivias' and the Third Bolivian Revolution." *Journal of Latin American Studies* 39, no. 1 (2007): 133–66.

Dutra e Silva, Sandro. "A Construção Simbólica do Oeste Brasileiro (1930–1940)." In Dutra e Silva et al. eds. *Vastos sertões: História e natureza na ciência e na literatura*. Rio de Janeiro: Mauad, 2015.

Epp, Mark. "Establishing New Agricultural Communities in the Tropical Lowlands: The San Julián Project in Bolivia." MA Thesis (M.P.S.), Cornell University, 1975.

Epp, Marlene. *Women Without Men: Mennonite Refugees of the Second World War*. Toronto: University of Toronto Press, 2000.

Espejo Ticoña, Juan. *Historia de Yapacaní*. El Alto: Ediciones Qhanañchwai, 2013.

Fabricant, Nicole, and Bret Gustafson, eds. *Remapping Bolivia: Resources, Territory, and Indigeneity in a Plurinational State*. Santa Fe: School for Advanced Research Press, 2011.

Fabricant, Nicole, and Nancy Postero. "Sacrificing Indigenous Bodies and Lands: The Political-Economic History of Lowland Bolivia in Light of the Recent TIPNIS Debate." *The Journal of Latin American and Caribbean Anthropology* 20, no. 3 (2015): 452–74.

Faguet, Jean-Paul. *Decentralization and Popular Democracy Governance from below in Bolivia*. Ann Arbor: University of Michigan Press, 2012.

Farrés Cavagnaro, Juan B., and Yolanda E. Borquez. *Estudio sociológico sobre los grupos migratorios de braceros Bolivianos en Mendoza*. Mendoza: Universidad Nacional de Cuyo, Escuela Superior de Estudios Políticos y Sociales, Instituto de Estudios Políticos y Sociales, 1962.

Farthing, Linda C., and Benjamin H. Kohl. *Evo's Bolivia: Continuity and Change*. Austin: University of Texas Press, 2014.

Field, Thomas. *From Development to Dictatorship: Bolivia and the Alliance for Progress in the Kennedy Era*. Ithaca: Cornell University Press, 2014.

———. "Ideology as Strategy: Military-Led Modernization and the Origins of the Alliance for Progress in Bolivia." *Diplomatic History* 36, no. 1 (2012): 147–83.

Fifer, Valerie. *Bolivia: Land, Location, and Politics since 1825.* Cambridge: Cambridge University Press, 1972.

———. "A Series of Small Successes: Frontiers of Settlement in Eastern Bolivia." *Journal of Latin American Studies* 14, no. 2 (November 1982): 407–32.

Fischer, Brodwyn. *A Poverty of Rights: Citizenship and Inequality in Twentieth-Century Rio de Janeiro.* Stanford: Stanford University Press, 2010.

Fitzpatrick-Behrens, Susan. "From Symbols of the Sacred to Symbols of Subversion to Simply Obscure: Maryknoll Women Religious in Guatemala, 1953 to 1967." *The Americas: A Quarterly Review of Latin American History* 61, no. 2 (October 2004): 189–216.

———. *The Maryknoll Catholic Mission in Peru, 1943–1989: Transnational Faith and Transformation.* Notre Dame, IN: University of Notre Dame Press, 2012.

———. "The Maya Catholic Cooperative Spirit of Capitalism in Guatemala: Civil-Religious Collaborations, 1943–1966." In *Local Church, Global Church: Catholic Activism in Latin America from Rerum Novarum to Vatican II*, edited by Stephen Andes. Washington, DC: The Catholic University of America Press, 2016

Fretz, J. Winfield. *Mennonite Colonization in Mexico: An Introduction.* Akron, PA: Mennonite Central Committee, 1945.

———. "A Visit to the Mennonites in Bolivia." *Mennonite Life* 15, no. 1 (January 1960): 13–17.

Gandarilla Guardia, Nino. *Cine y televisión en Santa Cruz.* Santa Cruz de la Sierra: Universidad Autónoma Gabriel René Moreno, 2011.

García Taboada, Mariano. *San Julián, la colonizacion y su fortaleza.* La Paz: Fondo Editorial de los Diputados, 2008.

Garfield, Seth. *In Search of the Amazon: Brazil, the United States, and the Nature of a Region.* Durham: Duke University Press, 2013.

Garitano, Carmela. *African Video Movies and Global Desires: A Ghanaian History.* Athens: Ohio University Press, 2013.

Garrard-Burnett, Virginia. *Protestantism in Guatemala: Living in the New Jerusalem.* Austin: University of Texas Press, 1998.

Geidel, Molly. "The Point of the Lance: Gender, Development, and the 1960's Peace Corps," PhD diss., Boston University, 2011.

Geyer, Robert. *Death Trails in Bolivia to Faith Triumphant.* New York: Vantage Press, 1963.

Gildner, Matthew. "Indomestizo Modernism: National Development and Indigenous Integration in Postrevolutionary Bolivia, 1952–1964." PhD diss., University of Texas at Austin, 2012.

Gill, Lesley. "Religious Mobility and the Many Words of God in La Paz, Bolivia." In *Rethinking Protestantism in Latin America*, edited by Virginia Garrard-Burnett and David Stoll. Philadelphia: Temple University Press, 1993.

———. *Peasants, Entrepreneurs, and Social Change: Frontier Development in Lowland Bolivia.* Boulder: Westview Press, 1987.

González Navarro, Moisés. *Xenofobia y xenofilia en la historia de México, siglos XIX y XX*, SEGOB. Instituto Nacional de Migración, 2006.

Gordillo, Gastón. *Rubble: The Afterlife of Destruction.* Durham: Duke University Press, 2015.

Gotkowitz, Laura. *A Revolution for Our Rights: Indigenous Struggles for Land and Justice in Bolivia, 1880–1952.* Durham: Duke University Press, 2007.

Grandin, Greg. *The Blood of Guatemala: A History of Race and Nation.* Durham: Duke University Press, 2000.

———. *Fordlandia: The Rise and Fall of Henry Ford's Forgotten Jungle City.* London: Icon Books, 2010.

Gruzinski, Serge, and Heather MacLean. *Images at War: Mexico from Columbus to Blade Runner, 1492–2019.* Durham: Duke University Press, 2001.

Guano, Emanuela. "The Denial of Citizenship: 'Barbaric' Buenos Aires and the Middle-Class Imaginary." *City & Society: Journal of the Society for Urban Anthropology* 16, no. 1 (2004): 69–97.

Guevara, Ernesto. *The Bolivian Diary.* Melbourne, Ocean Press, 2006.

———. *Our America and Theirs: Kennedy and the Alliance for Progress: The Debate at Punto Del Este.* Melbourne: Ocean Press, 2006.

Guevara Arze, Wálter. *Plan inmediato de política económica del gobierno de la revolución nacional.* La Paz: Editorial. Letras, 1955.

Gunder Frank, Andre. *Chile; el desarrollo del subdesarrollo.* Santiago: Editorial Prensa Latinoamericana, 1967.

Guridy, Frank Andre. *Forging Diaspora Afro-Cubans and African Americans in a World of Empire and Jim Crow.* Chapel Hill: University of North Carolina Press, 2010.

Gustafson, Bret. "By Means Legal and Otherwise: The Bolivian Right Regroups." North American Congress on Latin America. Accessed August 6, 2017. https://nacla.org /article/means-legal-and-otherwise-bolivian-right-regroups.

———. "Spectacles of Autonomy and Crisis: Or, What Bulls and Beauty Queens Have to Do with Regionalism in Eastern Bolivia." *Journal of Latin American Anthropology* 11, no. 2 (2006): 351–79.

Hajdarpasic, Edin. *Whose Bosnia?: Nationalism and Political Imagination in the Balkans, 1840–1914.* Ithaca: Cornell University Press, 2015.

Hamilton, Susan. *An Unsettling Experience: Women's Migration to the San Julián Colonization Project.* Binghamton, NY: Institute for Development Anthropology, 1986.

Hartch, Todd. *Missionaries of the State: The Summer Institute of Linguistics, State Formation, and Indigenous Mexico, 1935–1985.* Tuscaloosa: University of Alabama Press, 2006.

Harvey, Penelope. "The Materiality of State Effects: An Ethnography of a Road in the Peruvian Andes." In *State Formation: Anthropological Perspectives*, edited by Christian Krohn-Hansen and Knut G. Nustad. London; Ann Arbor: Pluto Press, 2005.

Healey, Mark Alan. *The Ruins of the New Argentina: Peronism and the Remaking of San Juan after the 1944 Earthquake.* Durham: Duke University Press, 2011.

Healy, Kevin. "The Boom within the Crisis: Some Recent Effects of Foreign Cocaine Markets on Bolivian Rural Society and Economy." In *The Coca Leaf and its Derivatives*, 101–144. Ithaca: Cornell University, 1985.

Hecht, Susanna, and Charles Mann. "How Brazil Outfarmed the American Farmer." *Fortune* 157, no. 1 (2008): 92–106.

Hein, Laura Elizabeth, and Mark Selden. *Islands of Discontent: Okinawan Responses to Japanese and American Power.* Lanham: Rowman & Littlefield Publishers, 2003.

Hetherington, Kregg. *Guerrilla Auditors: The Politics of Transparency in Neoliberal Paraguay.* Durham: Duke University Press, 2011.

Hines, Sarah. "The Power and Ethics of Vernacular Modernism: The Misicuni Dam Project in Cochabamba, Bolivia, 1944–2017." *Hispanic American Historical Review* 98, no. 2 (2018): 223–56.

Hoey, Lesli. "'No Monuments, No Heroes' How Accidental Planners Established Bolivia's Flagship Land Reform Project through Spatial, Facilitated, and Adaptive Strategies." *Journal of Planning Education and Research* 36, no. 4 (2016): 400–13.

Hollweg, Mario Gabriel. *Alemanes en el Oriente Boliviano, Vol. 2.* Santa Cruz de la Sierra: Ed. Sirena, 1997.

Iacobelli, Pedro. *Postwar Emigration to South America from Japan and the Ryukyu Islands.* London: Bloomsbury Academic, 2017.

Imaizumi, Shichiro, Alfredo Clementelli, and Raúl Terrazas. "Estudio sobre las situaciones reales de Mecanización Agrícola en el Departamento de Santa Cruz. (2) Mecanización Agrícola de la Colonia San Juan." Santa Cruz de la Sierra: JICA-CIAT, 1991.

Instituto de Capacitación Política. *Lecciones de propagada, organización y agitación.* La Paz: SPIC, 1953.

Instituto Nacional de Colonización. *Proyecto de las zonas de colonización: Puerto Villarroel, km. 21, Chané-Piray, ampliación de San Julián.* La Paz: INC, 1974.

Instituto Nacional de Estadísticas. *Censo agropecuario 2013.* La Paz: INE, 2016.

International Fishmeal and Fish Oil Organisation. "The production of fishmeal and fish oil from Peruvian Anchovy." Accessed February 12, 2016. https://www.iffo.net/system /files/67_0.pdf.

James, Daniel. *Resistance and Integration: Peronism and the Argentine Working Class, 1946–1976.* Cambridge: Cambridge University Press, 1993.

Janzen, Rebecca. "Media Portrayals of Low German Mennonite 'Ghost Rapes' Challenge the Bolivian Plurinational State." *A ContraCorriente* 13, no. 3 (Spring 2016): 246–62.

Jemio, L. *Debt Crisis and Reform in Bolivia: Biting the Bullet.* New York: Palgrave, 2001.

Jemio, Miriam Telma. "Bolivia's Forest Fires and the Rise of Beef Exports." *Diálogo Chino.* Accessed October 25, 2019. https://dialogochino.net/29896-bolivias-forest-fires-and -the-rise-of-beef-exports/

Joseph, Gilbert M., Anne Rubenstein, and Eric Zolov. *Fragments of a Golden Age: The Politics of Culture in Mexico since 1940.* Durham: Duke University Press, 2001.

Karam, John. *Another Arabesque: Syrian-Lebanese Ethnicity in Neoliberal Brazil.* Philadelphia: Temple University Press, 2007.

Kerr, George. *Okinawa: History of an Island People.* Ruttland, VT: Tuttle Publishing, 2018.

King, John. *Magical Reels: A History of Cinema in Latin America.* New York: Verso, 1990.

Knight, Alan. "Racism Revolution and Indigenismo, Mexico." In *The Idea of Race in Latin America, 1870–1940,* edited by Richard Graham, 71–87. Austin: University of Texas Press, 1990.

Kniss, Fred. *Disquiet in the Land: Cultural Conflict in American Mennonite Communities.* New Brunswick: Rutgers University Press, 1997.

Kohl, Benjamin, and Linda C. Farthing. *Impasse in Bolivia: Neoliberal Hegemony and Popular Resistance.* London: Zed Books, 2006.

Kopp, Adalberto. *Las colonias menonitas en Bolivia: antecedentes, asentamientos y propuestas para un diálogo.* Santa Cruz de la Sierra: Fundación Tierra, 2015.

Kraybill, Donald B., and James P. Hurd. *Horse-and-Buggy Mennonites: Hoofbeats of Humility in a Postmodern World.* University Park: Pennsylvania State University Press, 2006.

Kruse Kevin. *White Flight: Atlanta and the Making of Modern Conservatism.* Princeton, N.J.: Princeton University Press, 2005.

Lai, Symbol. "Southern Encounters: Ethnic Nationalism and Okinawa Military Migration to Latin America." Paper presented at the American Historical Association Conference, Washington, D.C., January 2018.

Land, Lucas M. "Two Kingdoms: Low German Mennonites in Charagua, Bolivia." Wordpress. Accessed April 15, 2016. https://wwje.wordpress.com/2011/08/16/two -kingdoms-low-german-mennonites-in-charagua-bolivia/.

Langer, Erick. "Bringing the Economic Back In: Andean Indians and the Construction of the Nation-State in Nineteenth-Century Bolivia." *Journal of Latin American Studies* 41, no. 3 (2009): 527–51.

———. *Expecting Pears from an Elm Tree: Franciscan Missions on the Chiriguano Frontier in the Heart of South America, 1830–1949.* Durham: Duke University Press, 2009.

Lapenga, Pablo. *Soybeans and Power: Genetically Modified Crops, Environmental Politics and Social Movements in Argentina.* New York: Oxford University Press, 2016.

Larson, Brooke. "Capturing Indians Bodies, Hearths, and Minds: The Gendered Politics of Rural School Reform in Bolivia, 1910–52." In *Proclaiming Revolution: Bolivia in Comparative Perspective,* edited by Merilee Grindle and Pilar Domingo, 183–212. London: Institute of Latin American Studies, 2003.

———. *Trials of Nation Making: Liberalism, Race, and Ethnicity in the Andes, 1810–1910.* Cambridge: Cambridge University Press, 2004.

Latta, Alex, and Hannah Wittman, eds. *Environment and Citizenship in Latin America: Natures, Subjects and Struggles.* New York: Berghahn Books, 2012.

Laws, Edward A. *El Niño and the Peruvian Anchovy Fishery.* Sausalito, CA: University Science Books, 1997.

Lazar, Sian. *El Alto, Rebel City: Self and Citizenship in Andean Bolivia.* Durham: Duke University Press, 2008.

Lesser, Jeffrey. *Negotiating National Identity: Immigrants, Minorities, and the Struggle for Ethnicity in Brazil.* Durham: Duke University Press, 1999.

———. *Welcoming the Undesirables: Brazil and the Jewish Question.* Berkeley: University of California Press, 1995.

Li, Tania. *The Will to Improve: Governmentality, Development, and the Practice of Politics.* Durham: Duke University Press, 2007.

Liberman, Jacobo Z. *Bolivia: 10 años de revolución.* La Paz: Dirección Nacional de Informaciones, 1962.

Lim, Julian. *Porous Borders: Multiracial Migrations and the Law in the U.S.-Mexico Borderlands.* Chapel Hill: University of North Carolina Press, 2017.

Limerick, Patricia Nelson. *The Legacy of Conquest: The Unbroken Past of the American West.* New York: Norton, 2011.

Linares, Adolfo. *Problemas de colonización.* La Paz: CBF, 1964.

Loewen, Royden. *Diaspora in the Countryside: Two Mennonite Communities and Mid-Twentieth Century Rural Disjuncture.* Toronto: University of Toronto Press, 2015.
———. *Horse-and-Buggy Genius: Listening to Mennonites Contest the Modern World.*
———. *Village among Nations: "Canadian" Mennonites in a Transnational World, 1916–2006.* Toronto: University of Toronto Press, 2013.
Lomnitz, Claudio. *Deep Mexico, Silent Mexico: An Anthropology of Nationalism.* Minneapolis: University of Minnesota Press, 2011.
Mackey, Lee. "Legitimating Foreignization in Bolivia: Brazilian Agriculture and the Relations of Conflict and Consent in Santa Cruz, Bolivia." Paper presented at the International Conference on Global Land Grabbing, University of Sussex, Brighton, UK, April 2011.
McKay, B. and G. Colque. "Bolivia's Soy Complex: The Development of 'Productive Exclusion.'" *Journal of Peasant Studies* 43, no. 2 (2016): 583–610.
Malitsky, Joshua. *Post-Revolution Nonfiction Film: Building the Soviet and Cuban Nations.* Bloomington: Indiana University Press, 2013.
Malloy, James. *Bolivia: The Uncompleted Revolution.* Pittsburgh: University of Pittsburgh Press, 1970.
Malloy, James, and Richard Thorn. *Beyond the Revolution: Bolivia since 1952.* Pittsburgh: University of Pittsburgh Press, 1971.
Manshard, Walther, and William B. Morgan, eds. *Agricultural Expansion and Pioneer Settlements in the Humid Tropics: Selected Papers Presented at a Workshop Held in Kuala Lumpur, 17–21 September 1985.* Tokyo: United Nations University Press, 1988.
Masterson, Daniel, and Sayaka Funada-Classen. *The Japanese in Latin America.* Urbana: University of Illinois Press, 2004.
Mayer, Enrique. *Ugly Stories of the Peruvian Agrarian Reform.* Durham: Duke University Press, 2009.
McKeown, Adam. "Conceptualizing Chinese Diasporas, 1842 to 1949." *The Journal of Asian Studies* 58, no. 2 (1999): 306–37.
McNeill, J. R. *Something New Under the Sun: An Environmental History of the Twentieth-Century World.* New York: Norton, 2000.
Méndez, Cecilia. *The Plebeian Republic: The Huanta Rebellion and the Making of the Peruvian State, 1820–1850.* Durham: Duke University Press Books, 2005.
Mesa Gisbert, Carlos D. *La aventura del cine Boliviano, 1952–1985.* La Paz: Editorial Gisbert, 1985.
Métraux, Alfred. "Croyances Et Pratiques Magiques Dans La Vallée De Marbial, Haïti." *Journal De La Société Des Américanistes* 42, no. 1 (1953): 135–98.
Míguez Bonino, José. *Doing Theology in a Revolutionary Situation.* Philadelphia: Fortress Press, 1975.
Mihaylova, Dimitrina. "Reopened and Renegotiated Borders: Pomak Identities at the Frontier between Bulgaria and Greece." In *Culture and Power at the Edges of the State: National Support and Subversion in European Border Regions*, edited by Thomas Wilson and Hastings Donnan, 155–90. Piscataway: Transaction, 2005.
Minchin, J. B. "Eastern Bolivia and the Gran Chaco." *Proceedings of the Royal Geographical Society* 3, no. 7 (July 1881): 401–20.

Mize, Ronald L., and Alicia C. S. Swords. *Consuming Mexican Labor: From the Bracero Program to NAFTA.* Toronto: University of Toronto Press, 2010.

Montero Hoyos, Sixto, and Constantino Montero Hoyos. *Territorios ignorados: Sobre una visita a la serranía aurífera de San Simón y un estudio agropecuario de las provincias de ñuflo de chávez y velasco del ingeniero constantino montero hoyos.* Santa Cruz de la Sierra, Bolivia: Editorial Nicolás Ortíz, 1936.

Morales Coca, Gonzalo. "Efectos de labranza profunda y fertilizacion en el cultivo de soya, Gylcine max." MA Thesis, Universidad Autónoma Gabriel René Moreno University, 1991.

Morales, Waltraud Q. "From Revolution to Revolution: Bolivia's National Revolution and the 'Re-founding' Revolution of Evo Morales." *The Latin Americanist* 55, no. 1 (2011), 131–44.

Moreno Tejada, Jaime, and Bradley Tatar, eds. *Transnational Frontiers of Asia and Latin America since 1800.* New York: Routledge, 2017.

Mosse, David. *Cultivating Development: An Ethnography of Aid Policy and Practice.* London: Pluto Press, 2005.

Moya, José C. *Cousins and Strangers: Spanish Immigrants in Buenos Aires, 1850–1930.* Berkeley: University of California Press, 1998.

Nacify, Hamid. *A Social History of Iranian Cinema, Vol. 2.* Durham: Duke University Press, 2011.

Nash, Linda. *Inescapable Ecologies: A History of Environment, Disease, and Knowledge.* Berkeley: University of California Press, 2006.

Nelson, Michael. *The Development of Tropical Lands: Policy Issues in Latin America.* Baltimore: Johns Hopkins University Press, 1973.

Nobbs-Thiessen, Ben. "Channeling Modernity: Nature, Patriotic Engineering, and the Chaco War." In *The Chaco War: Environment, Ethnicity, and Nationalism,* edited by Bridget Chesterton. London: Bloomsbury Academic, 2016.

Ochoa Maldonado, Felipe. *Algodón: Información agro-económica del cultivo en Bolivia.* La Paz: Banco Agrícola Bolivia, 1973.

O'Donnell, Guillermo. *Modernization and Bureaucratic Authoritarianism.* Berkeley: University of California Pres, 1973.

Oliveira, Gustavo, and Susanna Hecht. "Sacred Groves, Sacrifice Zones and Soy Production: Globalization, Intensification and Neo-Nature in South America." *Journal of Peasant Studies* 43, no. 2 (2016): 251–85.

Oliven, Rúben George. *Tradition Matters: Modern Gaúcho Identity in Brazil.* New York: Columbia University Press, 1996.

Owensby, Brian. *Empire of Law and Indian Justice in Colonial Mexico.* Stanford: Stanford University Press, 2011.

Ozawa, Terutomo. *Multinationalism, Japanese Style: The Political Economy of Outward Dependency.* Princeton: Princeton University Press, 2014.

Pacino, Nicole. "Prescription for a Nation: Public Health in Post-Revolutionary Bolivia, 1952–1964." PhD diss., University of California, Santa Barbara, 2013.

Palmer, Jim. *Red Poncho and Big Boots: The Life of Murray Dickson.* Nashville: Abingdon Press, 1969.

Parker, Jason. *Hearts, Minds, Voices: US Cold War Public Diplomacy and the Formation of the Third World.* New York: Oxford University Press, 2016.

Peña, Paula. *La permanente construcción de lo cruceño: Un estudio sobre la identidad en Santa Cruz de la Sierra.* La Paz: Fundación PIEB, 2003.

Piper, Karen Lynnea. *Cartographic Fictions: Maps, Race, and Identity.* New Brunswick: Rutgers University Press, 2002.

Platt, Tristan. "Calendars and Market Interventions." In *Ethnicity, Markets, and Migration in the Andes: At the Crossroads of History and Anthropology,* edited by Brook Larson. Durham: Duke University Press, 1995.

Postero, Nancy Grey. *Now We are Citizens: Indigenous Politics in Postmulticultural Bolivia.* Stanford: Stanford University Press, 2007.

Pruden, Hernán. "Cruceños into Cambas: Regionalism and Revolutionary Nationalism in Santa Cruz de la Sierra, Bolivia, 1935–1939." PhD diss., State University of New York, Stony Brook, 2012.

Putnam, Lara. *The Company They Kept: Migrants and the Politics of Gender in Caribbean Costa Rica, 1870–1960.* Chapel Hill: University of North Carolina Press, 2002.

———. *Radical Moves: Caribbean Migrants and the Politics of Race in the Jazz Age.* Chapel Hill: University of North Carolina Press, 2013.

Quispe, Delfín E. *Historia de la Iglesia Evangélica Metodista en Bolivia 1906–2006: Una iglesia evangélica inculturada.* La Paz: Centro de Historia y Archivo Metodista, 2006.

Raffles, Hugh. *In Amazonia: A Natural History.* Princeton: Princeton University Press, 2002.

Rankin, Monica A. *¡México, la patria! Propaganda and Production during World War II.* Lincoln: University of Nebraska Press, 2009.

Redekop, Calvin. *The Old Colony Mennonites: Dilemmas of Ethnic Minority Life.* Baltimore: Johns Hopkins University Press, 1969.

Regehr, T. D. *Mennonites in Canada, 1939–1970: A People Transformed.* Toronto: University of Toronto Press, 1996.

Rein, Raanan. *Argentine Jews or Jewish Argentines? Essays on Ethnicity, Identity, and Diaspora.* Leiden: Brill, 2010.

Renique, Gerardo. "Race, Region, and Nation: Sonora's Anti-Chinese Racism and Mexico's Postrevolutionary Nationalism, 1920s–1930s." In *Race and Nation in Modern Latin America,* edited by Nancy Appelbaum, et al. Chapel Hill: University of North Carolina Press, 2003.

Rivera Cusicanqui, Silvia. *Oprimidos pero no vencidos: luchas del campesinado Aymara y Quechwa, 1900–1980.* La Paz, Bolivia: Hisbol, 1984.

Romero, Robert Chao. *The Chinese in Mexico, 1882–1940.* Tucson: University of Arizona Press, 2010.

"La ruta del desarrollo: Del altiplano al trópico." La Paz: CBF-BID, n.d.

Sadlier, Darlene. *Brazil Imagined: 1500-Present.* Austin: University of Texas Press, 2008.

Salmon, Hugo Alfonso, and Jorge Ruiz. "Un Transplante Humano: El Proyecto de Colonización Alto Beni." La Paz: Ministerio de economía nacional, oficina de proyectos especiales, 1965.

San Julián: Bloqueos campesinos camioneros: Testimonio de la lucha de un pueblo. Pamphlet. La Paz: Asamblea Permanente de Derechos Humanos de Bolivia, 1984.

Sánchez-H., José. *The Art and Politics of Bolivian Cinema.* Lanham: Scarecrow Press, 1999.

Sandoval Arenas, Carmen. *Santa Cruz, economía y poder, 1952–1993.* La Paz: Fundación PIEB, 2003.

Sawatzky, Harry Leonard. *They Sought a Country: Mennonite Colonization in Mexico. With an Appendix on Mennonite Colonization in British Honduras.* Berkeley: University of California Press, 1971.

Schoultz, Lars. *Beneath the United States: A History of U.S. Policy toward Latin America.* Cambridge: Harvard University Press, 1998.

Schroeder, William, and Helmut T. Huebert. *Mennonite historical atlas.* 2nd edition. Winnipeg: Springfield Publishers, 1996.

Schumacher E. F. *Small Is Beautiful; Economics as If People Mattered.* New York: Harper & Row; 1973.

Schumann, Debra A, and William L Partridge. *The Human Ecology of Tropical Land Settlement in Latin America.* Boulder: Westview Press, 1989.

Scott, James C. *Seeing Like a State: How Certain Schemes to Improve the Human Condition Have Failed.* New Haven: Yale University Press, 1998.

Scudder, Thayer. *The Development Potential of New Lands Settlement in the Tropics and Subtropics.* Washington, DC: U.S. Agency for International Development, 1984.

"Seminario de ideas y proyectos específicos." In *Santa Cruz y el desarrollo.* Santa Cruz de la Sierra: Comité de Obras Públicas, 1972.

Sharon, Tucker. "Inscribed in the Margins: Envisioning Road Colonization in Peru's Age of Development." PhD diss., University of British Columbia, 2017.

Shaw, Lisa. "Vargas on Film: From the Newsreel to the Chanchada." In *Vargas and Brazil: New Perspectives,* edited by Jens Hentschke. New York: Palgrave Macmillan, 2006.

Shesko, Elizabeth. "Constructing Roads, Washing Feet, and Cutting Cane for the 'Patria': Building Bolivia with Military Labor, 1900–1975." *International Labor and Working-Class History* 80, no. 1 (Fall 2011): 6–28.

Siekmeier, James F. *The Bolivian Revolution and the United States, 1952 to the Present* University Park: Pennsylvania State University Press, 2011.

Simón, Brigitte et al. *I Sold Myself, I Was Bought: A Socio-Economic Analysis Based on Interviews with Sugar-Cane Harvesters in Santa Cruz de La Sierra, Bolivia.* Copenhagen: IWGIA, 1980.

Sivak, Martín. *Evo Morales: The Extraordinary Rise of the First Indigenous President of Bolivia.* New York: Palgrave Macmillan, 2010.

Skar, Sarah Lund. *Lives Together-Worlds Apart: Quechua Colonization in Jungle and City.* New York: Oxford University Press, 1994.

Skidmore, Thomas. *The Politics of Military Rule in Brazil.* Oxford: Oxford University Press, 1988.

Smith, Rosaleen. "Film as Instrument of Modernization and Social Change in Africa: The Long View." In *Modernization as Spectacle in Africa,* edited by Peter Bloom, Stephen Miescher, and Takyiwaa Manuh. Bloomington: Indiana University Press, 2014.

Solem, Richard, Richard J. Green, David W. Hess, Carol Bradford Ward, and Peter Leigh Taylor. "Bolivia: An Integrated Rural Development in a Colonization Setting." USAID Project Impact Evaluation Report, no. 57, 1985.

Soliz, Carmen. "Fields of Revolution: Agrarian Reform and Rural State Formation in Bolivia, 1936–1971," PhD diss., New York University, 2014.

———. "Land to the Original Owners": Rethinking the Indigenous Politics of the Bolivian Agrarian Reform." *Hispanic American Historical Review* 97, no. 2 (2017): 259–96.

Sommer, Doris. *Foundational Fictions: The National Romances of Latin America.* Berkeley: University of California Press, 1991.

Soria, Oscar, and Jorge Sanjínes. *Qué es el plan decenal?* La Paz: Burillo, 1962.

Soruco, Ximena. "De la Goma a la Soya: El Proyecto Histórico de la Élite Cruceña." In *Los barones del oriente: El poder en Santa Cruz ayer y hoy*, edited by Ximena Soruco, Wilfredo Plata, and Gustavo Medeiros. Santa Cruz de la Sierra: Fundación Tierra, 2008.

Spitzer, Leo. *Hotel Bolivia: The Culture of Memory in a Refuge from Nazism.* New York: Hill and Wang, 1998.

Stearman, Allyn MacLean. *Camba and Kolla: Migration and Development in Santa Cruz, Bolivia.* Orlando: University of Central Florida Press, 1985.

Stefanoni, Pablo. *Qué hacer con los indios: Y otros traumas irresueltos de la colonialidad.* La Paz: Plural Editores, 2010.

Stepan, Nancy. *The Hour of Eugenics: Race, Gender, and Nation in Latin America.* Ithaca: Cornell University Press, 1991.

Sturtevant, Chuck. "Missions, Unions, and Indigenous Organization in the Bolivian Amazon: Placing the Formation of an Indigenous Organization in Its Context." *Latin American Research Review* 53, no. 4 (2018): 770–84.

Sutter, Paul. "The World With Us: The State of American Environmental History." *Journal of American History* 100, no. 1 (2013), 94–119.

Suzuki, Taku. *Embodying Belonging: Racializing Okinawan Diaspora in Bolivia and Japan.* Honolulu: University of Hawaii Press, 2010.

Taffet, Jeffrey. *Foreign Aid as Foreign Policy: The Alliance for Progress in Latin America.* New York: Routledge, 2012.

Tanji, Miyume. *Myth Protest and Struggle in Okinawa.* London: Routledge, 2006.

Thomson, Sinclair et al. eds. *The Bolivia Reader: History, Culture, Politics.* Durham: Duke University Press, 2018.

Tigner, James Lawrence. "The Okinawans in Latin America: Investigations of Okinawan Communities in Latin America, with Exploration of Settlement Possibilities." Washington, DC: Pacific Science Board, National Research Council, 1954.

Tinsman, Heidi. *Buying into the Regime: Grapes and Consumption in Cold War Chile and the United States.* Durham: Duke University Press, 2013.

———. *Partners in Conflict: The Politics of Gender, Sexuality, and Labor in the Chilean Agrarian Reform, 1950–1973.* Durham: Duke University Press, 2002.

Truett, Samuel. *Fugitive Landscapes: The Forgotten History of the U.S.-Mexico Borderlands.* New Haven: Yale University Press, 2008.

Tsing, Anna Lowenhaupt. *In the Realm of the Diamond Queen: Marginality in an Out-of-the-Way Place.* Princeton: Princeton University Press, 1993.

Turits, Richard Lee. *Foundations of Despotism: Peasants, the Trujillo Regime, and Modernity in Dominican History.* Stanford: Stanford University Press, 2003.

Uberto Barbieri, Sante. "That Strange Land Called Bolivia." *Lands of Witness and Decision.* Methodist Church Joint Section of Education and Cultivation. New York: Methodist Church, 1957.

"The United Republic of Soybeans: Take Two," *GRAIN*, June 2013. Accessed January 8, 2019. https://www.grain.org/article/entries/4749-the-united-republic-of-soybeans -take-two.

Urenda Díaz, Juan Carlos. *Autonomías departamentales: La alternativa al centralismo Boliviano*. La Paz: Editorial Los Amigos del Libro, 1987.

Urquidi, Wilde, Donald H. Lee, and Julio Tumiri Javier. "Principios de cooperativismo." *Servicio agrícola interamericano*, 1958.

Valdivia, José Antonio. *Testigo de la realidad: Jorge Ruiz, memorias del cine documental Boliviano*. La Paz: CONACINE, 1998.

Wachtel, Nathan. *Gods and Vampires: Return to Chipaya*. Chicago: University of Chicago Press, 1994.

Wade, Peter. *Race and Ethnicity in Latin America*. London: Pluto Press, 2010.

Walker, Charles F. *Shaky Colonialism: The 1746 Earthquake-Tsunami in Lima, Peru, and its Long Aftermath*. Durham: Duke University Press, 2008.

Wang, Zhuoyi. *Revolutionary Cycles in Chinese Cinema, 1951–1979*. New York: Palgrave Macmillan, 2014.

Welch, Cliff. *The Seed Was Planted: The Saõ Paulo Roots of Brazil's Rural Labor Movement, 1924–1964*. University Park: Pennsylvania State University Press, 1999.

White, Richard. *Railroaded: The Transcontinentals and the Making of Modern America*. New York: Norton, 2011.

Whiteford, Scott. *Workers from the North: Plantations, Bolivian Labor, and the City in Northwest Argentina*. Austin: University of Texas Press, 1981.

Wilcox, Robert W. *Cattle in the Backlands: Mato Grosso and the Evolution of Ranching in the Brazilian Tropics*. Austin: University of Texas Press, 2017.

———. "Ranching Modernization in Tropical Brazil: Foreign Investment and Environment in Mato Grosso, 1900–1950." *Agricultural History* 82, no. 3 (2008): 366–92.

———. "Zebu's Elbows: Cattle Breeding and the Environment in Central Brazil, 1890–1960." In *Territories, Commodities and Knowledges: Latin American Environmental History in the Nineteenth and Twentieth Centuries*, Christian Brannstrom et al., eds. London: Institute for the Study of the Americas, 2004.

Will, Martina E. "The Old Colony Mennonite Colonization of Chihuahua and the Obregón Administration's Vision for the Nation." MA thesis, University of California, San Diego, 1993.

Winder, Gordon M., and Andreas Dix, eds. *Trading Environments: Frontiers, Commercial Knowledge and Environmental Transformation, 1750–1990*. New York: Routledge, 2015.

Wolfe, Joe. *Watering the Revolution: An Environmental and Technological History of Agrarian Reform in Mexico*. Durham: Duke University Press, 2017.

"The World's Fastest Growing Cities and Urban Areas from 2006 to 2020." CityMayors Statistics. Accessed January 12, 2019. http://www.citymayors.com/statistics/urban _growth1.html.

Wright, Angus, and Wendy Wolford. *To Inherit the Earth: The Landless Movement and the Struggle for a New Brazil*. Oakland: Food First Books, 2003.

Young, Elliott. *Alien Nation: Chinese Migration in the Americas from the Coolie Era through World War II*. Chapel Hill: University of North Carolina Press, 2014.

Yu, Henry. "Los Angeles and American Studies in a Pacific World of Migrations." *American Quarterly* 56, no. 3 (2004): 531–43.

———. *Thinking Orientals: Migration, Contact, and Exoticism in Modern America.* New York: Oxford University Press, 2002.

Zulawski, Ann. *Unequal Cures: Public Health and Political Change in Bolivia, 1900–1950.* Durham: Duke University Press, 2007.

Index

Note: All topics, locations, and organizations refer to Bolivia unless otherwise noted. Page numbers followed by *f* indicate figures. Page numbers followed by n indicate notes.

abandonment, 119–20; appeals from settlers, 22–23, 105, 119–36; autonomy vs., 241; *braceros* and threat of emigration, 115–17; dependency and, 123–25, 150–51; emigration as, 1–2; lowland discourse on, 21, 37, 41–42, 237; Methodists and, 150–51; parallels with other regions and time periods, 15, 259n43; petitions for abandoned land, 109–10, 119; settler attrition, 119–24, 128–29, 225–26; Sylvain's study of settlers, 121–25

Acción Rural, 102, 104–5

Aceite Rico, 224–25

Africa, resettlement and villagization in, 5, 24, 176

agrarian citizenship, 11–15; Andean, 11–15, 101; Mennonite, in Bolivia, 11–15, 100, 193–95, 209, 217, 240, 252; Mennonite, in Mexico, 91; miners's appeals for, 113–14; Okinawan, in Okinawa, 73; Okinawan and Japanese, in Bolivia, 11–15, 21–22, 66, 81–84, 100–101, 254

agrarian nationalism, 102–3, 112, 232–33

agrarian radicalism, in Yapacaní, 160–67

Agrarian Reform Law, 44, 65–66, 194–95, 252

agrarian unions, 104, 126, 134

Agricultural Expansion Zone, 159, 168. *See also* San Julián Project

agriculture: and citizenship, 11–15; coca eradication programs, 185, 218–19, 240, 243–44; critique of Mennonite practices, 227–28; environmental history and migration, 11–15; farmers targeted for relocation, 29–30; farming as

"mining," 15, 259n42; food security, 4, 12, 25, 71, 211–12, 232–33, 239–40; gendered notions of citizenship in, 11; Green Revolution, 6, 12, 13–14, 152–53; growth and transformation of Santa Cruz, 8–9; mechanization by Mennonites, 206–7; model crop for Mennonites, 190, 209–11; National Day of Corn, 232–33; socio-economic function, 65, 194; soybean agribusiness, Mennonite, 24–25, 190–91, 211–28; soybean boom and migration, 2–3, 14–15, 212; soybean production, Brazilian, 5; transnational class of practitioners, 6; USAID policy, 152–53, 185, 218, 240, 243–44. *See also specific products*

agroenvironmental history, 12

Akamine, José, 66, 67, 71, 80, 81, 99

Alba López, José, 121

Alex, Gary, 177

Algodonera, agreement with Mennonites, 86–88

Alliance for Progress: funding for Bolivia, 6, 57–58, 117; Guevara's criticism of, 52–53, 57–58, 236; Punta del Este meeting, 52–53, 57; Ruiz's documentation of, 55–56; support of Guatemalan colonization program, 56

alternative cropping, 185, 218–19, 240, 243–44

Amazonian expansion, Brazilian. *See* March to the West

ANAPO. *See* Asociacíon Nacional de Productores de Oleaginosas

anchovy crisis, Peruvian, 214–15

Andeans: agrarian nationalism of, 102–3, 112; Bravo's story of migration, 1–2, 7; citizenship, agrarian, 11–15, 101; citizenship, indigenous people, 106; defining and categorizing, 105–6; diaspora of, 118–19; engagement with logic of March to the East, 102; farmers targeted for relocation, 29–30; INC cynicism about, 173; as *interculturales*, 253–54; as "invaders" of lowlands, 13, 25, 181, 223, 240, 244–45; lingering culture in new land, 123; migration, comparison with Mennonites and Okinawans, 101, 102, 111–12, 118; migration to Santa Cruz, 7, 258n14; migratory strategies in neoliberal Brazil, 225; Morales's (Evo) story of mobility, 7–8, 25, 218, 241, 242–45; pamphlet images of, 58–62, 61f; petitions (letters) from, 22–23, 102–38, 238; petitions (letters) from, appeals to settle frontier, 105, 106–19; petitions (letters) from, complaints from settlers, 105, 119–36; racialized reaction to, 8, 13, 26–27, 37, 41–45, 80–81, 237–38; settler attrition, 119–24; transnational labor of, 81, 235

Andrade, Victor, 77–78, 84–85

Antezana, Fernando, 115, 117

Arakaki, Yoei, 82

Araus, Alejandro, 172

Arazi, Shai, 169–70

Árbenz, Jacobo, 53, 55

Argentina: Bolivian workers in, 3, 9, 10, 114–19, 225; racism in, 117–18; rail links with Santa Cruz, 37–42, 38f; soybean exports to, 217

Arias, Mortimer, 169, 175, 184

Asociacíon Nacional de Productores de Oleaginosas (ANAPO): annual reports, 212, 220, 252; critique of Mennonite agricultural practices, 227; debt crisis intervention, 224, 227; national, racial, and ethnic categories of, 219, 252; Okinawan and Japanese participation in, 253; world markets opened to, 217

attrition, settler, 119–24, 225–26

autonomy, 241, 242–55; departmental, national referendum on, 245; gendered ideas of, 246, 247–48; indigenous groups, 25, 244, 249–55; liberation theology and, 153; lowland elites (*cruceños*), 8, 20, 25, 45–51, 244–48, 249; Mennonite model, 153, 249–52, 284n20; Mennonites and regional symbolism, 247–48

Aviles Randolph, Jorge, 1–2

Ayaviri Villca, Eustaquio, 120

Aymara: continued use of language, 122–23; defining Andeans, 106; first indigenous president, 7, 242; Methodist recruitment of, 148–49; migration of, 1–2, 118, 144; "Our Indian" editorial *vs.*, 44–45; *Plan immediato* dismissal of, 42; settler attrition, 122–23. *See also* Bravo, Jaime; Morales Ayma, Juan "Evo"

Ayoreo, 25, 248–49

Baldivieso, Ruben, 157, 162–64

Balkanization, 47

Banzer, Hugo, 240; agricultural policy, 212; Banzerato, 143, 164; faith-based development under, 23, 137–38, 143, 164–86, 239; kibbutz system proposal, 169–70, 216; military coup, 23, 143, 164; overthrow of, 177–78; Public Works Conference (1972), 169–70, 181–82; regime's reprisals against Methodists, 165–67; threat to Mennonite exemptions, 203, 252

Barbieri, Sante Uberto, 143–44, 147

Barrientos, René, 56, 104, 240; death of, 162; disaster response (1968 flooding), 154–56, 275n59; overthrow of Paz Estenssoro, 137; petitions to, 124, 126, 127–28, 133–34; on "titanic struggle," 157

beauty queens, and regional identity, 246, 247–48

Belaúnde Terry, Fernando, 6, 56, 58, 64

Belize: Mennonite emigration from, 203, 229; Mennonite settlement in, 93, 95 belonging: agrarian citizenship and, 12; race/ethnicity and, 13

Beltrán, Luis Ramiro, 54–57, 64

Belzu, Manuel Isidoro, 103–4

Bohan, Erwin, 17

Bolívar Menacho, Jesús, 207–9, 220, 250

Bolivia: balkanization of, 47; bureaucratic authoritarianism in, 143; conflicts and loss of territory, 17, 109–10; consensus on eastern expansion, 20; geography of, 3–5, 4*f*; as landlocked nation, 3–4; military coup and authoritarian rule in, 23–24; plurinationality in, 250–55; political periods of, 240–41; proliferation of institutions in, 269n14; relationship with Cuba, 57; spacial sensibility in, 29–30; special exemptions for Mennonites, 22, 24, 65–66, 87–89, 90, 202–3, 209, 239, 250–52; territoriality and sovereignty in, 15–18; U.S. funding for, 6, 58–59, 64, 77, 117, 263n106

Bolivian Development Corporation (CBF), 269n14; archives, 120; disbanding of, 182, 191; letters and petitions to, 105, 112, 134, 163; milk processing plant, 208; proposed Japanese investment, 84–86; regard for Okinawans, 78

Bolivian Institute for Foreign Commerce (IBCE), 253

Bolivian migration: citizenship notions in, 10–11; lowland ascendancy in, 8; lowlands as landscape of migration, 3; narratives of, 1–3; parallels in global south, 5; parallels in Latin America, 5–6; past and present of, 230–41; recasting Bolivia from the margins, 6–9; transnational revolution in, 9–11; U.S. funding for, 58–59. *See also* Andeans; frontier imaginary; March to the East; Mennonites; Okinawans

Bolsonaro, Jair, 20

Bonino, José Míguez, 160

Bonoli, Felipe, 80, 266n54

braceros (field workers): Antezana exposé on, 115, 117; Bolivian, in Argentina, 3, 9, 10, 114–19, 225; Guevara Arze article on, 115–16; invoked in petitions for land, 114–19; racism toward, 117–18; in Santa Cruz, orientation program for, 157–59

Bravo, Germán, 254

Bravo, Jaime: arrest and imprisonment, 165–67; career trajectory, 184–85; embrace of capacity building, 181–82; exile from Bolivia, 165–67, 178; Methodist faith and development work, 149–51, 158, 160–67; migration story, 1–2, 7, 149; parting with Methodist Church, 175; personal trajectory, 241; racial and civil rights discourse, 236, 254; radicalism, 160–67; view on secularization, 177

Brazil: Bolivian loss of territory to, 17, 109; bureaucratic authoritarianism in, 143; frontier imaginary of, 28–29, 32–33, 64; highway link with Santa Cruz, 38*f*; March to the West, 5, 29; Mennonite settlement in, 70; migrant-driven environmental change in, 13; Okinawans in, 68, 75; rail links with Santa Cruz, 38–42, 38*f*; soybean production, 5, 213–14

Brokensha, David, 176

buen vivir, 241

Buhler, Wilhelm, 226

bulls, and regional identity, 246, 247

bureaucratic authoritarianism, 143

Busch, Alberto Natusch, 199

Busch, Germán, 44

"By the New Highway" (Clouzet), 43–44

Café Filho, João, 38

CAICO (Cooperativa Agropecuaria Integral Colonias Okinawa Ltda.), 253

CAISY (Cooperativa Agropecuaria Integral San Juan de Yapacaní Ltda.), 253

Cajías, Eduardo, 125

California Colony, 252

Camba Nation, 245, 254

cambas, 8, 13, 26–27, 44, 106, 237, 245, 254–55

campesino, 106

Canada: Mennonite citizenship in, 94, 100;
  Mennonite emigration from, 86, 90, 229;
  Mennonite exemptions in, 252;
  Mennonite return to, 94–96, 97, 225–26;
  Mennonite settlement in, 69–70;
  Okinawan settlement in, 68
Canadiense I (colony), 88–90, 232, 233
Canadiense II (colony), 232, 233
CAO (Chamber of Eastern Agriculture), 253
capacity building, 170, 181–82
Capobiano, Guillermo, 162–63, 165
cattle: bulls and regional identity, 246, 247;
  Mennonite dairy industry, Bolivian, 204,
  207–10, 223, 226–27, 226f, 239–40;
  Mennonite dairy industry, Mexican, 91,
  96; Mennonite import of, 201, 204–5,
  207–8; in Santa Cruz, 8–9; Zebu, 8–9,
  33, 204–5, 214, 235, 247
Caulfield, Robert and Rosa, 147, 149
CBF. *See* Bolivian Development
  Corporation
CECOYA (Central de Colonizadores de
  Yapacaní), 162–63
Center for Tropical Agricultural
  Investigation (CIAT), 204
Central de Colonizadores de Yapacaní
  (CECOYA), 162–63
Cerruto, Walter, 32, 34, 51
Certificate of Identity, Okinawan, 75–77,
  100
Chaco War (1932–35), 17, 193
Chamber of Eastern Agriculture (CAO),
  253
Chávez, Hugo, 240
cheese, Mennonite: Bolivian, 199, 208–10,
  223, 226–27, 231, 239–40; Mexican, 91, 96
Chibana, Hiro, 100
Chicha, 43, 62, 122
Chile: agrarian reform and gendered
  notions of citizenship in, 10–11; Bolivian
  loss of territory to, 17, 109; grape
  production, 214
China, agrarian reform in, 52
Chiquitano, 25, 248–49
Chiriguano, 45

CIAT (Center for Tropical Agricultural
  Investigation), 204
CIDOB (Confederation of Indigenous
  Peoples of the Bolivian Oriente), 249–50
citizenship: environmental change and,
  5–6; gendered notions of, 11; indigenous
  people, 11, 106; lowland colonization
  and, 10–11; Mennonite, Canadian, 94,
  100; Mennonite, identification
  (resident) card and, 251–52; Mennonite,
  noncitizens by choice, 11, 87, 193–94;
  Okinawan postwar problems with, 11, 68,
  74, 75–77, 100. *See also* agrarian
  citizenship
CIU. *See* United Church Committee
"civic struggles," over oil royalties, 45–48
Clouzet, Ramón, 43–44
coca: association with criminality, 219;
  consumption, Andean, 122–23, 218;
  economic crisis and production of, 218;
  eradication programs, 185, 218–19, 240,
  243–44; Santa Cruz, 8; soybeans *vs.*,
  218–19, 240; union organization,
  Morales (Evo) and, 7, 243–44, 249
cocaine, 218–19, 243–44
Cochabamba: highway link with, 38f,
  43–44; rail link with, 38–42, 38f
Colombia, Mennonite colonization
  program in, 232
Colque, Gonzalo, 14–15
Colque, Moises, 108
COMIBOL (Corporación Minera de
  Bolivia), 121, 182, 269n14
Committee of Reception of Japanese
  Immigrants for Okinawa Colony, 82
Committee Pro-Santa Cruz, 46–47, 50, 169,
  245, 246
Communal Lands of Origen (TCOs), 249
Condori, Francisco, 117–18
Confederation of Indigenous Peoples of
  the Bolivian Oriente (CIDOB), 249–50
conscientious objectors, Mennonites
  as, 152
consolidation stage, of settlements, 176,
  178, 183

constitution, rewrite under Morales (Evo), 256–46, 249, 250

Cooperativa Agropecuaria Integral Colonias Okinawa Ltda. (CAICO), 253

Cooperativa Agropecuaria Integral San Juan de Yapacaní Ltda. (CAISY), 253

Cornell University, 176, 183, 218

Corporación Minera de Bolivia (COMIBOL), 121, 182, 269n14

Correa, Rafael, 240

corruption, 125, 172

Cossio, Angel, 112

Cotoca Colony, 120–27

Cotton, 85–88, 167, 169, 198, 210, 213, 249

La Crónica (newspaper), 192, 194, 197, 204

cruceños, 8

Cruceño Youth, 50, 245

Cuba: Alliance for Progress challenged by, 52–53, 57; Cuban Institute of Cinematography and Art, 31; guerrilla force in Bolivia, 187–89; relationship with Bolivia, 57

dairy industry, Mennonite: Bolivian, 204, 207–10, 223, 226–27, 226f, 239–40; Mexican, 91, 96

"Day of the Indian," 44

El Deber (newspaper): denunciation of dependency on highland markets, 42; flooding coverage (1968), 153–56; "Homage to Santa Cruz" editorials, 28; letter promoting dairy, 208; martial law protests described, 47–48; racial identity shaped in, 44–45; racism in, 78–83; railroad coverage, 39–42, 261n49; regionalist discourse and critique, 37; reporting on soybeans, 219, 221, 222; reporting on Yapacaní unrest, 161–62; response to "Day of the Indian," 44; review of *La Vertiente*, 48–49

debt cancellation, for Bolivia, 224

debt crisis, Mennonite, 221–28, 282n125; ANAPO intervention, 224, 227; MCC intervention, 223–24, 227; MCC report, 221

decentralization, 174

decolonization, language of, 240–41

Del Norte (colony), 216–17

dependency, 123–25, 150–51, 172–74, 178, 183, 234

development, faith-based. *See* faith-based development

development narrative, in films, 51–58

development studies: Nelson's stages of settlements, 176, 178, 183; San Julián Project, 141–42, 175–77, 179–80, 183

El Día (newspaper), 248

El Diario (newspaper): coverage of Mexican Mennonites, 191–94, 195, 197, 201, 204; denunciation of rail contraband, 41; "Santa Cruz: crucified," 46

Díaz Ordaz, Gustavo, 93

Dickson, Murray, 146, 150, 151, 185

Dietrich, William, 9, 208, 230

Dillon, Douglas, 53

directed settlers, spontaneous colonists *vs.*, 131

Doing Theology in a Revolutionary Setting (Bonino), 160

Domínguez, Jorge, 95–96

Doña Bárbara (Gallegos novel), 30

Driediger, Arthur, 153, 197

drought: Bolivian, and Mennonite debt crisis, 221–28, 282n125; Mennonite settlement design and wind-based erosion, 223–24; Mexican, 92–98

drought-resistant crops, 226

Dueck, Martin, 1–2, 9, 13, 99

Duguid, Julian, 16–17

Duranović, Jaime, 224–25, 253

Eastern Lowlands Project, 229

Ecuador, Ruiz–Beltrán film in, 55

egg production, 207–10, 220

El Salvador, U.S. policy toward, 284n9

Empresa Brasiliera de Pesquisa Agropecuária (EMBRAPA, Brazil), 214

environmental protests, 20, 249

Epp, Mark, 173–74, 176, 183

Espejo Ticoña, Juan, 160–61, 163, 165–66
eurocentrism, 79–80
Evangelical Mennonite Mission Church, 200
evangelism: Mennonite Central
    Committee, 168; Mennonite conflict
    over, 97–99, 200; Methodist "lands of
    decision" campaign, 6, 146–47, 153, 185;
    World Gospel Mission, 23, 144–46,
    238–39, 248, 274n22
"exodus" cooperative, 118
EXPOCRUZ, 246, 247

faith-based development, 18–19, 139–86,
    238–39; Banzer regime and, 23, 137–38,
    143, 164–86, 239; disaster response (1968
    flooding), 153–56; Mennonite Central
    Committee, 151–53; Methodist Church,
    146–51, 149f; Methodist radicalism,
    160–67; ministering *vs.* administering,
    139–40, 144, 168, 174–75, 181–82, 185–86,
    238–39; MNR welcoming of, 143–44; as
    proxy for state, 138, 142, 156, 167, 169,
    182–86, 239; recognition of local
    knowledge, 160; San Julián Project, 6,
    23–24, 139–43, 168–86; secularization of,
    177; settler orientation program, 139,
    157–59, 170–72; soybean promotion, 211;
    United Church Committee, 23, 139–43,
    155–59
Fajardo, Ana, 148–49, 150, 163, 166–67, 184
Falange Socialista Boliviana (FSB), 47
Federation of San Julián Colonists, 180
Fellman Velarde, José, 31
femininity: in frontier imaginary, 27, 30,
    33, 43, 64; in ideas of autonomy, 246,
    247–48
Fernández Prado, Luis, 126–27
FIDES (Inter-American Foundation for
    Development), 177, 178, 181, 278n155
films, 26–37; Anglo-Americans inspiration
    for, 31; fertility and femininity in, 27, 30,
    33; gender terms in, 26–27, 30, 34, 37, 49;
    influence of *Plan immediato* on, 29–30;
    lowlands as transformative space in, 27,
    29; mandated distribution of, 32;

Methodist, 149; MNR-produced
    propaganda, 7, 31–34; reception from
    imagined subjects, 21, 28, 64;
    revolutionary states and, 31; Ruiz, 21,
    26–31, 34–37, 236–37; Ruiz, *A Little Bit of
    Economic Diversification,* 26–31, 28f, 36,
    42, 63–64, 237, 243; Ruiz, *Little Johnny
    Can Read,* 44–45; Ruiz, *The Mountains
    Never Change,* 58–59, 83–84; Ruiz, *Los
    Primeros,* 35–36, 64; Ruiz, *La Vertiente,*
    34–36, 35f, 48–49, 51, 52, 64, 236; Ruiz,
    Vuelva Sebastiana, 44, 55; Ruiz, *Los
    Ximul,* 53–57, 236; self-awareness and
    representation in, 33–34; theaters for,
    Santa Cruz, 49–50; transnational topics
    and audiences, 51–58; *Traveling Through
    Our Land,* 32–34; visual distinction
    between east and west in, 21, 27
First National Congress of Colonizers, 163
fiscal shock program, 217–19, 243
flooding: Río Grande (1968), 143, 153–56,
    275n59; San Julián (1983), 178
Flores, Casimiro, 108
food security, 4, 12, 25, 71, 211–12, 232–33,
    239–40
foreign actors, as proxies for state, 15,
    259n44
Forestry Law, 227, 232
Frank, Andre Gunder, 123
Fretz, J. W., 89–90, 92, 151–52
*From Death Trails in Bolivia to Faith
    Triumphant* (Geyer), 144
frontier imaginary, 20, 21, 26–64, 236–37;
    Brazilian, 28–29, 32, 64; fertility in, 27,
    30, 33, 64; film, 7, 21, 26–36, 236–37;
    National Congress "Homage to Santa
    Cruz", 28; influence of *Plan immediato*
    on, 29–30; lowlands as transformative
    space in, 27, 29; MNR propaganda films,
    7, 31–34; pamphlet, 58–62, 60f, 61f, 237;
    in pamphlets, 58–62; petitions from
    Andeans, 106–19; reception from
    imagined subjects, 21, 28, 64; regional
    tropes in, 32, 33, 34, 43; rivers, roads, and
    rail in, 36–51; Ruiz films, 21, 26–31, 34–36,

42, 58–59; self-awareness and representation in, 33–34; territoriality and state sovereignty in, 15–18; transnational topics and audiences, 52–58, 64

Galarza, Ernesto, 115
García Linear, Álvaro, 232–33
García Meza, Luis, 177–78
Gazit, Y., 169–70
gender: and citizenship, 11; and colony terminology, 228, 283n148; and frontier imaginary in films, 26–27, 30, 33, 34–36, 49; and frontier imaginary in pamphlets, 59–62; and ideas of autonomy, 246, 247–48; and ownership rights, 131; and perception of Mennonites, 192; and settler attrition, 128–29
Geyer, Robert, 144–46, 248
"Ghost Rapes," 284n20
Goff, James and Margaret, 165–66
Gómez, Flora, 177
Good Neighbor Policy, 17, 55
Government of the Ryukyuan Islands (GRI), 72
Great Disjuncture, 206
green hell, 16
Green Revolution, 6, 12, 13–14, 140, 152–53
GRI (Government of the Ryukyuan Islands), 72
Grierson, John, 31
growth stage, of settlements, 176, 178, 183
Guaraní, 25, 45, 106, 144, 248–51
Guarayo, 144, 248–49
Guatemala, Ruiz documentary on colonization program in, 53–58
Guerrilla Warfare (Guevara), 189
Guevara, Ernesto "Che," 52–53, 57, 164, 187–89, 228, 236
Guevara Arze, Wálter: article on *braceros*, 115–16; occupation authorized by, 48; *Plan immediato*, 29–30, 35, 42, 84, 110–11, 113; regionalists *vs.*, 64; Tigner and, 71–72

Hardeman Camp/Colony, 154–56
Heifer Project International, 183

Henry, Walter, 172
Hess, David, 176, 177, 183–84
Hickman, John, 150, 151
Higa, Katoshi, 225
Higa, Yoshihide, 75, 81, 83–84
Highland Echoes (Methodist newsletter), 149–50
highlanders *(kollas)*, 8, 13, 26–27, 43–44, 105–6, 237, 245, 254
"Homage to Santa Cruz" (Bolivian Congress), 28
Hoover Institute, 71
Horowitz, Michael, 176
horse-and-buggy, as regional symbol, 247–48
Hoy Bolivia (news series), 57
Hughes, Evert C., 189
Hurtado Medina, Ricardo, 89
hyperinflation: and San Julián Project, 178, 179–80; and soybean production, 211–12, 215–20, 221

IBCE (Bolivian Institute for Foreign Commerce), 253
ICB. *See* Instituto Cinematográfico Boliviano
IDA. *See* Institute for Development Anthropology
IDB (Inter-American Development Bank), 58, 160
identification card, for Mennonites, 251–52
IICA (Inter-American Institute for Cooperation on Agriculture), 54–55
import duties, Mennonite exemption from, 202–3
import-substitution, 29–30
INC. *See* Instituto Nacional de Colonización
India, dam-building and resettlement in, 5
"Indian question," 104
indigenous people: autonomy movement, 25, 249–55; citizenship status of, 11, 106; "Day of the Indian," 44; evangelism and, 144–46, 248; extension of rights to, 3; "incorporate into national life," 105,

indigenous people (*continued*)
269n15; lowland *vs.* highland, 45; March
for Territory and Dignity, 249; Morales
(Evo) as first indigenous leader of
Bolivia, 7, 242; naïve monarchism of,
103; preservation of language and
culture, 249, 250; protected territories,
249; violent conflict with settlers, 144.
See also Andeans
indio, 106
Indonesia, transmigration in, 5
Industrias Oleaginosas (IOL, S.A.), 213,
246
infant industry critique, 151
Institute for Development Anthropology
(SUNY-Binghamton), 6, 24, 141–42, 176,
179–80, 183
Institute for Political Capacitation, 31
Instituto Cinematográfico Boliviano
(ICB), 31–35, 237; attempts to humanize
March to the East, 63; *The Mountains
Never Change*, 58–59; Ruiz and, 27, 34–36
Instituto Nacional de Colonización (INC),
269n14; *braceros* issue, 116, 271n53;
cynicism about Andeans, 173; Lemaitre's
defense of Japanese, 192–93; Lemaitre's
defense of Mennonites, 192–94, 195;
letters and petitions, 106, 116, 127, 271n53;
Mennonite negotiations with, 202;
Mennonite titles and land survey, 199;
ownership (title) issues, 129–31; Public
Works Conference (1972), 169–70;
reservations about Mexican
Mennonites, 195–97; San Julián contract,
23, 170; San Julián protest against,
180–82; San Julián report, 172–73
Inter-American Agricultural Service (SAI),
27, 55, 263n110
Inter-American Development Bank (IDB),
58, 160
Inter-American Economic and Social
Council, 52
Inter-American Foundation for
Development (FIDES), 177, 178, 181,
278n155

Inter-American Institute for Cooperation
on Agriculture (IICA), 54–55
interculturales, 253–54
Interdenominational Evangelical Faculty of
Theology (Buenos Aires), 160
International Development Bank, 163
Interoceanic Corridor, 232
"In the Trail of the Pioneers" (Brazilian
newsreel), 32
Isiboro-Sécure Indigenous Territory
(TIPNIS), 249
Italian migrants, 79–80, 266n54

Japan: economic miracle of, 22, 86, 100–101;
immigration policy for descendants, 225;
investment in Bolivia, 84–86, 230
Japanese community and migrants:
Bolivian congressman, 253, 254;
challenges to racism, 82–83; egg
production, 220; emigration to Bolivia,
75–76; past and present, 230–31;
petitioners opposed to, 112;
plurinationality and integration, 252–53;
return to Japan, 225–26; Uruma Society,
66, 67, 71, 77–78; xenophobia toward, 13,
19, 21, 75, 78–83, 266n60
Japanese International Cooperation
Agency (JICA), 19, 253
Japanese Social Center, 230–31
Juanito sabe leer (Ruiz film), 44

Keegan, Maureen, 140, 157, 158
Kennedy, John F., 52, 58
"The Ketchup Song," 94
Kibbutz model, 169–70, 216
knowledge, transplanted, 236
knowledge transfer, 61. *See also*
technicians
kollas, 8, 13, 26–27, 43–44, 105–6, 237
Kopp, Alfred, 198, 200, 208, 210
Kubitschek, Juscelino, 5

Labor Federation of Rural Workers, 111
"lands of decision" campaign, 6, 146–47,
153, 185

Latin America 3s volunteers (Methodist), 150, 183

Le Corbusier, 103

Lemaitre, Roberto, 116, 192–96, 199, 202–3, 228

Lemnitzer, Lyman, 76

letters. *See* petitions and letters

liberation theology, 153, 236

Limachi, Marcelino, 157–58

Linares, Adolfo, 84–85

Linsenmeyer, Dale, 165, 167, 169

A Little Bit of Economic Diversification (Ruiz film), 26–30, 28*f*, 36, 42, 64, 237, 243

Little Johnny Can Read (Ruiz film), 44

Litwiller, Nelson, 174

Loeppky, Johan, 91

Loewen, Daniel, 98–99

López, Mario, 190, 195–96

lost decade (1980s), 142, 177, 219

Loucks, Lynn, 167

lowland ascendancy, 8

lowlands: Eastern Lowlands Project, 229; exploration and territoriality, 15–18; geography of, 4–5, 4*f*; as green hell, 16; as land of migration, 3; past and present in, 230–41; as transformative space, 27, 29. *See also* March to the East; Santa Cruz; *specific colonies and settlers*

Luft, Murray, 174

Lutheran World Relief, 181–82

Machaca, Belberto, 132

Malaysian resettlement, 258n4

Maldonado, Hugh, 48–49

Mao Zedong, 52

La Máquina (Guatemala), 53–58

March for Territory and Dignity, 249

March to the East, 3–6; Andean engagement with logic of, 102; divergence between state and settler aims, 103; humanization of, 63; as migratory project, 37; as modernizing vision, 37; Morales's (Evo) personal trajectory in, 7–8, 25, 218, 241, 242–45; as

national romance, 21, 29–37; parallels in global south, 5; parallels in Latin America, 5–6; petitions referencing, 111; populations targeted for, 29–30; propaganda films promoting, 7; Ruiz films and, 21, 26–30, 34–37. *See also* frontier imaginary

March to the West (Brazil), 5, 29, 32; link to *bandeirantes*, 29

"Marginal Highway of the Jungle," 6, 57, 59, 64

Marinkovic, Branko, 246–47

Marinkovic, Silvio, 212–13, 215, 224–25, 246–47

Marinkovic, Tatiana, 246

market access, complaints about, 124, 133–36

Martens, Wilhelm, 205

martial law, 47–51

Martin Rivas (Blest Gana novel), 30

Marx, Karl, on rail, time, and space, 17–18

Maryknoll nuns: disaster response (1968 flooding), 154; faith-based development, 140; long-term community engagement, 156, 275n68; settler orientation program, 139, 157–58; targeted after military coup, 166; United Church Committee, 23, 139–43, 155–59, 168–86

MAS (Movement to Socialism), 241

Mason, John Charles, 16

Matto Grosso, Brazil, 5

Mayorga, Evorista, 110

McIntyre, Loren, 54

mechanization, Mennonite, 206–7

Médici, Emílio, 5

Mendoza-López, Deputy, 34, 51

Mennoblatt (newspaper), 65, 90

Mennonite Central Committee, 151–53; archives of, 18, 197; "Bolivia Mystique," 153; Bolivian activities of, 23; debt crisis intervention, 223–24; debt crisis report, 221; disengagement approach, 156, 275n68; expectations for being "Mennonite embassy," 197, 223; faith-based development, 18, 140; fellow

Mennonite Central Committee (*continued*)
brethren as initial goal, 151; founding of,
151; impact of military coup on, 165–67;
indigenous autonomy movement and,
249–50; Mexican activities of, 92, 97;
ministering *vs.* administering, 168,
174–75; North American volunteers,
152–53; as proxy for state, 167; Public
Works Conference (1972), 169–70;
relationship with immigrant Mexican
Mennonites, 197–200; Teacher Abroad
Program, 152, 167, 277n125; Tres Palmas
health clinic, 151–52; United Church
Committee, 23, 139–43, 155–59, 168–86
Mennonite Life (periodical), 90, 151
Mennonites: autonomy model, 153, 249–52,
284n20; citizenship, agrarian, in Bolivia,
11–15, 100, 193–95, 209, 217, 240, 252;
citizenship, agrarian, in Mexico, 91;
citizenship, Canadian, 94, 100;
citizenship, noncitizens by choice, 11, 87,
193–94; communication (letter writing),
90, 258n100; conflict over rubber tires,
22, 96–99, 227; diasporic history of, 2, 9,
66–67, 68–70, 69f, 70f, 88, 99, 235;
emigration from Belize, 203, 229;
emigration from Canada, 86, 90, 229;
emigration from Mexico, 1–2, 9, 13, 22,
66, 99 (*see also* Mennonites, Mexican);
emigration from Paraguay, 22, 65–67,
86–90, 229 (*see also* Mennonites,
Paraguayan); emigration from Russia,
65, 68–69; erosion of privileges, 251–52;
guerrilla vision/backwoods guerrilla
warfare, 189, 229; identification
(resident) card for, 251–52; Latin
American settlements, 70; map of lowland
colonies, 190f; migration, comparison
with Andeans, 101, 102, 111–12, 118;
migration, comparison with Okinawans,
22, 66–70, 86, 99–101; MNR and, 65–66,
87–89; race and attitudes toward, 88,
192–94; rape cases ("Ghost Rapes"),
284n20; regional symbolism and, 247–48;
sources for study, 18–19, 189–90, 197;

special exemptions for, 11, 68–69, 251–52;
special exemptions for, Bolivian, 22, 24,
65–66, 87–88, 90, 202–3, 209, 239, 250–52;
special exemptions for, Mexican, 91, 93,
250, 252; as "state within a state," 250;
struggle in Bolivia, Fretz's story of, 89–90;
term, explanation of, 67, 68–69; tradition
vs. modernity, 13, 24, 96–99, 227–28;
transnational revolution and, 9
Mennonites, in Mexico, 90–99; agrarian
citizenship, 91; conflict over rubber tires,
22, 96–99; drought impact on, 92–98;
emigration of, 1–2, 9, 13, 22, 66, 99
(*see also* Mennonites, Mexican);
evangelical conflict, 97–99; landlessness
problem, 92, 98; return to Canada,
94–96, 97; Social Security participation
and, 93–94; special exemptions for, 91,
93, 250, 252
Mennonites, Mexican (in Bolivia), 24–25,
66–97, 90, 99, 187–240; agrarian
citizenship of, 193–95, 209, 217, 240, 252;
agricultural practices, critique of,
227–28; autonomy of, 249–52; cattle
importing, 201, 204–5, 207–8; colony
expansion, 228, 229, 232;
cosmopolitanism of, 200–201; cultural
characteristics, 13, 193–94; dairy
industry, 204, 207–10, 223, 226–27, 226f,
239–40; diary of Wiebe (Johan), 201–4;
drought and debt crisis, 221–28, 282n125;
Dueck's migration story, 1–2, 9, 13, 99;
exemption from import duties, 202–3;
exploratory committee, 188–89;
gendered images of, 192; hyperinflation
and, 211–12, 215–20, 221; imprisonment
for debts, 221, 223, 282n125; INC
reservations about, 195–97; Kopp's
clandestine visit to, 198, 200; land prices
and parcel sizes for, 195–97, 201–2, 239,
280n55, 283n19; machinery import and
mechanization, 206–7; "Minimum Plan
of Work," 194–95, 199, 204; model crop,
search for, 190, 209–11; native economies
revolutionized by, 189, 207, 229;

negotiations with local and national governments, 200–204; paradox of isolation and economic success, 209, 229, 239, 250–51; population growth, 223, 228; press reception and perception of, 191–94, 197, 204; relationship with Mennonite Central Committee, 197–200; Santa Cruz street scenes and market day, 231, 231*f*; settlement design and wind-based erosion, 223–24; social-economic function of settlement, 194, 216; soybean agribusiness, 24–25, 190–91, 211–28, 239–40; theological justifications of, 202–3; transnational trade, 200–211

Mennonites, Paraguayan (in Bolivia), 22, 65–67, 86–90, 229; anniversary commemoration, 233; Canadiense I, 88–90, 232, 233; Canadiense II, 232, 233; comparison with Mexican Mennonites, 209; exploratory delegation in Santa Cruz, 65–66; September 24 Cooperative, 86–88

Mennonite Service Committee (MSC), 97–99

Mennonitische Post, 90, 258n100

Menses, Julio, 111, 112

Mercado, Orlando, 50

mestizo, 8, 27, 43, 45, 106

Methodist Mission Board, 6, 23, 146

Methodist Rural Institute, 147, 152, 157, 165

Methodists, 146–51; areas of operation, 147, 148*f*; career springboards, 183–85; Chané-Bedoya organizing, 160; disaster response (1968 flooding), 154–56; disengagement approach, 156, 275n68; establishment of Bolivian operations, 146; extension work in colonies, 159–64; films, 149; indigenous speakers recruited by, 148–49; "lands of decision" campaign, 6, 146–47, 153, 185; Latin America 3s volunteers, 150, 183; low cost and collaborative work of, 150–51; "out on their limb," 165, 167; political involvement, reflections on, 166–67;

Public Works Conference (1972), 169–70; radicalism, 160–67; reprisals against, following coup, 165–67, 168; schools and alumni, 146; transnational theological strain in, 159–60; United Church Committee, 23, 139–43, 155–59, 168–86; waning North American ties to Bolivia, 166; youth volunteers, 163, 276n107

Métraux, Alfred, 121

Mexico: drought impact on Mennonites, 92–98; Mennonite conflict over rubber tires, 22, 96–99; Mennonite emigration from, 1–2, 9, 13, 22, 66–67, 99; Mennonite landlessness problem in, 92, 98; Mennonites as agrarian citizens in, 91; Mennonite settlement in, 70, 90–99; participation in Social Security, 93–94; special exemptions for Mennonites, 91, 93, 250, 252

Miami International Airport, 230

military coup: against Banzer, 177–78; against NMR, 56, 137, 143, 164

Miller, Marty, 140, 157, 158, 169

Minchin, J. B., 15–16

miners: depiction in Ruiz films, 26, 29–30; "Exodus" cooperative, 118; targeted for relocation, 29–30

mini-dust bowl, in Mexico, 92

"Minimum Plan of Work" (Mexican Mennonites in Bolivia), 194–95, 199, 204

Miranda, Rogelio, 116, 271n53

missionaries, 6, 23. *See also* religion and religious organizations

MNR. *See* Nationalist Revolutionary Movement

Modernization initiatives, 103, 183

Moffat, Bob, 175

Mojeño, 45, 248–49

Molina, Max, 130, 131

Montero Hoyos, Constantino, 17

Moore, James E., 76

Morales, Marcelino, 163

Morales Ayma, Juan "Evo": autonomy movement challenging, 8, 20, 244–45;

Morales Ayma, Juan "Evo" (*continued*)
as Bolivia's first indigenous leader, 7,
242; colony support for, 253;
constitutional rewrite, 245–46, 249, 250;
election of, 240–41; environmentalism,
241; personal migration story of, 7–8, 25,
218, 241, 242–45; plot to assassinate, 246;
pro-lowland stance, 20; socialism of,
240–41; union organizing for coca
workers, 7, 243–44, 249
The Mountains Never Change (Ruiz film),
60, 83–84
Movement to Socialism (MAS), 241
El Mundo (newspaper), 117

"Nachos Mexicambas," 254–55
La Nación (newspaper), 31
El Nacional (newspaper), 91
NADEPA, San Julián Project, 170
Nagatani, Michiaki, 253, 254
naïve monarchism, 103, 132
National Bolivian Workers Congress, 160
National Congress of Bolivian Colonists,
276n104
National Day of Corn, 232–33
National Emergency Committee, 155
National Federation of Bolivian Colonists,
7–8, 163, 276n104
National Institute of Colonization. *See*
Instituto Nacional de Colonización
nationalism, agrarian, 102–3, 232–33
Nationalist Revolutionary Movement
(MNR): Andean petitions to, 22–23,
104–5, 110–13; anniversary and film,
58–59; conflict with Santa Cruz over oil
royalties, 45–47; March to the East, 3–6;
Mennonites and, 65–66, 87–89; military
coup against, 56, 137, 143, 164;
modernization initiatives, 103;
paternalistic views of indigenous people,
5, 32, 45; *Plan immediato*, 29–30, 42, 71,
84, 110–11, 113; propaganda films produced
by, 7, 31–34; rise to power, 65; state–
citizen exchange under, 137; terminology
for indigenous people, 105–6

nature's metropolis (Cronon), Santa
Cruz as, 8
Nelson, Michael, 139, 175–76, 177, 178, 183
neo-extractivism, 15, 25
neoliberalism, 20, 63, 142, 182, 191, 240–44;
agricultural impact (soybeans), 14, 25,
211–20; election of Morales (Evo) as
break with, 240–41, 244; fiscal shock
program, 217–19; grassroots resistance
to, 181; migratory strategies under, 225
New Economic Policy (NEP), 219–21
NGOs, 177. *See also specific organizations*
Niemeyer, Oscar, 103
Nishikawa, Toshimichi, 82
Nkrumah, Kwame, 5, 31
no-till technology, 227–28
Nueva Esparanza (colony), 203
Nuñez del Prado, Deputy, 46–48
Nyerere, Julius, 5, 103

Obregón, Álvaro, 91, 252
O'Donnell, Guillermo, 143
oil industry: conflict over royalties, 45–47;
in *Los Primeros* (Ruiz film), 35–36, 45
Okinawa: history of, 67–68; Japanese
economic miracle and, 22, 86, 100–101;
paranoid ideological climate of, 72;
postwar, agrarian and militarized
landscapes in, 71–77; protests against
United States, 72–75, 264n18;
resettlement within, 74; U.S. rule and
displacement on, 2, 21, 66, 68, 71–77;
U.S. "thinning policy" for, 74–75
Okinawan People's Party, 73–74
Okinawans: anniversary commemorations,
230, 31, 233–34; Certificate of Identity,
75–77, 100; challenges to racism, 82–83;
Christmas card list of accomplishments,
83; citizenship, agrarian, in Bolivia, 11–15,
21–22, 66, 81–84, 100–101, 254; citizenship,
agrarian, in Okinawa, 73; citizenship,
postwar problems with, 11, 68, 74, 75–77,
100; disparagement from Japanese
settlers, 81; emigration history of, 66, 68,
99–100, 235; expense per emigrant to

Bolivia, 74; failure of early settlement ("Uruma disease"), 78; journey to Bolivia, 75–76; migration, comparison with Andeans, 101, 102, 111–12, 118; migration, comparison with Mennonites, 22, 66–70, 86, 99–101; in *The Mountains Never Change* (Ruiz film), 83–84; plurinationality and integration, 252–53; restricted markets for, 83–84; return to Japan, 225–26; Santa Cruz settlement, 71, 75, 77–86; sources for study, 18–19; spatial positioning of, 81, 83–84; term, explanation of, 67–68, 264n8; Tigner's report on, 1–2, 9, 71–73, 74, 92; Uruma Society and, 66, 67, 77–78, 81; U.S. support for emigration of, 71, 74–78, 86, 100; xenophobia toward, 13, 19, 21, 75, 78–83, 266n60

Old Colony Mennonites, 68, 189, 191. *See also* Mennonites, in Mexico; Mennonites, Mexican

orientation program, for settlers, 139, 157–59, 170–72

Oroza, Felix, 111, 112

Ortega Estrada, Andrés, 97

Orton, James, 15–16

Ovando Candía, Alfredo, 56, 137, 165, 203

ownership of land: Andean complaints over titles, 124, 128–33, 162; Mennonite colonies as title holders, 224; titles for Mexican Mennonites, 199; titles withheld, 11, 259n31

oxfam grant, for San Julián Project, 174

Pagura, Federico José, 160

Painter, Michael, 179–80, 183

pamphlets: abandonment depicted in, 128–29; frontier imaginary in, 58–62; "How Will I Live and Work in My New Parcel?," 59, 61f; "A Human Transplant," 62, 237; images of Andean bodies in, 58–62, 61f; "Lessons of Propaganda, Organization and Agitation," 31; "The Route to Development," 128; soybean promotion, 211–12; tropic motif in, 59,

60f, 62; "We Will Form Our Cooperative," 59; "What is the 10 Year Plan?," 59, 60f

Paraguay: Bolivian loss of territory to, 109; Mennonite emigration from, 22, 65–67, 86–90, 229; Mennonite settlement in, 72; property rights of settlers in, 129

Paraguay River, 17

Partridge, William, 179

paternalism: citizenship and, 11; Cotoca Colony, 123; deficient, complaints about, 134; dependency and, 123–25, 150–51, 172–74; "Indian question," 104; in *Little Johnny Can Read* (Ruiz film), 44; MNR and, 5, 31–32, 44–45; as obstacle to success, claim of, 172–73; petitions and, 123–25, 238; San Julián Project and, 172–74, 183; wariness of, 156

PAX program, 152

Paz Estenssoro, Víctor, 3, 14; agreement with Mennonites, 87; faith-based development welcomed by, 143–44, 147; first interview as president, 65; fiscal shock program, 217–19, 243; founding of ICB, 32; inauguration of railroad, 38–39; overthrow of, 137; petitions to, 104, 127, 134–35; privatization overseen by, 191

PCM (Peasant–Military Pact), 137

Peace Corps, 154

Peacock, Harry: background and beginning of work in Bolivia, 147–50, 185; Baldivieso and, 162, 163–64; on capacity building, 170; career trajectory, 185; choice to work with Banzer regime, 168–69; departure from Bolivia, 178; disaster response (1968 flooding), 154–56; ministering *vs.* administering, 139–40, 174–75, 185; personal trajectory, 241; Public Works Conference (1972), 169–70; return to Bolivia, 187; San Julián Project, 139, 168–72, 174–78; settler orientation program, 139, 157, 170–72; USAID recruitment and posts, 139, 174–75, 178, 185, 218, 243

peanuts, 226

Peasant–Military pact (PCM), 137
Pereyra, Mary de Porres, 140, 157, 158
Pérez Baldivieso, Felix, 88–90
permanent resident card, for Mennonites, 251–52
Perón, Eva and Juan, 104, 114
Peru: environmental (fishery) collapse, 212, 214–15; frontier elites, 37; Okinawans in, 68; Ruiz's work in, 56; transnational highway imagined in, 6
petitions and letters, Andean, 22–23, 102–38, 238; appeals to settle frontier, 105, 106–19; *braceros* invoked in, 114–19; characterization of homelands, 107, 108; complaints about technical advice, 124–28, 162; complaints about titles to land, 124, 128–33, 162; complaints about transport and market access, 124, 133–36, 162; complaints from settlers, 105, 119–36; diversity of petitioners, 106–7; expectations for new land, 109; miner petitions for land, 112–14; motifs of hope and failure in, 136; naïve monarchism, 103, 132; opposition to foreigners, 111–12; political climate and, 137–38; postrevolutionary, 104–5; "speaking to a state" 103; resonance with revolutionary aims, 110–13; subaltern practices, 103–5, 238
petitions and letters, Mexican Mennonite, 202–4, 216–17
physiocracy, 12
PIL Andina, S. A., 227
pink tide, 240
Pinto Parada, Melchor, 46, 48
pioneering stage, of settlements, 176
Plan 3000, 254
Plan Bohan, 17
plan inmediato de politica economica del gobierno de la revolución nacional, 29–30, 42, 71, 84, 110–11, 113
plurinationality, 250–55
Point Four Program, 6, 51–52, 147
populism, 104, 137

Un poquito de diversificación económica (Ruiz film), 26–31, 28f, 36, 42, 63, 237, 243
Porras, Emilio, 87
Prebisch, Raúl, 123
predatory lending, and Mennonite debt crisis, 221–28
Presencia (newspaper), 115–16, 270n50
Los Primeros (Ruiz film), 35–36, 45, 64
privatization, 7, 182, 191
Privilegium, for Mennonites: in Bolivia, 87–88, 90, 202–3, 209; in Mexico, 91, 93
propaganda: MNR-produced films, 7. *See also* frontier imaginary
property rights, complaints and disputes over, 128–33, 162
Protestant Boom, 142–44, 260n64
Public Works Conference (1972), 169–70, 181–82
Punta del Este (Uruguay) meeting, 53, 57

Quechua: continued use of language, 122–23; defining Andeans, 106; Methodist recruitment of, 148–49; migration stories of, 2, 118, 144; "Our Indian" editorial *vs.,* 44–45; *Plan immediato* dismissal of, 42; settler attrition, 122–23
Los que nunca fueron (Ruiz–Beltrán film), 55

race: Argentinian view of *braceros,* 117–18; attitudes toward Mennonites, 88, 192–94; belonging complicated by, 13; eurocentrism, 79–80; highlanders *(kollas)* vs. lowlanders *(cambas),* 8, 13, 26–27, 43–44, 106, 245, 254; Morales's (Evo) antiracism campaign, 240–41; Okinawan and Japanese challenges to racism, 82–83; "Our Indian" editorial, 44–45, 48, 106; overcoming divides, 254–55; petitions opposing foreigners, 111–12; racialized reaction to Andeans, 8, 13, 26–27, 37, 42–45, 80–81, 237–38; racism in *El Deber,* 78–83; Santa Cruz reaction to outsiders, 37, 42, 78–81; soybean production by, 219; union in *mestizo*, 8, 27, 43, 45, 106; "whitening," 79, 192–93, 265n48; xenophobia toward Japanese and Okinawans, 13, 19, 21, 75, 78–83, 266n60

radicalism: Methodist, in Yapacaní, 160–67; San Julián roadblock, 180–82

railroads: accidents and deaths, 39, 41, 261n49; connection to Brazilian commerce, 41; contraband and smuggling via, 41–42; frontier imaginary, 36–51; frustration with, 39–42; grueling saga of travel, 40–41; inaugural celebration, 38–39; Marx on, 17–18; past and present, 231–32; pilgrimage to Cotoca, 40; river crossing and drownings, 39–40; Santa Cruz links with Cochabamba, Brazil, and Argentina, 38–41, 38*f*; surplus passengers on open decks, 40; transnational agreements, 17

"Railway or Nothing" movement, 261n41

rape cases, Mennonite, 284n20

religion and religious organizations, 23, 238–39; Banzer regime and, 23, 137–38, 143, 164–86, 239; disaster response (1968 flooding), 153–56; evangelical conflict among Mennonites, 97–99; interconnections, 142; liberation theology, 153; Methodist "lands of decision" campaign, 6, 146–47, 153, 185; ministering *vs.* administering, 139–40, 144, 168, 174–75, 181–82, 185–86, 238–39; Protestant Boom, 142–44, 260n64; as proxy for state, 138, 142, 156, 167, 169, 182–86, 239; San Julián Project, 6, 23–24, 139–43, 168–86; United Church Committee, 23, 155–59. *See also specific organizations*

Revista Oeste (West Magazine), 29

Reyes, Ulrich, 169

Reynolds, Earle, 100, 269n146

Ribera, Guillermo, 215, 217

Río Grande flooding (1968), 143, 153–56

Rios, Epifanio, 195–96

Riva Palacio (colony), 191, 194–95; agricultural practices, 227; dairy industry, 207–10; diary of Wiebe (Johan), 201–4; drought and debt crisis, 221–28; expansion of, 228, 232; Kopp's clandestine visit to, 198, 200; mechanization of, 206–7; soybean production, 212–20

Rivera Prado, Oscar, 189, 191, 194

Rivero Mercado, Pedro, 155, 162

roadblocks: grassroots resistance to neoliberalism, 181; San Julián, 180–82

roads: cautionary tale ("By the New Highway"), 43–44; frontier imaginary, 36–51; indigenous territory, protests against, 249; linking Santa Cruz and Cochabamba, 38*f*, 43–46; linking Santa Cruz and highlands, 17, 26, 42–51; "Marginal Highway of the Jungle," 6, 57, 58, 64; Marxian equation, 17–18; past and present, 230–32; Second Ring, 230; settler complaints about, 124, 133–36, 162

Roca, Augusto, 54–55

Roca, Jorge, 50

Roca, José Luis, 202

Rodríguez, Miguel, 17

Roque Chuquimia, Celestino, 132

rubber tires, Mennonite conflict over, 22, 96–99, 227

Ruiz, Jorge, 21; attempts to humanize March to the East, 63; directorship of ICB, 34–36; interview with Paz Estenssoro, 67; itinerary and repertoire of, 51–52; news series (*Hoy Bolivia*), 57; pamphlet produced by, 62, 237; personal trajectory, 241; tour with Grierson, 31; U.S. support for, 64; work for Guatemalan regime, 53–58. *See also* Ruiz films

Ruiz Cortines, Adolfo, 93, 94, 96

Ruiz films, 26–30, 34–36, 236–37; diffusion of economic policy in, 111; early works, 44; gendered terms in, 26–27, 30, 34–36, 49; influence of *Plan immediato* on, 29–30; *A Little Bit of Economic*

Ruiz films (*continued*)
Diversification, 26–30, 28f, 36, 42, 64, 237, 243; *Little Johnny Can Read,* 44; lowlands as transformative space in, 27, 29; *The Mountains Never Change,* 58–59, 83–84; paternalism toward indigenous people in, 44–45; *Los Primeros,* 35–36, 45, 64; transnational topics and audiences, 51–58, 64; *La Vertiente,* 34–36, 35f, 48–49, 50, 52, 64, 236; visual distinction between east and west in, 21, 27; *Vuelva Sebastiana,* 44, 55; *Los Ximul,* 53–57, 236
Russia: Mennonite emigration from, 65, 68–69; Russification policy, 69

Sachs, Jeffrey, 217, 219, 243
Sagrado Corazón, 232–33
SAI (Inter-American Agricultural Service), 27, 54–55, 263n110
Salas López, Daniel, 93, 64, 96–97
Sanabría Fernández, Hernando, 45, 47–48
Sánchez de Lozado, Gonzalo, 244
San Francisco Peace Treaty of 1951, 71
San Julián Project, 6, 23–24, 139–43, 168–86; academic study of, 141–42, 175–77, 179–80, 183; anniversary, 234; capacity-building goals, 170; as career springboard, 183–85; challenges to development, 177–82; consolidation stage of, 176, 178, 183; expansion of, 232; flooding (1983), 178; funding for, 174–77; growth stage of, 178, 183; hyperinflation and, 178, 179–80; INC report, 172–73; Mexican Mennonite settlement *vs.,* 216; NADEPAs of, 170; paternalism/ dependency theory and, 172–74, 183; political instability and, 177–78; Public Works Conference (1972) and, 169–70, 181–82; reputation of, 142; roadblock and radicalism, 180–82; secularization and FIDES shift, 177; socioeconomic forces threatening, 180; spatial design of, 140, 141f, 170, 171f; titles for land withheld, 259n31; USAID progress report on, 178–80

Santa Cruz: Andean migration to, 7, 258n14; autonomy movement, 8, 20, 25, 45–51, 244–48, 249; balkanization of, 47; case studies of transnational migration history, 234–35; response to representation, 37; conflict with MNR over oil royalties, 45–47; dependency on highland markets, 42; ecotone of, 4–5, 16, 187; editorials promoting, 29; exploration and territoriality, 15–18; Green Revolution and rural modernization in, 6; growth and transformation of, 8–9, 233, 234; Guevara's guerrilla force in, 187–89, 228; highway linking to highlands (Cochabamba), 17, 26, 38f, 42–51; infrastructure projects converging on (1950s), 38f; in *A Little Bit of Economic Diversification* (Ruiz film), 26–28, 42; martial law, occupation, and protests, 47–51; Mennonite exploration, Mexican, 188–89; Mennonite exploration, Paraguayan, 65–66; Mennonite settlement in, 86–90, 99; migrant-friendly motto, 234; in *The Mountains Never Change* (Ruiz film), 58–59; national images and narratives of, 46; nativist posturing, 79, 265n50; Okinawan settlement in, 71, 75, 77–86; past and present in, 230–41; Public Works Conference (1972), 169–70, 181–82; racialized reaction to Andeans, 8, 13, 26–27, 37, 42, 45, 80–81, 237–38; racialized reaction to Japanese and Okinawans, 80–83; railroads, 17–18, 38–42, 38f; regionalism, 37, 45–46, 64, 237, 244–48, 262n63; settler orientation program in, 139, 157–59; street scene and Mennonite market day, 231, 231f; theaters in, 49–50; transnational revolution and, 9–11; in *Traveling Through Our Land* (short films), 32–33; Viru Viru International Airport, 230. *See also specific settlers and settlements*

Santa Cruz–Corumbá railway, 38*f. See also* railroads

Santa Cruz de la Sierra, 8–9. *See also* Santa Cruz

Santa Cruz Department, 4–5, 4*f. See also* Santa Cruz

Santa Cruz–Yacuíba railway, 38*f*

Santa Rita (colony), 227, 251

Saucedo Sevilla, Lucas, 37

Schoutten, Bastiaan, 139, 175, 185

Schrock, Elwood, 154, 197

Scott, James, 22, 133, 237

Scudder, Thayer, 24, 176, 179, 273n127

Secretary of Press, Information, and Culture (SPIC), 31–33

"seeing like a state," 22, 103, 238

Selich Chop, Andrés, 164

Senaga, Kamejiro, 73–74

September 24th Cooperative, 86–88

sesame production, 226

settler attrition, 119–24, 128–29, 225–26. *See also* abandonment.

sharecropping agreement, Mennonite, 86–88

Shelley, Andrew, 98

Shōkō, Ahagon, 73

Siemel, Alexander, 16

Siglo XX mine, 112–13

Siles, Luis Adolfo, 162

Siles Zuazo, Hernán, 14, 46–47, 84, 104, 114

Silva, Lula da, 240

Simón García, Luis, 80

sirionó, 25

socialism, Morales (Evo) and, 240–41

social message films, 31–32. *See also* films

Solem, Richard, 180–82

Sora agroindustrial cooperative, 110

Soria, Oscar, 30, 55, 59

Soviet Union, film propaganda in, 31

soy and soybeans: ANAPO annual reports, 212*f*, 220, 252; boom and migration of, 2–3, 14–15, 212, 235–36; Brazilian production, 5, 213–14; coca production *vs.*, 218–19; conditions influencing shift to, 212–20; early attempts at growing,

213; farming as "mining," 15, 259n42; hyperinflation and, 215–20, 221; Mennonite agribusiness, 24–25, 211–28, 239–40; Mennonite agricultural practices, critique of, 227; Mennonite debt crisis and drought, 221–28, 282n125; national, racial, and ethnic categories of production, 219, 252; Okinawan and Japanese involvement in, 253; pamphlet promoting, 211–12; personal accounts of production, 220, 222, 222*f*, 223; Peruvian environmental collapse and, 212, 214–15; "United Republic of Soy," 212, 235

"soylandia," 212

soyscape, 14

Special Federation of San Julián Colonists, 181

SPIC (Secretary of Press, Information, and Culture), 31–33

spontaneous colonists, 131

The Spring (Ruiz film), 34–36, 35*f*, 48–49, 51, 64, 236

stages of settlements, 176, 178, 183

Stalin, Josef, 103

State Broadcasting and Entertainment Service (SODRE, Uruguay), 31

state sovereignty, 15–18

Stauffer, Russ, 152

Stoesz, Edgar, 167, 198

"struggle for the eleven percent," 45–48

subalterns, 103–5, 238, 273n129

subimperialism, 68, 70

Sukarno, 5

Supreme Decree 3342, 32

Swift Current (colony), 191, 227, 254

Sylvain, Jeanne, 121–25

syngenta, 2, 14, 235

Takata, Minuro, 82

Tamplin, Carroll, 144–46, 248

TAP (Teacher Abroad Program), 152, 167, 277n125

Taylor, Brooks, 161, 163, 165

Taylor, Peter Leigh, 183

TCOs (Communal Lands of Origen), 249

Teacher Abroad Program (TAP), 152, 167, 277n125

technicians: complaints about technical advice, 124–28, 162; faith-based development, 18–19, 139–43; as proxies for state, 121, 142

Terrazas, Juan, 180

territoriality, 15–18

theaters, in Santa Cruz, 49–50

Tigner, James, 1–2, 9, 71–73, 74, 92, 241

TIPNIS (Isiboro-Sécure Indigenous Territory), 249

titles to land: Andean settler complaints about, 124, 128–33, 162; Mennonite colonies as holders of, 224; Mexican Mennonite, 199; withheld, 11, 259n31

Torres, Juan José "J. J.," 137, 155, 164, 165

Torres, Mario, 113

Torrico, Armando, 196–97

"transnationalism through parochialism," 67

transnational revolution, in migration, 9–11

transport: past and present, 230–32; settler complaints about, 124, 133–36, 162. *See also* railroads; roads

Traveling Through Our Land (short films), 32–34

Tsukui, Heizo, 84

Union of Poor Campesinos (UCAPO), 160, 163–65

United Church Committee (CIU), 23, 139–43, 155–59; absence from colony narrative, 234; career springboards, 183–85; disaster response (1968 flooding), 153–56; Peacock's departure from, 174; Public Works Conference (1972), 169–70; San Julián Project, 6, 23–24, 139–43, 168–86, 234; secularization of, 177, 278n155; settler orientation program, 139, 157–59, 170–72

United Nations Andean Mission, 120–21

United Nations World Food Program, 118

United States: bipolar logic of foreign policy, 243, 284n9; cultural ambassadors from, 31; development agenda of, 51–52;

drug policy, 218–19, 243–44; funding for Bolivia, 6, 57–58, 63–64, 77, 117, 263n106; Good Neighbor Policy, 17, 55; Okinawan emigration support, 71, 74–78, 86, 100; Okinawan protests against, 72–75; Okinawan rule and displacement by, 2, 21, 66, 68, 71–77; promotion of Green Revolution, 6; purchase of old machinery from, 206–7. *See also specific agencies*

United States Agency for International Development. *See* USAID

United States Civil Administration of the Ryukyu Islands (USCAR), 18, 68, 72–78, 83, 86, 118

United States Information Service (USIS), 52

Únzaga de la Vega, Óscar, 47–48

Urquidi, Ricardo, 85

Urriolagoitía, Mamerto, 16–17

"uruma disease," 78

Uruma Society, 66, 67, 71, 77–78, 81

Los urus (Ruiz film), 44

USAID: agricultural policy, 152–53, 185, 218, 240, 243; career trajectories, 183–85; colonization pamphlet, 128; colonization partnership, 58, 120; Mennonite Central Committee compared with, 167; Morales expulsions of, 240; Peacock recruitment and posts, 139, 174–75, 178, 185, 218, 243; reliance on non-state agencies, 183; road construction, 133, 135; San Julián funding, 174–77; San Julián Project progress report, 178–80

USCAR. *See* United States Civil Administration of the Ryukyu Islands

USIS (United States Information Service), 52

Vaca Flor, Claudina, 261n49

Vaca Pereira, Raúl, 50

Valdés, Luisa, 132

Valdivia, Augusto, 85, 86

Valenzuela, Néstor, 195–96

Valverde, Carlos, 50

Vargas, Getúlio, 5, 106
La Vertiente (Ruiz film), 34–36, 35f, 48–49, 51, 52, 64, 236
villagization, in Africa, 5, 24
Viru Viru International Airport, 230
Vuelva Sebastiana (Ruiz film), 44, 55

War of the Pacific (1878–83), 17
WGM. *See* World Gospel Mission
Whitaker, Milton, 185, 218, 243–44
"whitening," racial discourse and migration, 79, 192–93, 265n48
widows and orphans fund *(Waisenamt)*, 87, 93, 215–16
Wiebe, Johan, 201–4, 208, 241, 247, 252

World Gospel Mission (WGM), 23, 144–46, 238–39, 248, 274n22
Wuhl, Jack, 208

Los Ximul (Ruiz film), 53–57, 236

Yapacaní (colony), Methodists and agrarian radicalism in, 160–67
Ydígoras, Miguel, 54
El Yeso society, 111

Zabalaga, Virgilio, 134
Zambrana, Carlos, 196, 201, 202
Zebu cattle, 8–9, 33, 204–5, 214, 235, 247
Zelada, Victor Hugo, 113

CPSIA information can be obtained
at www.ICGtesting.com
Printed in the USA
LVHW092219100320
649693LV00003B/667